Micro-Electro Discharge Machining: Principles, Recent Advancements and Applications

Micro-Electro Discharge Machining: Principles, Recent Advancements and Applications

Editors

Irene Fassi
Francesco Modica

MDPI • Basel • Beijing • Wuhan • Barcelona • Belgrade • Manchester • Tokyo • Cluj • Tianjin

Editors
Irene Fassi
STIIMA, Institute of Intelligent
Industrial Technologies and
Systems for Advanced
Manufacturing
CNR, National Research Council
of Italy
Milano
Italy

Francesco Modica
STIIMA, Institute of Intelligent
Industrial Technologies and
Systems for Advanced
Manufacturing
CNR, National Research Council
of Italy
Bari
Italy

Editorial Office
MDPI
St. Alban-Anlage 66
4052 Basel, Switzerland

This is a reprint of articles from the Special Issue published online in the open access journal *Micromachines* (ISSN 2072-666X) (available at: www.mdpi.com/journal/micromachines/special_issues/Micro-Electro_Discharge_Machining).

For citation purposes, cite each article independently as indicated on the article page online and as indicated below:

LastName, A.A.; LastName, B.B.; LastName, C.C. Article Title. *Journal Name* **Year**, *Volume Number*, Page Range.

ISBN 978-3-0365-1933-3 (Hbk)
ISBN 978-3-0365-1932-6 (PDF)

Contents

About the Editors

Irene Fassi

Irene Fassi is currently Research Director at CNR-STIIMA where she leads the research group MEDIS, performing research activities in micro and meso scale manufacturing and robotics. She gives courses and seminars in robotics, precision engineering and advanced manufacturing systems, at University of Brescia and Politecnico di Milano. She is member of several Expert Evaluators Panel: European Commission, Agence Nationale de la Recherche, Norwegian Research Council, Eureka, MiSE, MIUR. She has been general or scientific chair of several international Conferences in the field of micromanufacturing, microsystems, and robotics, including the World Congress on Micro and Nano Manufacturing. She is a member of the Executive Board of SIRI (Italian Robotics and Automation Association), the Italian Association of Manufacturing Technologies (AITeM), the ASME Technical Committee on Micro and Nano Manufacturing, and currently serves as President of the International Institution for Micro Manufacturing

Francesco Modica

Francesco Modica is full time researcher at STIIMA-CNR, where he is currently researching mainly on micro-manufacturing and micro-engineering at the STIIMA-CNR MicroManufacturing laboratory in Bari. From 2000 to 2009, he has conducted research activities in mechatronic analysis and simulation of machine tools. From 2005 to 2009, he has been involved as researcher and project manager in collaboration with Sintesi s.p.a. developing mechatronic solutions (devices and simulations) for improving dynamic performance of machine tools. In 2009 he was visiting researcher at Katholieke Universiteit Leuven applying on micro-electrical discharge machining. Successively he has led different research activities concerning micro-manufacturing processes (in particular related to micro-EDM, micro injection moulding and Additive Manufacturing for micro applications) and design and fabrication of micro-components and devices.

micromachines

Editorial

Editorial for the Special Issue on Micro-Electro Discharge Machining: Principles, Recent Advancements and Applications

Irene Fassi [1,*] and **Francesco Modica** [2,*]

1   STIIMA CNR, Institute of Intelligent Industrial Technologies and Systems for Advanced Manufacturing, National Research Council, Via A. Corti 12, 20133 Milan, Italy
2   STIIMA CNR, Institute of Intelligent Industrial Technologies and Systems for Advanced Manufacturing, National Research Council, Via P. Lembo 38/F, 70124 Bari, Italy
*   Correspondence: Irene.Fassi@stiima.cnr.it (I.F.); Francesco.Modica@stiima.cnr.it (F.M.)

Citation: Fassi, I.; Modica, F. Editorial for the Special Issue on Micro-Electro Discharge Machining: Principles, Recent Advancements and Applications. *Micromachines* **2021**, *12*, 554. https://doi.org/10.3390/mi12050554

Received: 6 May 2021
Accepted: 7 May 2021
Published: 13 May 2021

Publisher's Note: MDPI stays neutral with regard to jurisdictional claims in published maps and institutional affiliations.

Micro Electrical Discharge Machining (micro-EDM) is a thermo-electric and contactless process most suited for micro-manufacturing and high-precision machining, especially when difficult-to-cut materials, such as super alloys, composites, and electro conductive ceramics, are processed. Many industrial domains exploit this technology to fabricate highly demanding components, such as high-aspect-ratio micro holes for fuel injectors, high-precision molds, and biomedical parts.

Moreover, the continuous trend towards miniaturization and high precision functional components boosted the development of control strategies and optimization methodologies specifically suited to address the challenges in micro- and nano-scale fabrication.

This Special Issue showcases 12 research papers and a review article focusing on novel methodological developments on several aspects of Micro Electrical Discharge Machining: machinability studies of hard materials (TiNi Shape Memory Alloys, Si_3N_4–TiN ceramic composite, ZrB_2-Based Ceramics Reinforced with SiC Fibers and Whiskers, Tungsten-cemented carbide, Ti-6Al-4V alloy, Duplex Stainless Steel and Cubic boron nitride), process optimization adopting different dielectrics or electrodes, characterization of mechanical performance of processed surface, process analysis and optimization via discharge pulse-type discrimination, hybrid processes, fabrication of molds for inflatable soft microactuators, and implementation of low-cost desktop micro-EDM system.

In further detail, Zhao et al. [1] have investigated Laser-Assisted Electrochemical Discharge Machining for micro-processing of insulating and hard-brittle materials typified by glass. This study compared morphological features obtained by means of single processing and hybrid processing methods concerning the fabrication of microgrooves on quartz workpiece. The combination of these two methods allows to transform gradually V-shaped microgrooves into U-shaped microgrooves. Micro Electrical Discharge Machining has been characterized by Zhu et al. [2] for Surface Modification of TiNi Shape Memory Alloys using a TiC Powder Dielectric. The study reveals that TiC powder addition with a concentration of 5 g/L had a positive effect on increasing the electro-discharge frequency and MRR, reducing the surface roughness. The machined surface presents a recast layer with good adhesion and high hardness, due to metallurgical bonding, and contains CuO_2, TiO_2, and TiC phases, contributing to an increase in the surface microhardness from 258.5 to 438.7 HV, which could be beneficial for wear resistance in biomedical orthodontic applications. In accordance with potential effects on mechanical properties and oxidation performance, a study on ZrB_2-Based Ceramics reinforced with SiC Fibers and whiskers has been performed by Quarto et al. [3] By means of the pulse-type characterization, results indicated how reinforcement shapes affect the energy efficiency of the process and change the surface aspect. The microEDM milling machinability of the Si_3N_4–TiN ceramic composite was investigated through the discharge pulse-type discrimination algorithm evaluating the material removal rate, tool wear ratio and surface quality in [4] by Marrocco et al. The analysis shows that MRR was sensitive only to normal pulses, while the occurrence of arcs and delayed pulses induced unexpected improvements in tool wear. Additionally, the inspection of the surface, performed by SEM and EDS analyses, showed the presence of re-solidified droplets and micro-cracks, which modified the chemical composition and the consequent surface quality of the machined micro-features. Ablyaz et al. [5] have developed a Surface Characterization and a Tribological Performance Analysis of Electric Discharge Machined Duplex Stainless Steel considering the main process parameters and three variants of electrode material (Graphite, Copper–Tungsten and Tungsten). The results revealed that the machined surface at high spark energy in EDM oil portrayed porosity, oxide formation, and intermetallic compounds, and then exhibited 70% superior wear resistance compared to the un-machined sample and contextually improving surface wettability.

The machinability via Electrodischarge Drilling of microholes with 410 μm in diameter and 1 mm deep in difficult-to-cut cubic boron nitride (c-BN) material was studied by Wyszynski et al. [6]. Tests were executed using tungsten electrode tool in Exxol80 (hydrocarbon oil) and varying pulse on and off times, pulse frequency and working current. Results show that relatively high working current and short pulses improve machining quality. Wire electrical discharge grinding (WEDG) is adopted by Milana et al. [7] for shaping several molds to cast rubber and create complex void for inflatable soft microactuators. All microactuators that have a cylindrical shape with a length of 8 mm and a diameter of 0.8 mm have been characterized and tested showing that complex deformation patterns, such as curling, differential bending or multi-points bending, can be achieved.

In order to deal with the high wear ratio bottleneck of micro-electrode tool, Huang et al. [8] have investigated the use of a liquid droplet suspended at the end of a capillary nozzle as an electrode. The liquid supply, the geometry and shape control of the tip were analyzed both via simulation and experimentation showing the feasibility of the method. The feasibility of a novel filling method to fabricate a composite 3D microelectrode for processing 3D microstructures via micro-EDM without steps is analyzed theoretically and experimentally by Lei et al. [9]. Wear and corrosion resistance of the Ti-6Al-4V alloy processed by EDM using multi-walled carbon nanotubes (MWCNTs) mixed with dielectric were experimentally investigated by Singh et al. [10]. The study reveals that the formation of intermetallic compounds, oxides, and carbides allows a dramatic enhancement of the wear-resistant and the corrosion protection efficiency up to 95%, and 96.63%, respectively, in comparison to using plan dielectric. Xu et al. [11] have proposed foil queue microelectrodes with tapered structures in order to solve the typical step effect issue when these microelectrodes are used for eroding 3D microstructure. The influence of the taper angle and the number of microelectrodes on the step effect were investigated and tested, reporting that the step effect on the 3D microstructure's surface became less evident when the taper angle and the number of foils–microelectrodes increase.

Wu et al. [12] have presented a low-cost desktop micro-EDM system for fabricating micro-holes in tungsten-cemented carbide materials reporting experimental test of micro-holes with a diameter of 0.07 mm and thickness of 1.0 mm drilled with a cut-side shaped micro-tool.

Finally, with the aim of helping researchers for selecting suitable powders and identifying different aspects of powder-mixed dielectric fluid of EDM, a paper review is presented by Abdudeen et al. [13]. The paper discusses and compares studies about various powders used for the process focusing on achieving a more efficient metal removal rate, reduction in tool wear, and improved surface quality of the powder-mixed EDM process.

We thank all the authors who submitted their papers to this Special Issue "Micro-Electro Discharge Machining: Principles, Recent Advancements and Applications". We would also like to acknowledge the reviewers who carefully reviewed all the submitted papers dedicating their time and helping to improve the quality of this Special Issue.

Conflicts of Interest: The author declares no conflict of interest.

References

1. Zhao, D.; Zhang, Z.; Zhu, H.; Cao, Z.; Xu, K. An Investigation into Laser-Assisted Electrochemical Discharge Machining of Transparent Insulating Hard-Brittle Material. *Micromachines* **2021**, *12*, 22. [CrossRef] [PubMed]
2. Zhu, Z.; Guo, D.; Xu, J.; Lin, J.; Lei, J.; Xu, B.; Wu, X.; Wang, X. Processing Characteristics of Micro Electrical Discharge Machining for Surface Modification of TiNi Shape Memory Alloys Using a TiC Powder Dielectric. *Micromachines* **2020**, *11*, 1018. [CrossRef] [PubMed]
3. Quarto, M.; Bissacco, G.; D'Urso, G. Study on ZrB2-Based Ceramics Reinforced with SiC Fibers or Whiskers Machined by Micro-Electrical Discharge Machining. *Micromachines* **2020**, *11*, 959. [CrossRef] [PubMed]
4. Marrocco, V.; Modica, F.; Bellantone, V.; Medri, V.; Fassi, I. Pulse-Type Influence on the Micro-EDM Milling Machinability of Si3N4–TiN Workpieces. *Micromachines* **2020**, *11*, 932. [CrossRef] [PubMed]
5. Ablyaz, T.R.; Shlykov, E.S.; Muratov, K.R.; Mahajan, A.; Singh, G.; Devgan, S.; Sidhu, S.S. Surface Characterization and Tribological Performance Analysis of Electric Discharge Machined Duplex Stainless Steel. *Micromachines* **2020**, *11*, 926. [CrossRef] [PubMed]
6. Wyszynski, D.; Bizon, W.; Miernik, K. Electrodischarge Drilling of Microholes in c-BN. *Micromachines* **2020**, *11*, 179. [CrossRef] [PubMed]
7. Milana, E.; Bellotti, M.; Gorissen, B.; Qian, J.; De Volder, M.; Reynaerts, D. Shaping Soft Robotic Microactuators by Wire Electrical Discharge Grinding. *Micromachines* **2020**, *11*, 661. [CrossRef] [PubMed]
8. Huang, R.; Yi, Y.; Zhu, E.; Xiong, X. Investigation of a Liquid-Phase Electrode for Micro-Electro-Discharge Machining. *Micromachines* **2020**, *11*, 935. [CrossRef] [PubMed]
9. Lei, J.; Jiang, K.; Wu, X.; Zhao, H.; Xu, B. Surface Quality Improvement of 3D Microstructures Fabricated by Micro-EDM with a Composite 3D Microelectrode. *Micromachines* **2020**, *11*, 868. [CrossRef] [PubMed]
10. Singh, G.; Ablyaz, T.R.; Shlykov, E.S.; Muratov, K.R.; Bhui, A.S.; Sidhu, S.S. Enhancing Corrosion and Wear Resistance of Ti6Al4V Alloy Using CNTs Mixed Electro-Discharge Process. *Micromachines* **2020**, *11*, 850. [CrossRef] [PubMed]
11. Xu, B.; Guo, K.; Zhu, L.; Wu, X.; Lei, J. Applying Foil Queue Microelectrode with Tapered Structure in Micro-EDM to Eliminate the Step Effect on the 3D Microstructure's Surface. *Micromachines* **2020**, *11*, 335. [CrossRef] [PubMed]
12. Wu, Y.-Y.; Huang, T.-W.; Sheu, D.-Y. Desktop Micro-EDM System for High-Aspect Ratio Micro-Hole Drilling in Tungsten Cemented Carbide by Cut-Side Micro-Tool. *Micromachines* **2020**, *11*, 675. [CrossRef] [PubMed]
13. Abdudeen, A.; Abu Qudeiri, J.E.; Kareem, A.; Ahammed, T.; Ziout, A. Recent Advances and Perceptive Insights into Powder-Mixed Dielectric Fluid of EDM. *Micromachines* **2020**, *11*, 754. [CrossRef] [PubMed]

 *micromachines* 

Article

An Investigation into Laser-Assisted Electrochemical Discharge Machining of Transparent Insulating Hard-Brittle Material

Douyan Zhao [1], Zhaoyang Zhang [1,*], Hao Zhu [1], Zenghui Cao [2] and Kun Xu [1]

[1]   Laser Technology Institute, School of Mechanical Engineering, Jiangsu University, Zhenjiang 212013, China; 2211803033@stmail.ujs.edu.cn (D.Z.); haozhu@ujs.edu.cn (H.Z.); xukun@ujs.edu.cn (K.X.)
[2]   Fengjiang Intelligent Technology Co., Ltd., Changzhou 213100, China; 2211803060@stmail.ujs.edu.cn
*   Correspondence: zhaoyanz@ujs.edu.cn; Tel.: +86-0511-8878-0171

Abstract: Electrochemical discharge machining (ECDM) and laser machining are emerging nontraditional machining technologies suitable for micro-processing of insulating and hard-brittle materials typified by glass. However, poor machinability of glass is a major constraint, which remains to be solved. For the micro-grooves processed by ECDM, the bottom surface is usually uneven and associated with protrusion structures, while the edges are not straight with obvious wave-shaped heat-affected zones (HAZs) and over-cutting. Besides, the cross section of the micro-grooves processed by the laser is V-shape with a large taper. To solve these problems, this study proposed the laser-assisted ECDM for glass micro-grooving, which combines ECDM and laser machining. This study compared morphological features of the single processing method and the hybrid processing method. The results show that ECDM caused cylindrical protrusions at the bottom of the microgrooves. After processing these micro-grooves by laser, the cylindrical protrusions were removed. However, the edge quality of the microgrooves was still poor. Therefore, we used the laser to get microgrooves first, so we got micro-grooves with better edge quality. Then we use ECDM to improve the taper of microgrooves and the cross-sectional shape of the microgrooves transformed from a V-shape to a U-shape.

Keywords: electrochemical discharge machining; laser machining; glass; micro-groove

Citation: Zhao, D.; Zhang, Z.; Zhu, H.; Cao, Z.; Xu, K. An Investigation into Laser-Assisted Electrochemical Discharge Machining of Transparent Insulating Hard-Brittle Material. *Micromachines* **2021**, *12*, 22. https://dx.doi.org/10.3390/mi12010022

Received: 27 October 2020
Accepted: 24 December 2020
Published: 27 December 2020

Publisher's Note: MDPI stays neutral with regard to jurisdictional claims in published maps and institutional affiliations.

1. Introduction

Insulating and hard-brittle materials typified by glass are used widely in many fields such as chemical engineering, electronics, communications, instrumentation, and nuclear engineering. Especially in some complex, harsh, or extreme working environments, the parts made of glass materials show excellent corrosion resistance and high temperature stability. Although glass possesses many excellent characteristics, these materials are typically hard and brittle, and are therefore difficult to machine by conventional approaches, which affects the machining quality and limits its further application [1].

Traditional methods that have been reported for glass machining, including diamond grooving, mechanical grinding, micro-abrasive air jet [2] and water jet [3], in which material removed because of brittle crack formation caused by the cutting tool squeeze or jet impingement. Defects such as cracks and chipping easily produced during processing, leading to poor processing accuracy and surface quality.

For higher machining efficiency, better processing accuracy and surface morphology, nontraditional technologies have proposed for glass in recent years [4], and the typical methods are laser machining [5–7] and electrochemical discharge machining (ECDM) [8–12], both of which are noncontact methods and removing materials in a stress-less manner. Therefore, noncontact methods can avoid the stress damage, and reduce the chipping and cracks [13] significantly improving the processing accuracy and surface quality.

Specifically, the two nontraditional methods have their own characteristics. For laser machining, the light beam focuses on a small area and the pulse energy emits within a short or ultrashort pulse duration, and therefore achieves extremely high energy intensity up to GW/cm^2 even PW/cm^2, leading to unique nonlinear light, heat, and force effects [14,15]. In the process of ultrashort laser processing, such as picosecond laser, the laser energy first absorbed by electrons in material, and the amount determined by the absorption rate of the material itself. The electrons in the material absorb photons and then migrate, and, then, the electrons excited to the conduction band further improve the energy absorption of the workpiece. When the electron density of the conduction band exceeds the critical value, a Coulomb explosion may occur, and helps increase material removal.

The ECDM shown in Figure 1, where the tool electrode and auxiliary electrode immersed in the solution. After the DC power turned on, a current loop formed. As the reaction continues, the bubbles would fuse into a gas film, insulating the tool electrode from the working fluid. When the potential difference between the tool electrode and the working fluid increased up to the discharge voltage that could cause breakdown in the gas film, a spark discharge occurred, releasing a huge amount of energy. The emitted discharge energy led to high temperature and high pressure within the tiny distance between the tool electrode and material, and therefore the material may melt or even vaporize, and is then thrown away by high-pressure impact. Meanwhile, the rise of temperature also promoted the chemical corrosion of the alkaline solution on the workpiece. The two work together to increase material removal [16–18].

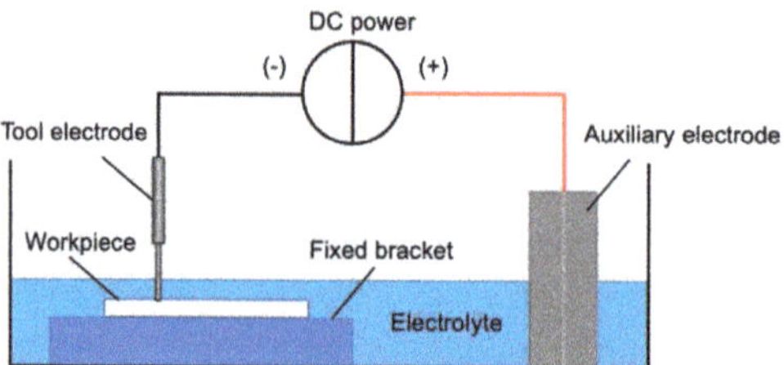

Figure 1. Schematic of the electrochemical discharge machining (ECDM).

In the existing literature, scholars from various countries have carried out research on the micro-machining technologies of insulating and hard-brittle materials, and a wide range of hybrid processing methods have been proposed.

A hybrid method of ECDM and micro grinding using polycrystalline diamond (PCD) tools explored by Xuan et al. [19], which had experimentally shown the grinding under PCD tools reduced the surface roughness of ECDM structures. The fundamental material removal mechanisms of hard-brittle materials under indentation and material removal mechanisms in the gap between the tool side and the hole wall have been studied by Nath et al. [20]. Ho et al. [21] explored the feasibility of ECDM with an aiding flow jet on nonconductive quartz glass materials. This research found that an aid nozzle improved the drilling depth by ensuring that an enough flow of electrolyte was available for spark generation and debris removal. Chen et al. [22] added a magnetic field to ECDM. The mechanism of the magnetohydrodynamic effect in ECDM was researched. By using this hybrid method, the ECDM micro-hole drilling performance was significantly improved.

However, much work is still in early exploration and the experimental research stage. The hybrid processing technology for insulating and hard-brittle materials is still based on one of laser or spark discharge, and then introduces rotation, vibration, inflation, or grinding for aid. However, there is still little research on etching glass by combining the two main energies of laser and spark discharge to improve the processing quality. Therefore, there are many theoretical and technical issues that need to be studied for the optoelectronic composite stepwise processing of glass.

This study focused on the step-by-step composite processing of glass using different sequences of laser and ECDM. According to the thermal effect of laser shock wave and the principle of ECDM effect, this study compared and analyzed the morphology characteristics of different machining methods. It also analyzed the influence of laser pre-processing on forming an ECDM gas film and the ablation of materials.

2. Experimental Design

2.1. Experimental Setup

As shown in Figure 2, an ECDM system has developed for this study, including pulse power supply (GKPT series, Shenzhen Shicheng Electronic Technology Co., Ltd. Shenzhen, China), current probe (CP8030B, Shenzhen Zhiyong Electronics Co., Ltd. Shenzhen, China), motion controller, four-axis linkage processing platform, microscope camera, tool electrode, auxiliary electrode, reaction tank, oscilloscope (TDS2012C, American Tek company, Beaverton, OR, USA), etc.

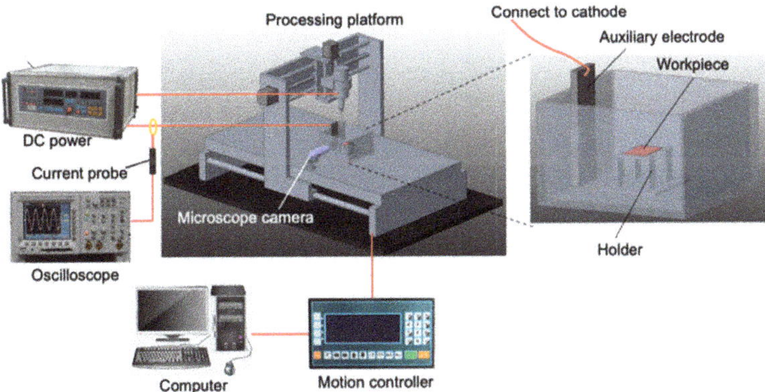

Figure 2. Schematic diagram of the experimental apparatus of ECDM.

The solution tank (size—160 × 120 × 80 mm^3) made of acrylic sheet fixed on the processing platform, which can precisely move along X–Y axes via a computer-controlled motion controller unit. The tool electrode clamped onto a vertical chuck rotated by a motor, and was moved up and down by the main spin. The negative pole of the DC pulse power supply connected to the tool electrode, and the positive pole connected to the auxiliary electrode. The high frequency current probe employed to monitor the current during ECDM, and the oscilloscope connected to it was used to display and store the collected current signal.

Tool electrodes and auxiliary electrodes are important in ECDM [23–25], and the material choice for them should be carefully considered. The tool electrode should not only have good electrical conductivity, but also needs to withstand the high temperature of electric spark discharge and strong alkali corrosion. Therefore, tungsten carbide (WC) was selected to make the tool electrode for this study. In addition, since the spiral thread electrode is more effective in expelling machining waste and renewing the electrolyte near the machining position compared to the smooth electrode, this experiment used a cylindrical tool electrode with a thread on the sidewall. The auxiliary electrode was made of block graphite with strong alkali resistance and good conductivity (size—80 × 20 × 15 mm^3).

Figure 3 shows the schematic of picosecond laser machining system, which was mainly composed of a picosecond laser, computer control system, optical path system, four-axes precision motion system, optical measurement system (charge coupled device camera, (CCD camera)), and other auxiliary equipment. A Nd: YVO4 laser (Edgewave PX100-1-GM) was employed in this study, which was operating in Gaussian mode at 1064 nm wavelength with a 12 ps pulse duration and the beam quality is M2 < 1.3. The maximum output power was about 70 W, while the pulse repetition frequency can change from 0.2 to 1 MHz, resulting in a maximum pulse energy of up to 260 µJ at the frequency of 0.2 MHz. The output power was adjusted by changing the high voltage (HV) level working on the HV modulator, which is superior to controlling the pump current as the laser spot size not affected by the varied HV level in this laser machine. In addition, higher HV level results in more energy emission and therefore the increased laser intensity. The laser beam emitting from the generator first expanded by a beam expander (Eoptics VE-532–1064, JENOPTIK, Jena, Germany), then directed to a galvanometer (IntelliSCAN 14–1064, SCANLAB, Munchen, Germany) with a focal length of 100 mm and a typical marking speed of 2 m/s, and finally arrived at the target specimen surface. A mechanical beam blocker also included in the light path for a protection purpose. To keep the laser system in a stable operating stage, deionized and filtered water was looping inside the laser machine to remove the generated heat and hence maintain a constant temperature. By using this optical setup, a focusing laser spot of ∼20 μm in diameter achieved in the focal plane.

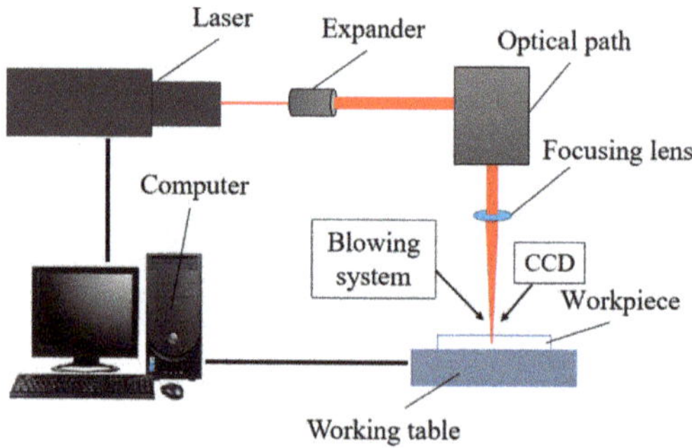

Figure 3. Schematic of picosecond laser machining system.

2.2. Machining Procedures

In this experimental study, the laser machining and ECDM was first carried out separately. In the ECDM experiment, the workpiece was a square piece of quartz with a size of $20 \times 20 \times 1$ mm^3, which was fixed to the bottom of the solution tank by a bracket. The electrolyte was a 30 wt.% NaOH solution, and the electrolyte level was adjusted to 2 mm above the workpiece. The ECDM parameters selected in this experiment are shown in Table 1. In the experiment, the layer was fed 3 times, and each feeding amount was 0.2 mm.

In this ultrashort pulsed laser micromachining process, the laser used to heat the material may cause melting, vaporization, plasma formation, or direct phase explosion as in the femtosecond laser. The workpiece with a size of $20 \times 20 \times 1$ mm^3 was fixed to the worktable of the laser system, and the CCD camera that comes with the laser, which is used for accurate focusing, such that the laser focused on the upper surface of the workpiece to process microgrooves. In the experiment, the scanning path designed by the computer in the laser processing system was a parallel line, the scanning distance between two adjacent parallel lines was 10 μm, the number of scanning lines was 20, and the scanning length was 2 mm. After each time the laser scanned a layer, the laser focus fed down by 20 μm to scan again, for 10 feeds. Because the laser focus diameter was 20 μm, the scanning distance between two adjacent parallel lines was 10 μm, and the scanning lines were 20, microgrooves with a width of around 200 μm were processed by this method. The laser machining parameters selected in this experiment are shown in Table 2. The scanning velocity and repetition times can be set by controlling the galvanometer. After the laser machining was completed, the workpiece was immersed in an alcohol solution for ultrasonic cleaning for 10 min and dried to remove the attached processing products.

Table 1. Parameters for ECDM.

Factors	Parameters
Tool electrode diameter (μm)	200
Voltage (V)	25
Peak current (A)	0.3
Duty factors (%)	50
Frequency (Hz)	1000
Rotating speed (RPM)	3600
Feed speed (μm/s)	10

Table 2. Parameters for laser machining.

Factors	Parameters
Layer	10
Number of elements	20
Scanning speed (mm/s)	300
Power (W)	12
Frequency (Hz)	3×10^5

Figure 4 was taken by a microscopic camera, showing the cross-sectional morphology of the microgrooves, respectively processed by laser and ECDM. The results show that although the edge profile of microgrooves processed by laser is relatively flat and neat, the cross-sectional taper is relatively large, showing a clear V-shape. On the contrary, the microgrooves processed by ECDM have almost no taper, but the edges have obvious wave-shaped heat-affected zones.

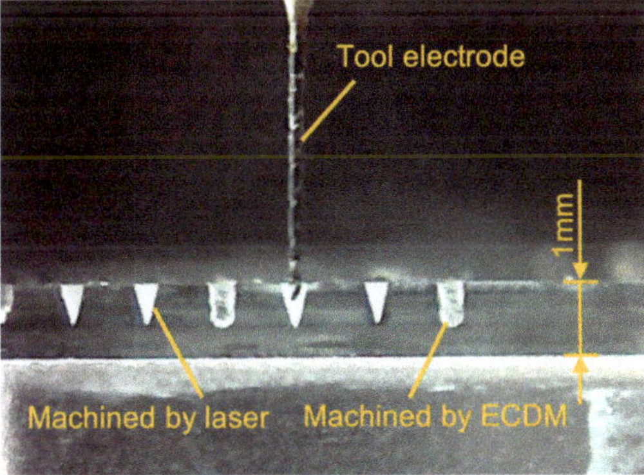

Figure 4. Cross-sectional morphology of the microgrooves respectively processed by laser and ECDM.

In order to obtain microgrooves with a small taper and good surface quality, the laser machining and ECDM carried out in a certain sequence on the same position to form a hybrid machining method, which combined the advantages of the two methods.

Figure 5 illustrates the processing sequence. Figure 5a shows ECDM pre-groove processing. The first step is to process the microgrooves on the workpiece by ECDM. Then the workpiece immersed in alcohol for ultrasonic cleaning for 10 min and dried to remove the attached processing products. The second step is to fix the workpiece to the laser processing platform. Then, use the CCD camera to observe the position of the microgroove, and use the software in the computer to focus on the center of a microgroove. In this way, the microgrooves processed by laser can completely overlap with the microgrooves processed by ECDM in the first step. The parameters used in these two steps were the same as described above. Except that the number of layer feeds in this ECDM experiment was one and the rotating speed is 0 RPM. Figure 5b shows laser pre-groove processing.

The order of this processing method was reverse to that in Figure 5a. In the first step, a laser used to process microgrooves on the workpiece. Then the workpiece is immersed in alcohol for ultrasonic cleaning for 10 min and dried to remove the attached processing products. The second step is to fix it on the ECDM platform. Then observe the position of the microgroove with a microscopic camera, align the tool electrode at the center of a microgroove, and perform the ECDM. The parameters used in these two steps were the same as described above. Except that the rotating speed is 0 RPM.

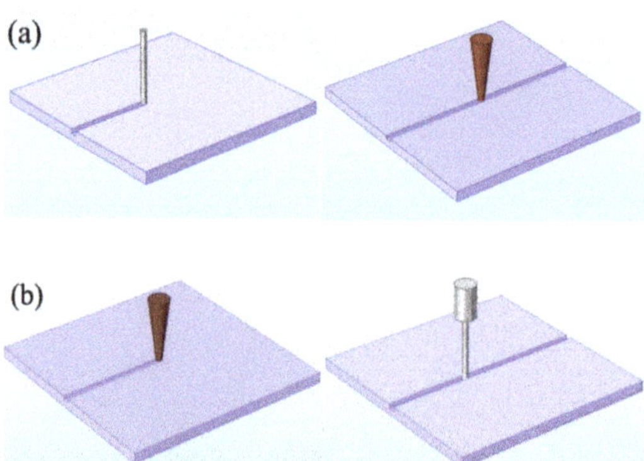

Figure 5. Processing diagram: (**a**) laser pre-groove processing and (**b**) ECDM pre-groove processing.

After processing, the glass was cleaned and dried. The scanning electron microscope (SEM, Hitachi S-3400N) was used to observe the surface morphology of the processed microgrooves. Because the glass itself is an insulator, it is necessary to spray a metal film or carbon film on the material to form a conductive layer before observation.

3. Results and Discussion

3.1. ECDM Preprocess

Figure 6 is the SEM image of a microgroove processed by ECDM. Taking the dotted line as the boundary in the figure, the left side shows the morphology of a microgroove produced by ECDM only, while the right side matches the result of post-processed by laser machining after ECDM. The ECDM parameters of Figure 6a were 1000 Hz and 0 RPM. Protrusions appear regularly at the bottom of the microgroove, and these protrusions are cylindrical with steep sidewalls. Further, the distance between the peaks of two adjacent protrusions is about 70 μm, while the bottoms not connected. Figure 6b shows the microgrooves processed with different ECDM parameters (800 Hz and 0 RPM). Compared with Figure 6a, the regularity of the protrusions' distribution decreases. The protrusions are nearly cone-shaped with a gentle sidewall. The distance between the tops of the adjacent two protrusions is about 30–80 μm, and the bottoms are connected.

The formation of this protrusion structure may be explained by two aspects. On the one hand, the discharge is more difficult to generate at the center of the tool electrode that is flat-bottomed. Specifically, Jiang et al. [26,27] conducted a finite element analysis on the current density of the tool electrode with this shape, as shown in Figure 7. Compared to the center of the cylindrical tool electrode, the current density near the edge of the contour is stronger. Therefore, if geometric defects are not considered, sparks tend to generate along the edge of the tool electrode, and it is more difficult to generate near the geometric center of the bottom, so the material in this part is more difficult to remove. Figure 8 is the picture of a crater created by the ECDM process using a cylindrical tool of 0.5-mm diameter with the discharging duration of 2 s, which illustrates that material near the rim of the cylindrical tool was removed, indicating release of sparks distributed around the rim, or the "fringing effect". On the other hand, during the ECDM, bubbles will produce at the bottom of the tool electrode. Because the solution has a high viscosity and the bubbles mainly stressed in the vertical direction, the bubbles will accumulate at the bottom of the electrode and are difficult to discharge, which hinders the electrolyte into the gap between the tool electrode and the workpiece, making this part difficult to produce electrochemical discharge [28], and the material is difficult to remove. As a result, the morphology with protrusions at the bottom is created.

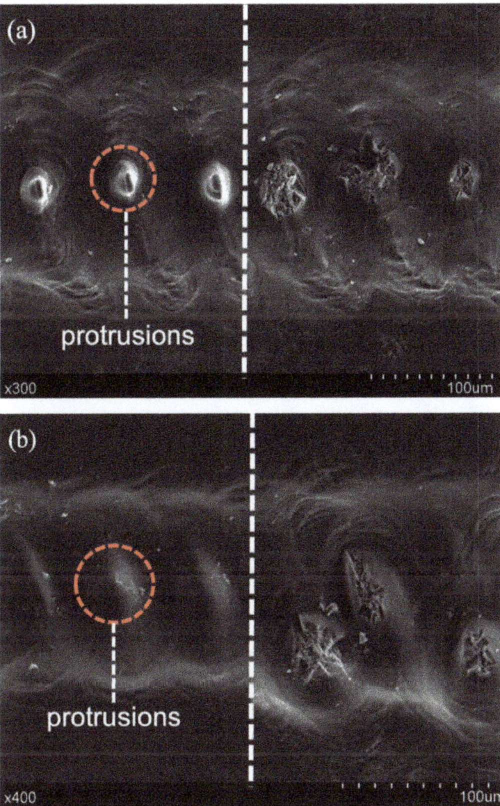

Figure 6. Morphology comparison of micro-grooves machined by ECDM only (left part) and hybrid method with laser processing. (**a**) 1000 Hz and (**b**) 800 Hz.

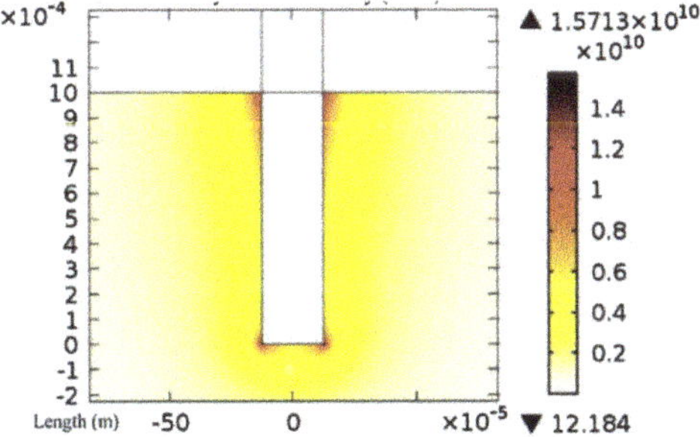

Figure 7. Finite element simulation of current density in electrochemical reaction on cylindrical electrode [26].

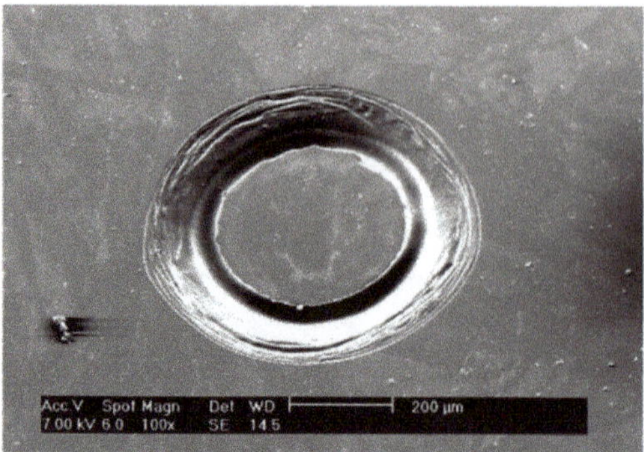

Figure 8. Crater created by a cylindrical tool electrode. Machined with 35-V electrode voltage and 2-s machining time [26].

It can be clearly seen from the topography on the right side of the dotted line in Figure 6 that the protrusions in the microgrooves were removed after laser machining, resulting in a smoother surface. Because ECDM preprocessing can promote the nonlinear absorption of the high-energy pulsed laser in the processing area and induce plasma generation. It can selectively remove the protrusion structures of different shapes formed in the ECDM. Besides, when the picosecond pulsed laser acts on a transparent medium, the high peak power can produce a strong nonlinear effect inside the transparent medium, and reduce the heat-affected zone effectively, which greatly improves the morphology quality.

But the drawbacks of this processing method are obvious. Although laser scanning can effectively remove the protrusion structures, the removed debris still existed at the bottom, and the surface roughness was not good. To further improve the morphology quality and composite effect, and to explore the influence of laser preprocessing on the gas film formation and material ablation in ECDM, the sequence of laser machining and ECDM was adjusted to form a laser-assisted ECDM microgroove machining method, as given in the details of the next section.

3.2. Laser Preprocess

In the laser preprocess experiments, the microgroove was processed by laser first, and then post-processed by ECDM. Specifically, the tool electrode with a suitable diameter was selected according to the width of the microgroove processed by laser pre-machining. In this experiment, the width W of the micro-groove was 200–240 μm, and the gas film thickness h was about 20 μm. The tool electrode diameter d_t is:

$$d_t = W - 2h \tag{1}$$

therefore, a tool electrode with a diameter of 200 μm was selected.

During the machining process, the relative position of the tool electrode tip and the pre-groove was monitored by connecting the micro camera to the computer to make the traces of ECDM and laser machining overlap accurately and completely. Figure 9 illustrates the change in the relative position of the tool electrode in the laser preprocessed groove during ECDM. First, the tool electrode descended to position A, where the tip of the tool electrode was slightly lower than the microgroove inlet level, but it was not in contact with the sidewall. ECDM was performed to remove burrs without forming a heat-affected zone on the edge of the microgroove. As the tool electrode continued to dive, it arrived at position B. At this time, there should be contact between the tool electrode and the sidewall of the microgroove. However, due to the discharge, this part formed a melting zone, which is further melted and eroded under the action of the electrolyte. As the tool electrode continued to dive, it finally reached position C, the bottom of the microgroove. As the

ECDM proceeded, the cross-section is changing gradually from V-shaped to U-shaped, and the taper decreased significantly.

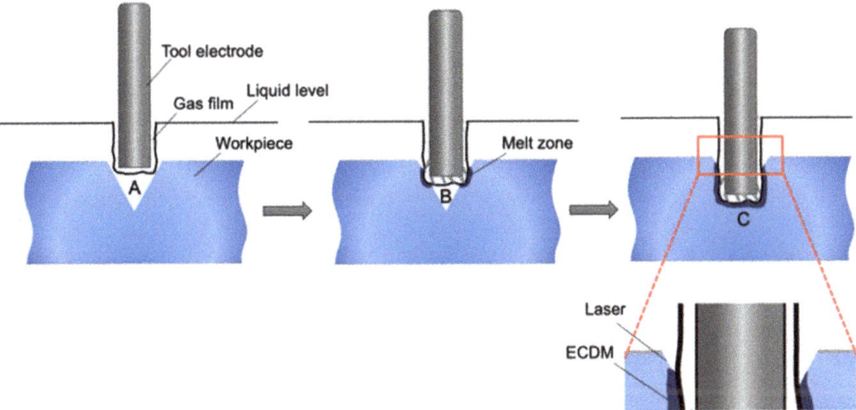

Figure 9. Schematic diagram of tool electrode position.

Figure 10 is a comparison diagram of the microgrooves with a width of 200 μm processed by combination method and ECDM only. Taking the dotted line in the figure as the boundary, the left side of the dotted line is combination, and the right side represents ECDM only.

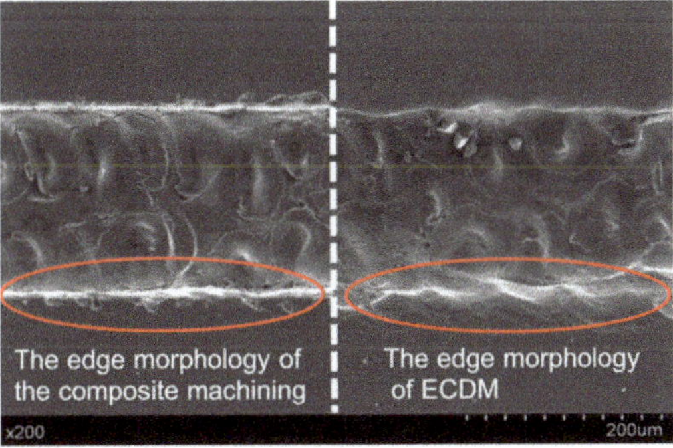

Figure 10. Comparison of microgrooves processed by combination method and ECDM only.

It can be clearly seen that the edges of the microgroove processed by ECDM only are tortuous. In addition, the processing quality is relatively poor, and there is a wave shaped heat-affected zone on the upper surface of the microgroove edge. This is because the molten material overflowed under the action of gas film agitation and thermal impact during the processing, and then cooled and solidified and accumulated on the edge of the microgroove, eventually forming a wave shape slightly higher than the surface of the material.

The edge of the microgroove processed by the combination method is flat, and the sidewall profile is more regular and smoother, and there is no visible wave-shaped heat-affected zone on the upper surface of the edge, showing the morphology of laser processing only.

Gas film is an important parameter which formed around the tool electrode and leads to discharge activities. The size and stability of the gas film significantly affected the material removal and accuracy of the microgrooves. A more stable gas film prepares a suitable condition for uniform discharges. The edge of the microgrooves presents the morphology of laser machining instead of ECDM, showing the microgrooves formed by laser preprocessing can reduce the thickness of the gas film, shorten the distance of electrical discharge erosion, and achieve better quality and less overcutting. This was because of the lack of material produced by laser processing increased the thickness of the liquid layer between the edge of the microgroove and the discharge area of the tool electrode, which increased the stability of the gas film. At the same time, the processed microgrooves squeezed the gas film on the tip of the tool electrode, which had a restrictive and standard effect on the ECDM energy. The dual action improved the localization of ECDM, reduced the amount of overcutting, and greatly improved the machining accuracy. Therefore, the edge of the laser preprocessed microgroove was not affected by the discharge, and still presents a neat edge of the preprocessed laser.

While the bottom surface presents the morphological characteristics of ECDM instead of laser machining. Since the microgrooves processed by the laser show a certain taper, the width of the microgrooves become narrower as the depth of the microgrooves increases. When the distance between the material and the tool electrode is less than the critical distance of electric discharge erosion (Figure 9, positions B and C), the material at that place will be eroded by ECDM. And since the final position of the tool electrode in the bottom of the groove reaches the effective processing range, the depth of ECDM is deeper than laser, so the bottom surface presents the morphological characteristics of ECDM instead of laser machining. As shown in the enlarged detail of Figure 9, laser-assisted ECDM can keep the laser processing morphology on the upper edge of the microgroove, and ECDM only affects the middle and lower regions. Therefore, the experiment can obtain high-quality microgrooves, which can have flat edges and small tapers. As shown in Figure 11, the width of the microgrooves of the combined machining is generally smaller than that of the ECDM. This study used different voltage parameters to verify that the combined method can improve the overcut phenomenon caused by ECDM, thereby improving the processing quality.

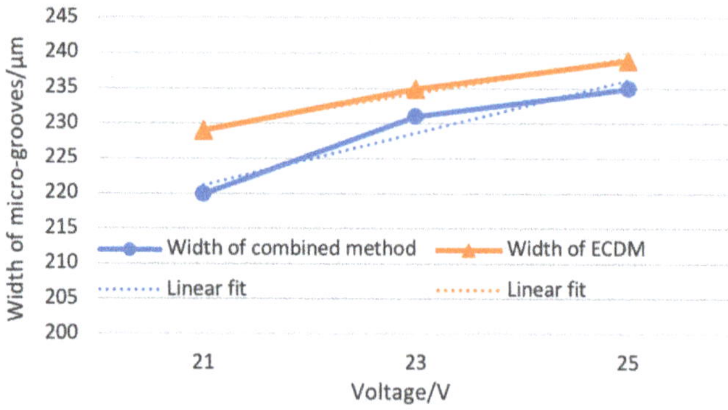

Figure 11. Comparison of width processed by combination method and ECDM only.

In summary, the microgroove first processed by laser, and then a tool electrode with an appropriate diameter was selected for ECDM according to the width of the groove. The combination of the two methods can effectively improve the edge quality of the microgrooves to obtain a more straight and neat edge morphology with less overcutting. At the same time, the taper of the microgroove can be effectively reduced, and the cross section is closer to a U-shape.

4. Conclusions

This study explored ECDM and laser processing of quartz, and compared the morphological differences between single processing methods and combined processing methods. The experimental results are summarized:

1. As a post-processing after ECDM, laser processing can effectively and selectively remove the protrusions at the bottom of the microgrooves. Because of the "cold processing" feature of the picosecond laser, there is no obvious heat-affected zone. Thus, the experiment obtained a more regular bottom structure of the microgrooves.
2. In the laser preprocessing, the pre-etched microgrooves can reduce the thickness of the gas film, shorten the distance of the electrical discharge erosion, and achieve better quality. The results show the bottom of the microgroove presented an ECDM morphology, while the upper surface of the edge presented a laser morphology and there is no visible wave-shaped heat-affected zone. The combined processing method can improve the overcut phenomenon caused by ECDM.
3. After the post-processing by ECDM, the V-shaped microgrooves gradually transform to the U-shape, and the taper of the microgrooves is reduced.

Author Contributions: Conceptualization, D.Z.; Investigation, Z.C.; Methodology, D.Z.; Resources, K.X.; Supervision, H.Z. and Z.Z.; Writing—original draft, D.Z.; Writing—review and editing, D.Z. and Z.Z. All authors have read and agreed to the published version of the manuscript.

Funding: This research was funded by National Natural Science Foundation of China, grant number 51675242, 51905226; Natural Science Research of Jiangsu Higher Education Institutions of China, grant number 18KJB460006, 18KJB460005; Natural Science Foundation of Jiangsu Province, grant number BK20180875.

Institutional Review Board Statement: Not applicable.

Informed Consent Statement: Not applicable.

Data Availability Statement: Data available on request due to restrictions e.g., privacy or ethical. The data presented in this study are available on request from the corresponding author. The data are not publicly available due to it was not agreed by all co-authors.

Conflicts of Interest: The authors declare no conflict of interest.

References

1. Lucas, H.; Jana, A.Z. Micro-Hole Drilling on Glass Substrates—A Review. *Micromachines* **2017**, *8*, 53. [CrossRef]
2. Sabahi, N.; Hajian, M.; Razfar, M.R. Experimental study on the heat-affected zone of glass substrate machined by electrochemical discharge machining (ECDM) process. *Int. J. Adv. Manuf. Technol.* **2018**, *97*, 1–8. [CrossRef]
3. Schwartzentruber, J.; Papini, M. Abrasive waterjet micro-piercing of borosilicate glass. *J. Mater. Proc. Technol.* **2015**, *219*, 143–154. [CrossRef]
4. Zhu, H.; Wang, J.; Yao, P.; Huang, C. Heat transfer and material ablation in hybrid laser-waterjet microgrooving of single crystalline germanium. *Int. J. Mach. Tools Manuf.* **2017**, *116*, 25–39. [CrossRef]
5. Xie, X.; Zhou, C.; Wei, X.; Hu, W.; Ren, Q. Laser machining of transparent brittle materials: From machining strategies to applications. *Opto Electron. Adv.* **2019**, *2*, 11–23. [CrossRef]
6. Zhu, H.; Zhang, Z.Y.; Xu, J.L.; Xu, K.; Ren, Y. An experimental study of micro-machining of hydroxyapatite using an ultrashort picosecond laser. *Precis. Eng.* **2018**, *54*, 154–162. [CrossRef]
7. Chichkov, B.N.; Momma, C.; Nolte, S.; Von Alvensleben, F.; Tunnermann, A. Femtosecond, picosecond and nanosecond laser ablation of solids. *Appl. Phys. A* **1996**, *63*, 109–115. [CrossRef]
8. Ghosh, A. Electrochemical discharge machining: Principle and possibilities. *Sadhana* **1997**, *22*, 435–447. [CrossRef]
9. Jain, V.K.; Dixit, P.M.; Pandey, P.M. On the analysis of the electrochemical spark machining process. *Int. J. Mach. Tools Manuf.* **1999**, *39*, 165–186. [CrossRef]
10. Bhattacharyya, B.; Doloi, B.N.; Sorkhel, S.K. Experimental investigations into electrochemical discharge machining (ECDM) of non-conductive ceramic materials. *J. Mater. Proc. Technol.* **1999**, *95*, 145–154. [CrossRef]
11. Lee, J.Y.; Lee, S.W.; Lee, S.K.; Park, J.H. Through-glass copper via using the glass reflow and seedless electroplating processes for wafer-level RF MEMS packaging. *J. Micromech. Microeng.* **2013**, *23*, 085012. [CrossRef]
12. Zhang, Z.Y.; Huang, L.; Jiang, Y.; Liu, G.; Nie, X.; Lu, H.; Zhuang, H. A study to explore the properties of electrochemical discharge effect based on pulse power supply. *Int. J. Adv. Manuf. Technol.* **2016**, *85*, 2107–2114. [CrossRef]
13. Ahmmed, K.; Colin, G.; Anne-Marie, K. Fabrication of Micro/Nano Structures on Metals by Femtosecond Laser Micromachining. *Micromachines* **2014**, *5*, 1219–1253. [CrossRef]
14. Kovalenko, A.F.; Vorobev, A.A. Energy-Conserving Regimes for Laser Machining of Glass and Ceramic Materials. *Glass Ceram.* **2014**, *71*, 198–200. [CrossRef]
15. Shin, J. Investigation of the surface morphology in glass scribing with a UV picosecond laser. *Opt. Laser Technol.* **2019**, *111*, 307–314. [CrossRef]

16. Jawalkar, C.S.; Sharma, A.K.; Kumar, P. Investigations on performance of ECDM process using NaOH and NaNO3 electrolytes while micro machining soda lime glass. *Int. J. Manuf. Technol. Manag.* **2014**, *28*, 80–93. [CrossRef]

17. Zheng, Z.P.; Cheng, W.H.; Huang, F.Y.; Yan, B.H. 3D microstructuring of Pyrex glass using the electrochemical discharge machining process. *J. Micromech. Microeng.* **2007**, *17*, 960–966. [CrossRef]

18. Pawar, P.; Ballav, R.; Kumar, A. Revolutionary developments in ECDM process: An overview. *Mater. Today Proc.* **2015**, *2*, 3188–3195. [CrossRef]

19. Xuan, D.C.; Bo, H.K.; Chong, N.C. Hybrid micromachining of glass using ECDM and micro grinding. *Int. J. Prec. Eng. Manuf.* **2013**, *14*, 5–10. [CrossRef]

20. Nath, C.; Lim, G.C.; Zheng, H.Y. Influence of the material removal mechanisms on hole integrity in ultrasonic machining of structural ceramics. *Ultrasonics* **2012**, *52*, 605–613. [CrossRef]

21. Ho, C.C.; Wu, D.S.; Chen, J.C. Flow-jet-assisted electrochemical discharge machining for quartz glass based on machine vision. *Measurement* **2018**, *128*, 71–83. [CrossRef]

22. Xu, Y.; Chen, J.; Jiang, B.; Liu, Y.; Ni, J. Experimental investigation of magnetohydrodynamic effect in electrochemical discharge machining. *Int. J. Mech. Sci.* **2018**, *142*, 86–96. [CrossRef]

23. Jain, V.K.; Adhikary, S. On the mechanism of material removal in electrochemical spark machining of quartz under different polarity conditions. *J. Mater. Proc. Tech.* **2008**, *200*, 460–470. [CrossRef]

24. Gaurav, S.; Alakesh, M. Amanpreet Singh Sethi, Investigations on performance of ECDM process using different tool electrode while machining e-glass fibre reinforced polymer composite. *Mater. Today Proc.* **2020**, *28*, 1622–1628. [CrossRef]

25. Kim, D.J.; Ahn, Y.; Lee, S.H.; Kim, Y. Voltage pulse frequency and duty ratio effects in an electrochemical discharge microdrilling process of Pyrex glass. *Int. J. Mach. Tools Manuf.* **2006**, *46*, 1064–1067. [CrossRef]

26. Jiang, B.; Lan, S.; Ni, J.; Zhang, Z. Experimental investigation of spark generation in electrochemical discharge machining of non-conducting materials. *J. Mater. Proc. Tech.* **2014**, *214*, 892–898. [CrossRef]

27. Jiang, B.; Lan, S.; Wilt, K. Modeling and experimental investigation of gas film in micro electrochemical discharge machining process. *Int. J. Mach. Tools Manuf.* **2015**, *90*, 8–15. [CrossRef]

28. Yang, C.K.; Wu, K.L.; Hung, J.C.; Lee, S.; Lin, J.; Yan, B. Enhancement of ECDM efficiency and accuracy by spherical tool electrode. *Int. J. Mach. Tools Manuf.* **2011**, *51*, 528–535. [CrossRef]

Article

Processing Characteristics of Micro Electrical Discharge Machining for Surface Modification of TiNi Shape Memory Alloys Using a TiC Powder Dielectric

Ziliang Zhu [1], Dengji Guo [1,2,*], Jiao Xu [1,*], Jianjun Lin [1], Jianguo Lei [2], Bin Xu [2], Xiaoyu Wu [2] and Xujin Wang [1]

[1] Institute of Semiconductor Manufacturing Research, Shenzhen University, Nan-hai Ave 3688, Shenzhen 518060, China; ziliang_zhu@163.com (Z.Z.); im.jianjun@szu.edu.cn (J.L.); wanghoshi@szu.edu.cn (X.W.)

[2] Guangdong Provincial Key Laboratory of Micro/Nano Optomechatronics Engineering, College of Mechatronics and Control Engineering, Shenzhen University, Nan-hai Ave 3688, Shenzhen 518060, China; ljg_sc111@163.com (J.L.); binxu@szu.edu.cn (B.X.); wuxy@szu.edu.cn (X.W.)

* Correspondence: guodj@szu.edu.cn (D.G.); xujiao@szu.edu.cn (J.X.)

Received: 29 October 2020; Accepted: 17 November 2020; Published: 20 November 2020

Abstract: Titanium-nickel shape memory alloy (SMA) has good biomedical application value as an implant. Alloy corrosion will promote the release of toxic nickel ions and cause allergies and poisoning of cells and tissues. With this background, surface modification of TiNi SMAs using TiC-powder-assisted micro-electrical discharge machining (EDM) was proposed. This aims to explore the effect of the electrical discharge machining (EDM) parameters and TiC powder concentration on the machining properties and surface characteristics of the TiNi SMA. It was found that the material removal rate (MRR), surface roughness, and thickness of the recast layer increased with an increase in the discharge energy. TiC powder's addition had a positive effect on increasing the electro-discharge frequency and MRR, reducing the surface roughness, and the maximum MRR and the minimum surface roughness occurred at a mixed powder concentration of 5 g/L. Moreover, the recast layer had good adhesion and high hardness due to metallurgical bonding. XRD analysis found that the machined surface contains CuO_2, TiO_2, and TiC phases, contributing to an increase in the surface microhardness from 258.5 to 438.7 HV, which could be beneficial for wear resistance in biomedical orthodontic applications.

Keywords: micro-EDM; TiNi shape memory alloy; TiC powder; surface modification; microhardness

1. Introduction

TiNi SMAs have broad application prospects in the aerospace, biomedical, and automobile fields due to their excellent biocompatibility, superelasticity, shape memory effect, and wear resistance [1]. Because the Young's modulus of titanium-nickel alloys is lower than that of other biomedical implant materials, they are widely used for medical implants [2]. In clinical medical applications, product safety and reliability are the primary requirements for long-term implants. However, amino acids and proteins in bodily fluids will accelerate metal corrosion, promoting the release of toxic nickel ions [3]. The release of metallic ions is detrimental to osseointegration and ultimately causes clinical failure [4]. Therefore, the surface modification of titanium-nickel alloys plays an important role in improving corrosion resistance and surface biocompatibility.

Previous studies have shown that a thin surface layer of titanium oxide (2–20 nm) will naturally form on the surface of TiNi alloys, and this layer can act as a barrier to human body corrosion and chemical reactions to limit the diffusion of nickel ions [5]. However, this film is unstable in the human body's complicated and volatile environment and can easily corrode and fall off the alloy material. Therefore, surface treatment techniques have been developed to treat TiNi alloys. Titanium oxide film has good blood compatibility [6] and is biologically inert [7], and it can effectively prevent the precipitation of nickel ions. Several surface treatment methods have been used commercially, such as anodic oxidation [8], plasma immersion ion implantation (PIII) [9], coating [10], and electrochemical polishing [11]. Qin et al. [12] used a glycerol electrolyte to obtain TiO_2 nanotubes on the surface of the TiNi alloy through anodic oxidation, which effectively improved the biocompatibility of the alloy.

Electrical discharge machining (EDM) is an unconventional machining technology that uses a series of pulse discharges between the tool and workpiece to process the workpiece [13]. It is mainly used for high-precision processing of difficult-to-cut materials. Wyszynski et al. [14] realized the high-precision micro-hole machining of cubic boron nitride and determined the optimal parameters. Wu et al. [15] developed a cut-side micro-tool suitable for the micro-EDM system and successfully realized the deep and high aspect ratio micro-holes machining on tungsten cemented carbide. In order to study the machining mechanism of micro-EDM, Liu et al. [16] analyzed the polarity effect of micro-EDM based on the movement characteristics of electrons and positive ions in the discharge plasma channel. Almacinha et al. [17] established an electro-thermal model for a single discharge of an electric discharge machining process based on the Joule heating effect theory. Roy et al. [18] analyzed the physical phenomenon behind occurrences of unusually high discharging points in reverse micro EDM by establishing a numerical model of ions and electrons' movement in the dielectric during machining. To realize the micro-EDM machining of the three-dimensional structure, Roy et al. [19] used reverse micro EDM to generate different shapes of protruded micro features, such as 3D hemispherical and 3D coni-spherical shapes. To study EDM's surface characteristics, Hsieh et al. [2] showed that the EDM process could successfully machine the ternary TiNiZr SMAs while ensuring its shape recovery ability. The recast layer generated on the machined surface can adhere to the substrate effectively by surface alloying, enhancing wear resistance [5]. Peng et al. [3] have reported that EDM can form a nanoporous biocompatible layer on the surface of Ti-6Al-4V, which is conducive to cell growth and proliferation.

To develop improved surface modification technologies, a new method of TiC-powder-assisted micro-EDM is proposed for the formation of a titanium oxide surface, and experiments were performed on a TiNi SMA. The effects of PMEDM parameters on the machining characteristics of TiNi SMA were investigated experimentally, and then, the surface roughness, surface morphology, and microhardness were characterized by relevant characterization techniques. Finally, the thickness and composition of the recast layer were studied deeply.

2. Principle and Mechanisms

Figure 1 illustrates the principle of the EDM process with TiC powder. By adding TiC-mixed powder particles with a particle size of 2 μm in deionized water, the tool electrode utilizes reciprocating movement to complete the powder-mixed EDM (PMEDM). To study the mechanism of the PMEDM, the following two assumptions were made: (1) the TiC-mixed powder particles are spherical, and (2) the electric field is an electrostatic field, shown as yellow circles and black lines in Figure 2a, respectively.

According to the principle of electronics [20], the conductive particles are polarized into bound charges under an electric field's action. When high voltage is applied between the electrode and workpiece, the TiC particles are polarized under the action of the electric field and become a bound charge. According to the electrostatic field theory [21], no matter how strong the electric field is applied to the conductor, the electrostatic balance of the conductor will make the internal field strength of the conductor zero. In order to achieve electrostatic equilibrium, the inside of the TiC particle will generate an electric field opposite to the uniform electric field, E0, as shown in Figure 2b. Therefore, the superposition of the electric field generated by the bound charge and the external electric field distorts the uniform electric field between the electrode and workpiece, as shown in Figure 2c. When the electric field's direction generated by the bound charge overlaps with the direction of the uniform electric field, the actual electric field intensity will reach the maximum value, i.e., points A and B in Figure 2c, and discharge breakdown takes priority here. The maximum value is given by the following [22]:

$$E_{max} = \left[1 + 2\left(\frac{\varepsilon_2 - \varepsilon_1}{\varepsilon_2 + 2\varepsilon_1}\right)\right]E_0, \tag{1}$$

where ε_1 is the dielectric coefficient of the dielectric fluid, ε_2 is the powder's dielectric coefficient, and E0 is the electric field strength of the uniform electric field.

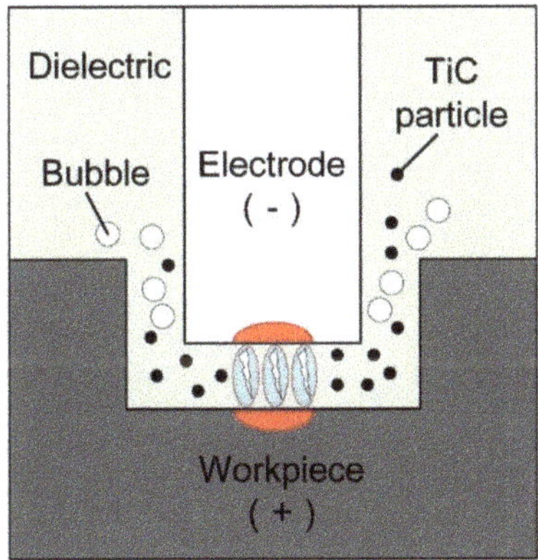

Figure 1. Principle of powder mixed EDM.

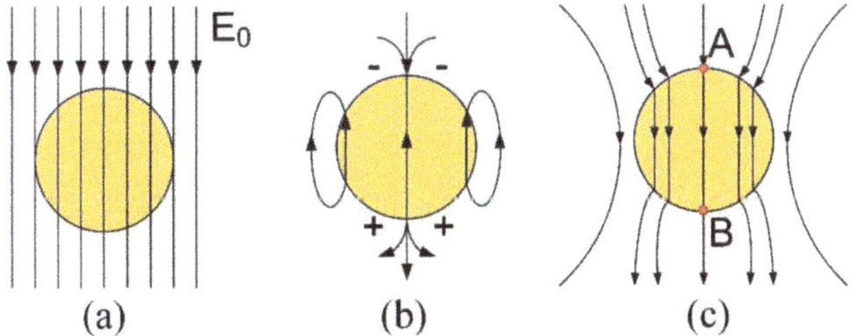

Figure 2. Illustration of electric fields: (**a**) uniform electric field, (**b**) electric field of the bound charge, (**c**) superimposed electric field.

The electrostatic field theory shows that the electric field intensity inside an ideal conductor is zero, and its relative dielectric constant is infinite [21]. Therefore the dielectric coefficient, ε_2, of the conductor tends toward infinity, and takes the limit based on Equation (1) to reach the maximum value, Emax, as follows:

$$E_{max} \approx 3E_0. \tag{2}$$

Therefore, the addition of the mixed powder increases the electric field strength between the electrode and the workpiece by a factor of three and expands the discharge gap by a factor of three, which promotes the removal of processing debris.

3. Experimental Methods

3.1. Experimental Procedure

Figure 3 illustrates a schematic diagram of the PMEDM equipment. During the experiment, the power supply unit included a high-voltage amplifier (Aigtek ATA-2021H, Aigtek, Xi'an, China) and a transistor-type pulse generator (Tektronix AFG3000C, Tektronix, Inc., Beaverton, OR, USA). The pulse waveform is generated by the waveform generator and amplified by a high-voltage amplifier to display the signal on the oscilloscope. The microelectrode movement in micro-EDM was realized by the three-axis micro-nano-motion platform (PI, Germany; M511.DD). A digital oscilloscope (Tektronix MDO 3000) was used to monitor the pulse signal in real-time during the machining process.

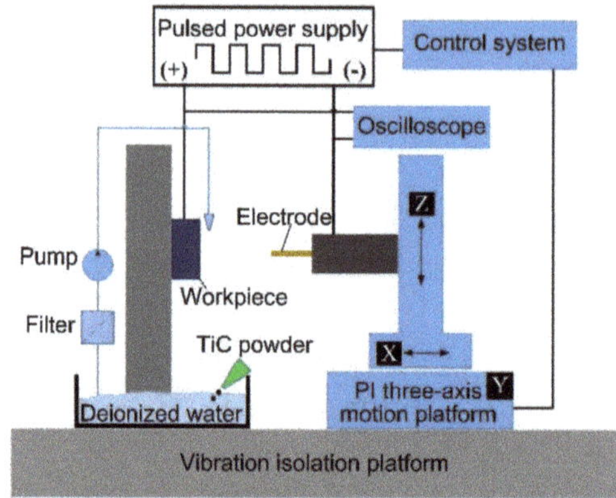

Figure 3. Experimental setup of powder-mixed EDM (PMEDM).

As demonstrated by Lin et al. [23], negative polarity processing can provide a larger MRR, while positive polarity processing can provide a thicker recast layer on the processed surface. Therefore, we chose to employ positive polarity processing in this study. A micropump was used to circulate and mix the dielectric fluid to ensure that the TiC powder was uniformly dispersed. The detailed experimental processing parameters are shown in Table 1.

Table 1. Experimental parameters.

Work Conditions	Description
Workpiece material	TiNi SMA
Electrode material	Brass (C2680)
Polarity	Positive
Dielectric fluid	Deionized water
Additive	TiC (2 μm)
Concentrations (g/L)	0, 3, 5, 7, 10
Duty (%)	50
Pulse durations (ms)	1, 4, 7, 10
Machining voltages (V)	60, 80, 100, 120
Cavity depth (mm)	100

3.2. Experimental Materials and Measurements

In the EDM process, the workpiece material used in the experiments was TiNi SMA (China Tai'zhou Cinoo Mental material Co., Ltd.). The as-received samples were a rectangular parallelepiped with a length of 100 mm, a width of 300 mm, and a thickness of 0.5 mm. Its elemental composition and main thermophysical properties are presented in Tables 2 and 3, respectively. Brass sheets of 1 mm thick were used as raw material for fabricating the microelectrodes. At present, deionized water is widely used as the dielectric fluid due to its weak electrical conductivity, which can also avoid carbon deposition in the spark oil. TiC powder was added to deionized water at different concentrations. Its physical properties are listed in Table 4. The fabrication of microelectrode and microcavity was shown in Figure 4, and the detailed process was as follows. First, according to the microcavity to be machined, a corresponding microelectrode model was designed (Figure 4a). Secondly, import model parameters into the low-speed wire EDM machine (LS-WEDM, Sodick, Japan; AQ250Ls) CNC System (Figure 4b) and start cutting (Figure 4c) to fabricate a single microelectrode with a size of 0.8 mm × 1 mm (Figure 4d). The microelectrode was then employed in the micro-EDM to process a microcavity with a depth of 100 μm (Figure 4e). Finally, a microcavity with a high-quality surface was obtained successfully (Figure 4f).

Table 2. Elemental composition of the workpiece material.

Element	Ni	Ti	Nb	C	O	Other
Wt.%	50.9	48.9	0.025	0.036	0.043	<0.025

Table 3. Physical and mechanical properties of the TiNi shape memory alloy (SMA).

Workpiece Material	TiNi SMA
Density (kg/m^3)	6450
Melting point (°C)	1310
Electrical resistivity (μΩ·m)	820
Modulus of elasticity (MPa)	42.3 × 103
Coefficient of thermal expansion (/°C)	11 × 10−6
Ultimate tensile strength (MPa)	880
Total elongation (%)	16

Table 4. Physical properties of TiC.

Property	Value
Density (kg/m^3)	4930
Melting point (°C)	3140
Thermal conductivity (W/m·K)	21

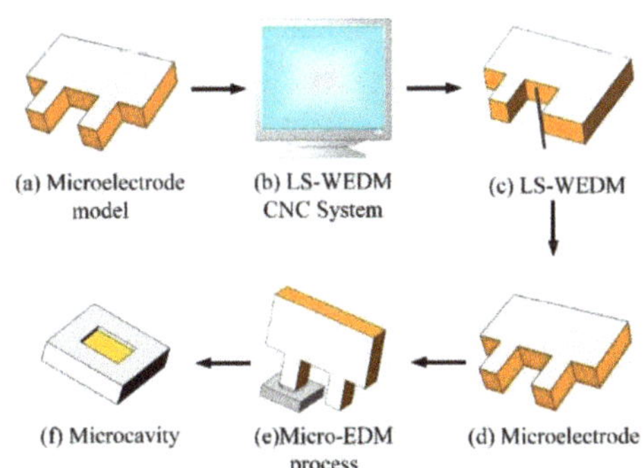

Figure 4. Fabrication process of microelectrode and microcavity.

A scanning electron microscope (SEM) manufactured by TESCAN, Czech Republic (model: LYRA3 XMH) was used to observe the surface morphology. A laser scanning confocal microscope (LSCM, Keyence, Japan; VK-X260K) was used to measure the surface roughness. Analyses of the EDM-treated surfaces were performed at room temperature using X-ray diffraction (XRD, MiniFlex600, Rigaku Corporation, Tokyo, Japan) at a 2θ scanning rate of 3° min −1. A microVickers hardness tester (MHV-1000A, HuaXing, Lai'zhou, China) was used to measure the surface hardness under a load of 100 g for 10 s. The average hardness value was taken from at least four test readings for each specimen.

4. Results and Discussion

4.1. Discharge Waveforms Comparison of EDM and PMEDM

A digital oscilloscope acquired the discharge voltage waveforms to study the effect of the addition of TiC powder on the electro-discharge behavior of the material. Processing parameters, including a pulse width of 4 µs, a duty of 50%, and a machining voltage of 80 V, were determined before micro-EDM processing. The voltage waveforms for the micro-EDM were acquired by an oscilloscope for both without TiC powder addition and with the addition of 5 g/L TiC powder, as shown in Figure 5. Taking the same time (8 µs) for comparison, the number of pulses for the discharge voltage in Figure 5b was significantly higher than in Figure 5a. This result indicates that the addition of TiC powder significantly improved the discharge characteristics of the TiNi SMA in micro-EDM. Moreover, a multiple discharging effect was observed within a single period in Figure 5b, which indicates that the TiC powder refined the discharging energy.

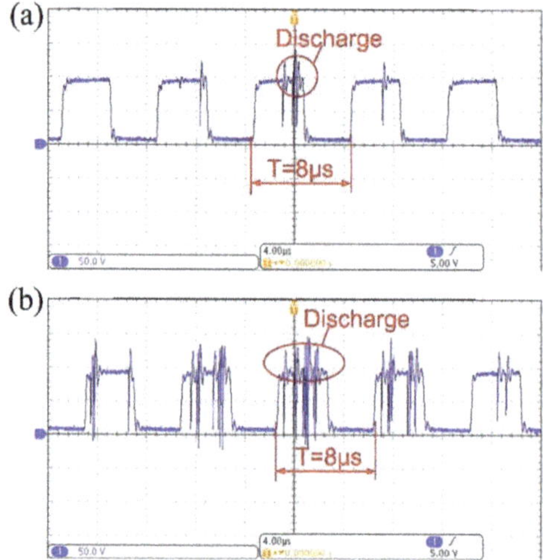

Figure 5. Discharge voltage waveform comparison: (**a**) without and (**b**) with the addition of TiC powder (5 g/L).

4.2. Influence of Machining Process Parameters on the Material Removal Rate

The MRR was calculated as the ratio of the volume of material removed from the workpiece to the processing time (mm 3/min). The volume of material removed was obtained by analyzing the three-dimensional surface topography scanned by LSCM.

Figure 6 shows the effect of the concentration of TiC powder on the MRR under different machining voltages and pulse widths. It is clear that the MRR increases with increasing concentration of TiC powder; regardless of the machining voltages and pulse widths, the maximum MRR is obtained at a concentration of 5 g/L. This result confirms that the discharge frequency is increased, and the discharge energy is improved by the addition of TiC powder to the dielectric fluid. In addition, when the concentration of TiC powder is greater than 5 g/L, the MRR tends to decrease. This trend agrees with that reported by Jahan et al. [24]. When the powder concentration is excessively high, the large number of conductive particles between the two poles cannot be removed easily and cause secondary sparking. Eventually, this leads to instability of the machining process and increases the machining time.

It is noted that the MMR increases with increasing machining voltage, as shown in Figure 6a. A high machining voltage can effectively increase the discharge channel's current density, which facilitates the melting and evaporation of materials. Figure 6b also shows an increase in the MRR with the pulse duration. The pulse duration determines the level of discharge energy, and high pulse durations can provide the necessary time to transmit discharge energy. Hence, a high MRR occurs at higher machining voltages and longer pulse widths in the micro-EDM process.

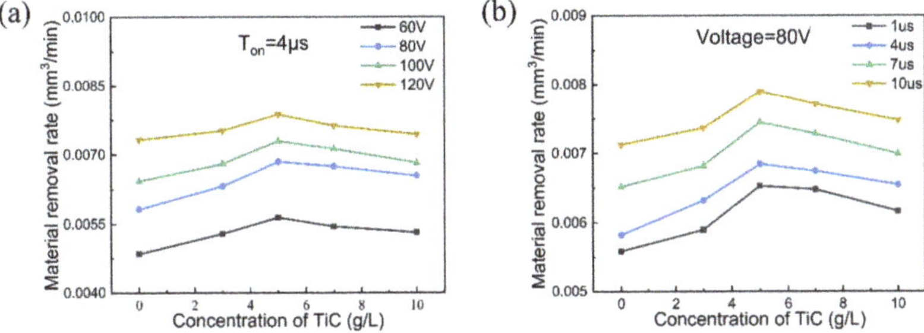

Figure 6. Material removal rates (MRRs) obtained with (**a**) different voltages and (**b**) different pulse durations in the TiC dielectric with concentrations of 0, 3, 5, 7, and 10 g/L.

4.3. Influence of Machining Process Parameters on the Surface Roughness

Figure 7 shows the effect of the concentration of TiC powder on the surface roughness under different machining voltages and pulse durations. In both cases, the surface roughness decreases with increasing TiC powder concentration up to 5 g/L; then, as the TiC powder concentration increases further, the surface roughness tends to increase. Liew et al. [25] recently showed that adding an appropriate amount of conductive powder to the dielectric fluid can uniformly disperse the discharge energy and reduce the craters' size, thereby improving the surface finish. Nevertheless, high-concentration TiC powder tends to accumulate on the workpiece's surface, which severely inhibits the transfer of discharge energy. Moreover, the deposited powder and melting material cannot be removed from the machining gap in time, which causes more frequent secondary sparking and circuiting. This effect will increase surface roughness.

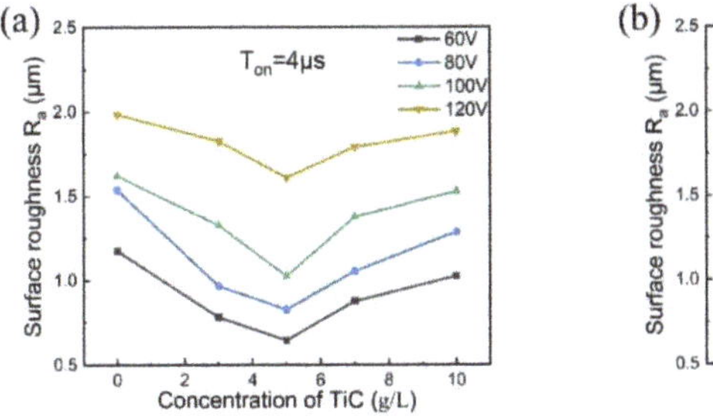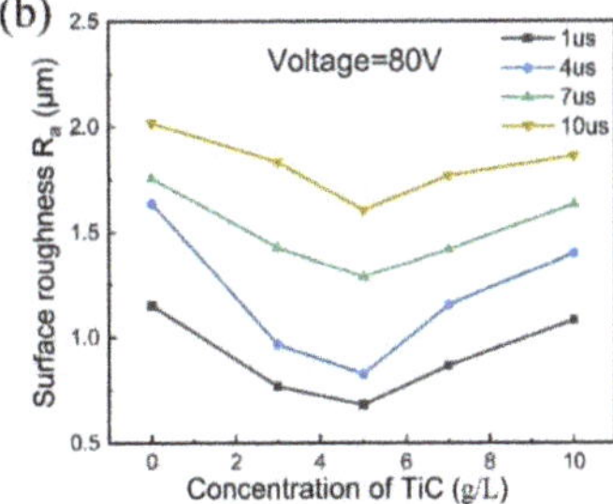

Figure 7. Surface roughness obtained with (**a**) different voltages and (**b**) different pulse durations in the TiC dielectric with TiC concentrations of 0, 3, 5, 7, and 10 g/L.

It is noted that the surface roughness increases with increasing voltage in Figure 7a and pulse width in Figure 7b. Under low machining voltages and short pulse widths, the pits on the machined surface were small and shallow, and the melting material was easily removed. When the machining voltage and pulse width were increased, the discharge time became longer, and the single pulse discharge energy increased. This caused the radius and depth of the discharge marks to increase, leading to an increase in the surface roughness.

4.4. Surface Morphology of the EDM-Treated TiNi SMA

In this study, the surface microtopography and surface roughness (Ra) was used to evaluate surface quality deviations. Five surface roughness measurements (Ra) made at different positions on the bottom of identical microcavity were averaged. Figure 8 shows SEM micrographs and surface roughness of the microcavities bottom variation with the machined voltage (60–20 V) increased under an applied pulse width of 4 µs and a TiC powder concentration of 5 g/L. When the machined voltage was 60 V, the machined surface was relatively smooth due to the low discharge energy, and the surface roughness was 0.645 µm. However, as the machined voltage increased, the surface quality gradually decreased. Until the machined voltage reaches 100 V, numerous discharge craters and melting drops were observed on the machined surface. Nevertheless, if the machined voltage increased to 120 V continuously, the machined surface became rougher, and the surface roughness up to 1.609 µm. This trend was consistent with that reported by Xu et al. [26]; the surface roughness (Ra) increases with greater machined voltage. This was because the electron flow in the channel had an enhanced bombardment effect on the anode under the condition of high voltage, and many melting drops, debris, micropores were observed on the surface.

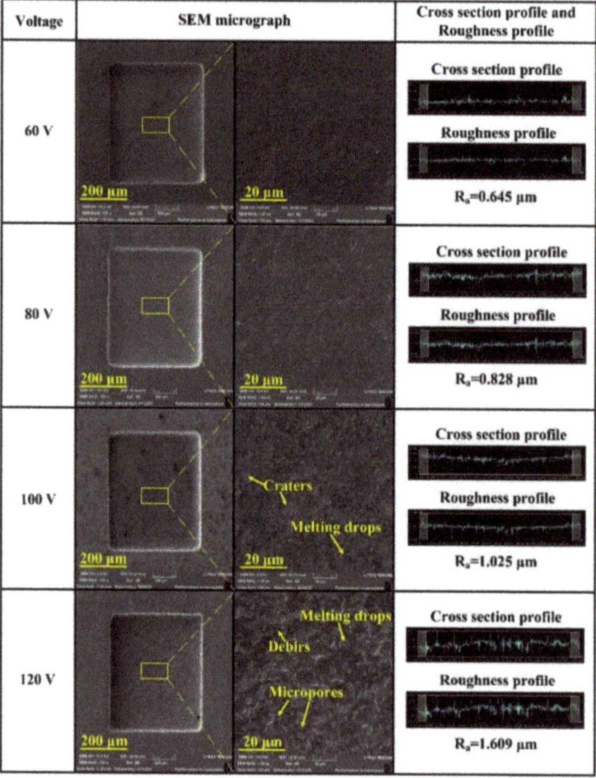

Figure 8. Effect of different machining voltages on surface roughness and scanning electron microscope (SEM) micrographs of the microcavities bottom.

Using a TiC powder concentration of 5 g/L, a machined voltage of 80 V, the influence of different pulse widths on SEM micrographs and bottom surface roughness of the microcavities were discussed in Figure 9. Similar to machined voltage, it can be found that the machined surface of a short pulse width contains shallower and smaller discharge craters compared to long pulse width; because lower pulse width has a smaller MRR, the TiC powder has enough time to refine the discharge energy at the same machining depth, resulting in a smooth bottom surface. As the pulse width increased to 10 µs, the surface roughness increased up to 1.628 µm. The analysis found that with the substantial increase of the pulse width, the current density in the discharge channel continued to increase, and the bombardment effect of charged particles was enhanced, which led to an increase in the radius and depth of the discharge marks. Moreover, the increase of discharge debris particles in long pulse width machining caused short circuits and arcing phenomenon and led to an unstable machining process [27]. Meanwhile, the melting material cools and solidifies on the surface of the workpiece during the deionization stage.

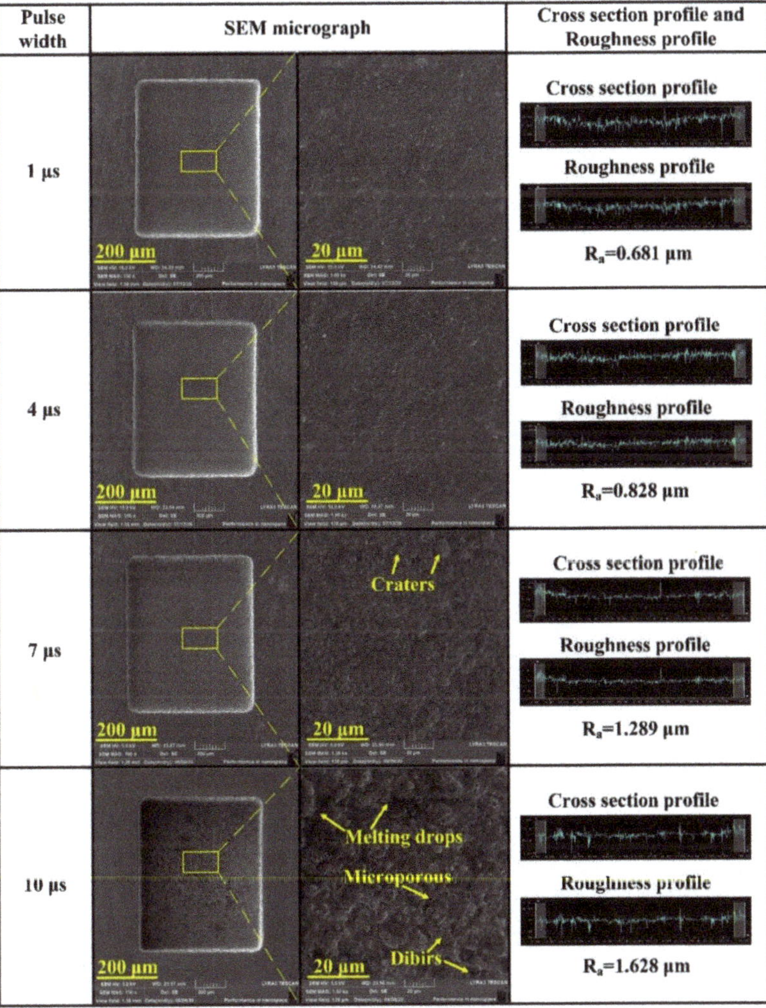

Figure 9. Effect of different pulse widths on surface roughness and SEM micrographs of the microcavities bottom.

The effect of TiC concentrations on SEM micrographs and surface roughness of the microcavities bottom under an applied machining voltage of 80 V and pulse width of 4 µs is given in Figure 10. Results indicate that the addition of TiC powder improved the surface quality, and the surface roughness was always lower than that of the material without the addition of mixed powder. When the TiC concentration was approximately 5 g/L, the minimum surface roughness was Ra 0.828 μm. According to Bui et al. [28], the addition of powder reduces the dielectric resistivity and increases the discharge gap. Conductive particles such as graphite [29], cobalt [30], and molybdenum [31] can form chains across the electrodes and enlarge the gap distance, which not only allows more working fluid to flow through but also lowers the single pulse explosion pressure, resulting in smaller and shallower craters. As mentioned in Section 4.1, TiC powder can disperse the discharge energy and increase the discharge gap. The discharge distribution becomes more uniform. Furthermore, many debris, micropores, and microcracks were observed on the machined surfaces with the pure dielectric fluid. According to the measurement results of surface roughness, when using higher TiC concentrations, such as 10 g/L, the sizes of the discharge craters are increased compared with those obtained with lower TiC concentrations.

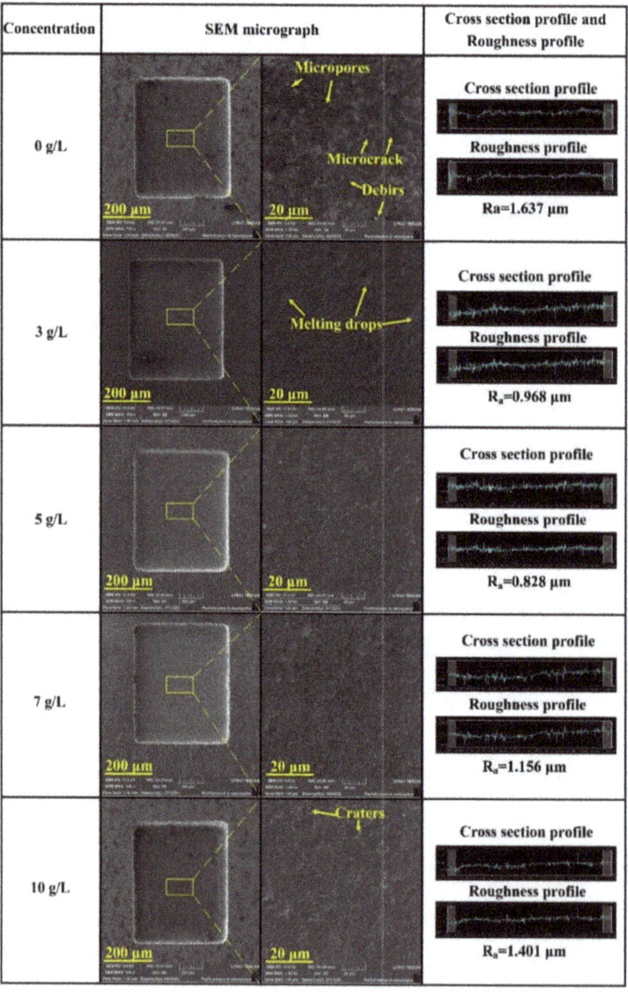

Figure 10. Effect of different TiC concentrations on surface roughness and SEM micrographs of the microcavities bottom.

4.5. Influence of Micro-EDM Parameters on the Recast Layer

The formation of the recast layer is affected by many factors, and the single pulse discharge energy is a key parameter for the formation of the recast layer [32]. Cross-section images of the recast layers in TiC-dielectric under different pulse widths are shown in Figure 11. As shown in Figure 11a–d, the thickness of the recast layer slightly increased with increasing pulse width. The discharge energy of micro-EDM was relatively low, which resulted in minimal changes to the thickness of the recast layer. The thickness of the recast layer measured in this experiment ranged from 1 to 3 µm, which was consistent with the research of Tan and Yeo [33]. The recast layer was composed of materials from the workpiece, electrode, and dielectric fluid. Increasing the pulse width caused more material to be melted and resolidified, thereby increasing the thickness of the recast layer.

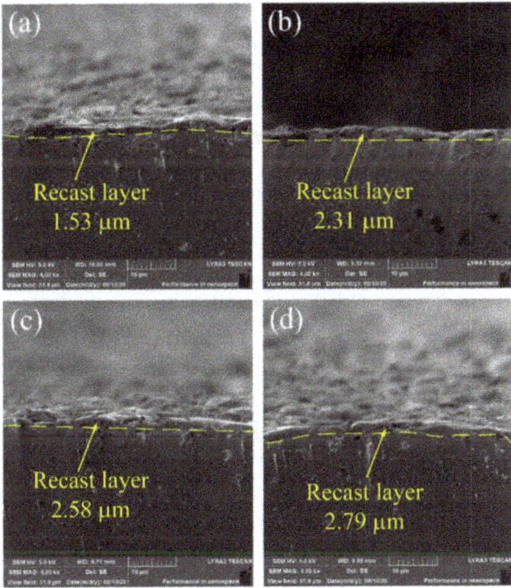

Figure 11. Cross-section images of EDM-treated TiNi SMA under pulse widths of (**a**) 1 µs, (**b**) 4 µs, (**c**) 7 µs, and (**d**) 10 µs.

Figure 12a–e show cross-section images of the recast layers under the TiC concentrations of 0, 3, 5, 7, and 10 g/L, respectively. The thickness of the recast layer decreased with the addition of up to 5 g/L TiC to the dielectric and then increased with further increases in the TiC concentration. The MRR reached a maximum at 5 g/L TiC, and the melting material and deposited particles could be effectively removed. Therefore, a thin recast layer was formed in the 5 g/L TiC dielectric. The high concentration of TiC powder caused frequent secondary discharge and short circuits, which will generate a large amount of heat. This heat accumulation on the surface of TiNi SMA is beneficial for the formation of the recast layer.

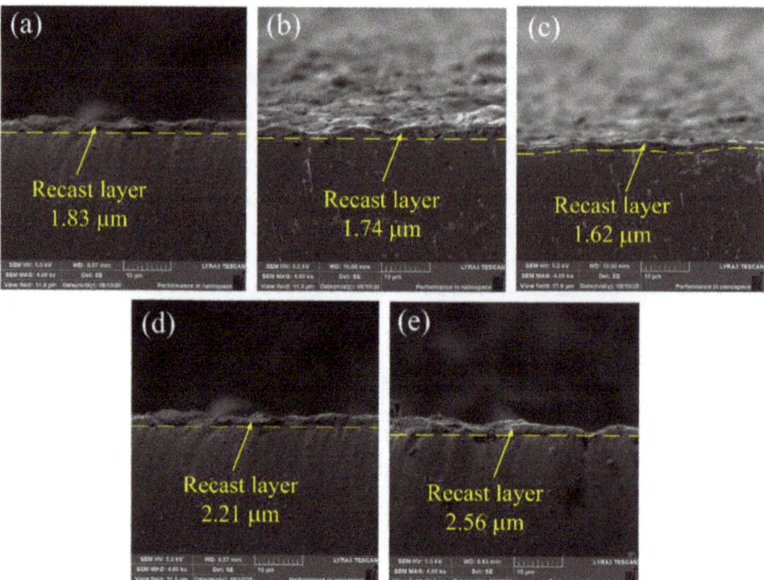

Figure 12. Cross section images of EDM-treated TiNi shape memory alloy (SMA) under the TiC concentrations of (**a**) 0 g/L, (**b**) 3 g/L, (**c**) 5 g/L, (**d**) 7 g/L, and (**e**) 10 g/L.

4.6. Effect of Compound Composition on Microhardness

Figure 13 shows the XRD pattern for the surface of the EDM-treated TiNi SMA with a pulse width of 4 μs, machining voltage of 80 V, and TiC powder concentration of 5 g/L. The results indicated that the EDM-treated surface layer consisted of TiO_2, Cu_2O, TiC, and TiNi phases. The presence of carbon and oxygen was associated with the decomposition of the TiC dielectric and the oxidation of molten metals. Cu electrode and TiC-dielectric medium were melted by EDM and deposited on the machined surface to form Cu_2O and TiC. The most important observation in the XRD analysis was the formation of Ti_2O. Jahan et al. [34] found that TiO_2 has good biocompatibility and can provide a protective coating for biomedical implant applications. Hence, micro-EDM can be used to modify the surface of TiNi SMA and improve the biocompatibility of the titanium–nickel alloy.

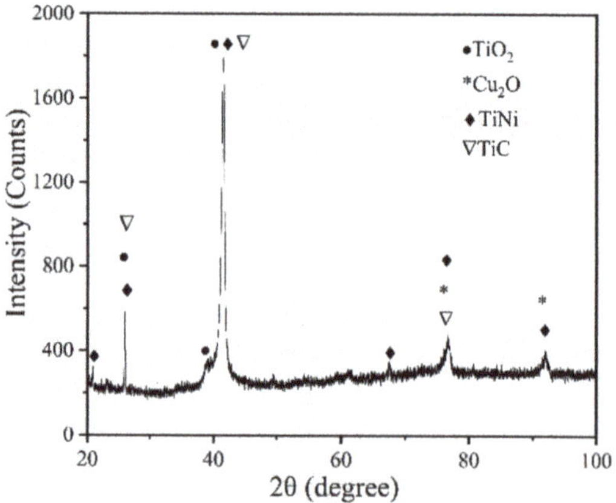

Figure 13. X-ray diffraction (XRD) pattern of TiNi SMA in 5 g/L TiC-dielectric.

The machined surfaces stand the continuous heating and cooling processes in EDM, which form a surface layer composed of the recast layer, heat affected zone, and base metal [34]. The recast layer has a great influence on the surface properties. Therefore, it is necessary to study the changes in surface microhardness. The microhardness curve at different distances from the center of the cavity and microhardness measurement of the substrate surface as shown in Figure 14a,b, respectively. The results show that the surface microhardness can reach 438.7 HV after micro-EDM, which is approximately 1.7 times the base material hardness. Chen et al. [35] recently showed that machined surfaces' hardening effect originates from the recast layer. Combined with Figure 13, the XRD analysis revealed that the machined surface was composed of TiC, Ti2O, Cu2O, and TiNi, which could improve the surface's microhardness.

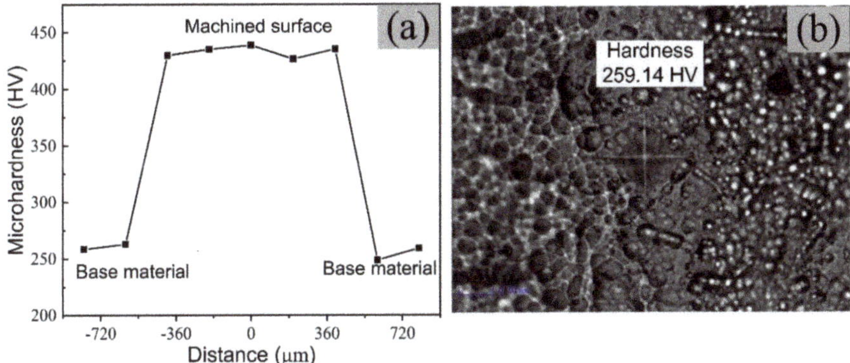

Figure 14. Measurement of surface microhardness of TiNi SMA: (**a**) microhardness curve of the surface and (**b**) microhardness measurement of the substrate surface.

5. Conclusions

The machining performance and feasibility of modifying the surface of a TiNi SMA through micro-EDM with the addition of TiC particles to the dielectric were discussed in this study. Discharge voltage waveforms demonstrated that the number of pulses in TiC-dielectric was significantly higher than in pure dielectric. The MRR, surface roughness, and thickness of the recast layer increased with an increase in discharge energy. MRR increased with an increase of TiC concentration, reaching a maximum at 5 g/L. Adding TiC particles to the dielectric can improve the surface finish by observing the surface morphology, and the machined surface in TiC-dielectric has smaller melting drops and craters compared to deionized water. The best surface finish occurred at a TiC concentration of 5 g/L. A layer ranging from 1 to 3 μm was obtained on the machined surface. The surface microhardness increased due to the formation of a recast layer containing TiC, Cu_2O, and Ti_2O; its hardness could reach 438.7 HV. Thus, this method can improve the wear resistance of the implant material, especially for orthodontic applications.

Author Contributions: Conceptualization, Z.Z. and D.G.; data curation, J.L. (Jianjun Lin) and B.X.; formal analysis, Z.Z. and J.X.; funding acquisition, D.G., X.W. (Xiaoyu Wu) and X.W. (Xujin Wang); investigation, Z.Z.; methodology, Z.Z. and J.X.; project administration, J.X.; resources, D.G., J.L. (Jianguo Lei) and X.W. (Xiaoyu Wu); software, J.L. (Jianjun Lin); supervision, Z.Z., D.G. and B.X.; validation, Z.Z. and J.X.; visualization, Z.Z. and J.X.; writing—original draft, Z.Z.; writing—review and editing, Z.Z. and D.G. All authors have read and agreed to the published version of the manuscript.

Funding: This work is supported by the innovation and entrepreneurship project for overseas high-level talents of Shenzhen (No. KQJSCX20180328095603847), the National Natural Science Foundation of China (No. 51975385, 51705503, 51805331), the Fundamental Research Free-exploring Project of Shenzhen (No. JCYJ20190809153205492).

Acknowledgments: The authors are grateful to their colleagues for their essential contribution to the work.

Conflicts of Interest: The authors declare no conflict of interest.

References

1. Jani, J.M.; Leary, M.; Subic, A.; Gibson, M.A. A review of shape memory alloy research, applications and opportunities. *Mater. Des.* **2014**, *56*, 1078–1113. [CrossRef]
2. Hsieh, S.F.; Hsue, A.W.J.; Chen, S.L.; Lin, M.H.; Ou, K.L.; Mao, P.L. Edm surface characteristics and shape recovery ability of Ti35.5Ni48.5Zr16 and Ni60Al24.5Fe15.5 ternary shape memory alloys. *J. Alloy. Compd.* **2013**, *571*, 63–68. [CrossRef]
3. Peng, P.W.; Ou, K.L.; Lin, H.C.; Pan, Y.N.; Wang, C.H. Effect of electrical-discharging on formation of nanoporous biocompatible layer on titanium. *J. Alloy. Compd.* **2010**, *492*, 625–630. [CrossRef]
4. McKay, G.C.; Macnair, R.; MacDonald, C.; Grant, M.H. Interactions of orthopaedic metals with an immortalized rat osteoblast cell line. *Biomaterials* **1996**, *17*, 1339–1344. [CrossRef]
5. Hsieh, S.F.; Lin, M.H.; Chen, S.L.; Ou, S.F.; Huang, T.S.; Zhou, X.Q. Surface modification and machining of TiNi/TiNb-based alloys by electrical discharge machining. *Int. J. Adv. Manuf. Technol.* **2016**, *86*, 1475–1485. [CrossRef]
6. Cheng, H.C.; Lee, S.Y.; Chen, C.C.; Shyng, Y.C.; Ou, K.L. Titanium nanostructural surface processing for improved biocompatibility. *Appl. Phys. Lett.* **2006**, *89*, 3. [CrossRef]
7. Shih, Y.H.; Lin, C.T.; Liu, C.M.; Chen, C.C.; Chen, C.S.; Ou, K.L. Effect of nano-titanium hydride on formation of multi-nanoporous TiO_2 film on ti. *Appl. Surf. Sci.* **2007**, *253*, 3678–3682. [CrossRef]
8. Ou, S.F.; Chou, H.H.; Lin, C.S.; Shih, C.J.; Wang, K.K.; Pan, Y.N. Effects of anodic oxidation and hydrothermal treatment on surface characteristics and biocompatibility of Ti-30Nb-1Fe-1Hf alloy. *Appl. Surface Sci.* **2012**, *258*, 6190–6198. [CrossRef]
9. Gurrappa, I.; Manova, D.; Gerlach, J.W.; Mandl, S.; Rauschenbach, B. Influence of nitrogen implantation on the high temperature oxidation of titanium-base alloys. *Surf. Coat. Technol.* **2006**, *201*, 3536–3546. [CrossRef]
10. Maleki-Ghaleh, H.; Khalili, V.; Khalil-Allafi, J.; Javidi, M. Hydroxyapatite coating on NiTi shape memory alloy by electrophoretic deposition process. *Surf. Coat. Technol.* **2012**, *208*, 57–63. [CrossRef]
11. Chu, C.L.; Wang, R.M.; Hu, T.; Yin, L.H.; Pu, Y.P.; Lin, P.H.; Wu, S.L.; Chung, C.Y.; Yeung, K.W.K.; Chu, P.K. Surface structure and biomedical properties of chemically polished and electropolished NiTi shape memory alloys. *Mater. Sci. Eng. C Biomimetic Supramol. Syst.* **2008**, *28*, 1430–1434. [CrossRef]
12. Qin, R.; Ding, D.Y.; Ning, C.Q.; Liu, H.G.; Zhu, B.S.; Li, M.; Mao, D.L. Ni-doped TiO_2 nanotube arrays on shape memory alloy. *Appl. Surface Sci.* **2011**, *257*, 6308–6313. [CrossRef]
13. Bilal, A.; Jahan, M.P.; Talamona, D.; Perveen, A. Electro-discharge machining of ceramics: A review. *Micromachines* **2019**, *10*, 10. [CrossRef] [PubMed]
14. Wyszynski, D.; Bizon, W.; Miernik, K. Electrodischarge drilling of microholes in c-BN. *Micromachines* **2020**, *10*, 179. [CrossRef]
15. Wu, Y.Y.; Huang, T.W.; Sheu, D.Y. Desktop Micro-EDM system for high-aspect ratio micro-hole drilling in tungsten cemented carbide by cut-side Micro-tool. *Micromachines* **2020**, *11*, 14. [CrossRef]
16. Liu, Q.Y.; Zhang, Q.H.; Zhang, M.; Yang, F.Z. Study on the discharge characteristics of single-pulse discharge in Micro-EDM. *Micromachines* **2020**, *11*, 55. [CrossRef]
17. Almacinha, J.A.; Lopes, A.M.; Rosa, P.; Marafona, J.D. How hydrogen dielectric strength forces the work voltage in the electric discharge machining. *Micromachines* **2018**, *9*, 240. [CrossRef]
18. Roy, T.; Balasubramaniam, R. Influence of ion-rich plasma discharge channel on unusually high discharging points in reverse micro electrical discharge machining. *Int. J. Adv. Manuf. Technol.* **2020**, *106*, 4467–4475. [CrossRef]
19. Roy, T.; Balasubramaniam, R. Effect of various factors influencing the generation of hemispherical micro features using non-conformal RMEDM. *J. Micromanuf.* **2019**, *2*, 110–122. [CrossRef]
20. Chen, S.L.; Yan, B.H.; Huang, F.Y. Influence of kerosene and distilled water as dielectrics on the electric discharge machining characteristics of Ti-6Al-4V. *J. Mater. Process. Technol.* **1999**, *87*, 107–111. [CrossRef]
21. Buckley, R.V. *Electrostatic Field Theory*; Palgrave: London, UK, 1981.
22. Wang, X.Z.; Yi, S.; Guo, H.; Li, C.J.; Ding, S.L. Erosion characteristics of electrical discharge machining using graphene powder in deionized water as dielectric. *Int. J. Adv. Manuf. Technol.* **2020**, *108*, 357–368. [CrossRef]
23. Lin, Y.C.; Yan, B.H.; Chang, Y.S. Machining characteristics of titanium alloy (Ti-6Al-4V) using a combination process of EDM with USM. *J. Mater. Process. Technol.* **2000**, *104*, 171–177. [CrossRef]

24. Jahan, M.P.; Rahman, M.; Wong, Y.S. Modelling and experimental investigation on the effect of nanopowder-mixed dielectric in micro-electrodischarge machining of tungsten carbide. *Proc. Inst. Mech. Eng. Part B J. Eng. Manuf.* **2010**, *224*, 1725–1739. [CrossRef]
25. Liew, P.J.; Yan, J.W.; Kuriyagawa, T. Carbon nanofiber assisted micro electro discharge machining of reaction-bonded silicon carbide. *J. Mater. Process. Technol.* **2013**, *213*, 1076–1087. [CrossRef]
26. Xu, B.; Guo, K.; Wu, X.Y.; Lei, J.G.; Liang, X.O.; Guo, D.J.; Ma, J.; Cheng, R. Applying a foil queue micro-electrode in micro-edm to fabricate a 3D micro-structure. *J. Micromech. Microeng.* **2018**, *28*, 11. [CrossRef]
27. Wu, Z.Z.; Wu, X.Y.; Xu, B.; Guo, D.J.; Tang, Y.; Diao, D.F. Reverse-polarity pmedm using self-welding bundled 3d-laminated microelectrodes. *J. Mater. Process. Technol.* **2019**, *273*, 8. [CrossRef]
28. Bui, V.D.; Mwangi, J.W.; Schubert, A. Powder mixed electrical discharge machining for antibacterial coating on titanium implant surfaces. *J. Manuf. Process.* **2019**, *44*, 261–270. [CrossRef]
29. Reddy, V.V.; Kumar, A.; Valli, P.M.; Reddy, C.S. Influence of surfactant and graphite powder concentration on electrical discharge machining of ph17-4 stainless steel. *J. Braz. Soc. Mech. Sci. Eng.* **2015**, *37*, 641–655. [CrossRef]
30. Tiwary, A.P.; Pradhan, B.B.; Bhattacharyya, B. Influence of various metal powder mixed dielectric on micro-edm characteristics of ti-6al-4v. *Mater. Manuf. Process.* **2019**, *34*, 1103–1119. [CrossRef]
31. Amorim, F.L.; Dalcin, V.A.; Soares, P.; Mendes, L.A. Surface modification of tool steel by electrical discharge machining with molybdenum powder mixed in dielectric fluid. *Int. J. Adv. Manuf. Technol.* **2017**, *91*, 341–350. [CrossRef]
32. Li, C.J.; Xu, X.; Li, Y.; Tong, H.; Ding, S.L.; Kong, Q.C.; Zhao, L.; Ding, J. Effects of dielectric fluids on surface integrity for the recast layer in high speed EDM drilling of nickel alloy. *J. Alloy. Compd.* **2019**, *783*, 95–102. [CrossRef]
33. Tan, P.C.; Yeo, S.H. Modeling of recast layer in Micro-electrical discharge machining. *J. Manuf. Sci. Eng. Trans. ASME* **2010**, *132*, 9. [CrossRef]
34. Jahan, M.P.; Mahbub, R.; Kirwin, R.; Alavi, F. Micro-EDM induced surface modification of titanium alloy for biocompatibility. *Int. J. Mach. Mach. Mater.* **2018**, *20*, 274.
35. Chen, S.L.; Hsieh, S.F.; Lin, H.C.; Lin, M.H.; Huang, J.S. Electrical discharge machining of TiNiCr and TiNiZr ternary shape memory alloys. *Mater. Sci. Eng. A Struct. Mater. Prop. Microstruct. Process.* **2007**, *445*, 486–492. [CrossRef]

Publisher's Note: MDPI stays neutral with regard to jurisdictional claims in published maps and institutional affiliations.

 *micromachines*

Article

Study on ZrB$_2$-Based Ceramics Reinforced with SiC Fibers or Whiskers Machined by Micro-Electrical Discharge Machining

Mariangela Quarto [1],*, Giuliano Bissacco [2] and Gianluca D'Urso [1]

[1] Department of Management, Information and Production Engineering, University of Bergamo, Via Pasubio 7/b, 24044 Dalmine, Italy; gianluca.d-urso@unibg.it

[2] Department of Mechanical Engineering, Technical University of Denmark, Produktionstorvet, Building 425, 2800 Kgs. Lyngby, Denmark; gibi@mek.dtu.dk

* Correspondence: mariangela.quarto@unibg.it; Tel.: +39-035-2052330

Received: 1 October 2020; Accepted: 26 October 2020; Published: 26 October 2020

Abstract: The effects of different reinforcement shapes on stability and repeatability of micro electrical discharge machining were experimentally investigated for ultra-high-temperature ceramics based on zirconium diboride (ZrB$_2$) doped by SiC. Two reinforcement shapes, namely SiC short fibers and SiC whiskers were selected in accordance with their potential effects on mechanical properties and oxidation performance. Specific sets of process parameters were defined minimizing the short circuits in order to identify the best combination for different pulse types. The obtained results were then correlated with the energy per single discharge and the discharges occurred for all the combinations of material and pulse type. The pulse characterization was performed by recording pulses data by means of an oscilloscope, while the surface characteristics were defined by a 3D reconstruction. The results indicated how reinforcement shapes affect the energy efficiency of the process and change the surface aspect.

Keywords: micro-EDM; Zirconium Boride; silicon carbide fibers; silicon carbide whiskers; advanced material

---

1. Introduction

Among the advanced ceramic materials, ultra-high-temperature ceramics (UHTCs) are characterized by excellent performances in extreme environments. This family of materials is based on borides (ZrB$_2$, HfB$_2$), carbides (ZrC, HfC, TaC), and nitrides (HfN), which are marked by high melting point, high hardness and good resistance to oxidation. In particular, ZrB$_2$-based materials are of particular interest because of their suitable properties combination and are considered attractive in applications, such as a component for the re-entry vehicles and devices [1–4].

The relative density of the base material ZrB$_2$ is usually about 85% because of the high level of porosity of the structure, furthermore, in the last years, researchers are focused on fabricating high-density composites characterized by good strength (500–1000 MPa). For these reasons, the use of single-phase materials is not sufficient for high-temperature structural applications. Many efforts have been done on ZrB$_2$-based composites in order to improve the mechanical properties, oxidation performances, and fracture toughness; however, the low fracture toughness remains one of the greatest limitations for the application of these materials under severe conditions [5–10]. Usually, the fracture toughness of ceramic materials can be improved by incorporating appropriate additives that activated toughening mechanisms such as phase transformation, crack pinning, and deflection. An example is the addition of SiC, in fact, it has been widely proved that its addition improves the fracture strength and the oxidation resistance of ZrB$_2$-based materials due to the grain refinement and the formation of a protective silica-based layer. Based on these aspects, recent works have focused on ZrB_2-based composites behavior generated by the addition of SiC with different shapes (e.g., whiskers or fibers). Specifically, it has been reported that the addition of whiskers or fibers gives promising results, improving the fracture toughness and this improvement could be justified by crack deflection [5,6,9,11,12]. The critical aspect of the reinforced process was the reaction or the degeneration of the reinforcement during the sintering process [13,14].

Despite all the studies that aim to improve mechanical properties and resistance, this group of materials is very difficult to machine by the traditional technologies, because of their high hardness and fragility. Only two groups of processes are effective in processing them: On one side the abrasive processes as grinding, ultrasonic machining, and waterjet, on the other side the thermal processes as laser and electrical discharges machining (EDM) [15–18].

In this work, ZrB_2 materials containing 20% vol. SiC whiskers or fibers produced by hot pressing were machined by the micro-EDM process; in particular, it would investigate the effect of the non-reactive additive shapes on the process performances, verifying if the process is stable and repeatable for advanced ceramics, and in particular for materials characterized by different geometry of the additive. The choice of 20% vol. is related to the evidence reported in some previous works [14,18,19], in which it has been shown that this fraction of additive has allowed generating the best combination of oxidation resistance and mechanical characteristics useful for obtaining better results in terms of process performances and dimensional accuracy for features machined by micro-EDM.

2. Materials and Methods

The following ZrB_2-based composites, provided by ISTEC-CNR of Faenza (Consiglio Nazionale delle Ricerche—Istituto di Scienza e Tecnologia dei Materiali Ceramici, Faenza (RA)—Italy), have been selected for evaluating the influence of additive shape on the process performances and geometrical of micro-slots machined by micro-EDM technology:

- ZrB_2 + 20% SiC short fibers, labelled as ZrB20f
- ZrB_2 + 20% SiC whiskers, labelled ZrB20w.

Such as reported in [14], commercial powders were used to prepare the ceramic composites: ZrB_2 Grade B (H.C. Starck, Goslar, Germany), SiC HI Nicalon-chopped short fibers, Si:C:O = 62:37:0.5, characterized by 15 μm diameter and 300 μm length or SiC whiskers characterized by average diameter 1 μm and average length 30 μm.

The powder mixtures were ball milled for 24 h in pure ethanol using silicon carbide media. Subsequently, the slurries have been dried in a rotary evaporator. Hot-pressing cycles were conducted in low vacuum (100 Pa) using an induction-heated graphite die with a uniaxial pressure of 30 MPa during the heating and were increased up to 50 MPa at 1700 °C (T_{MAX}), for the material containing fibers, and at 1650 °C (T_{MAX}) for the composites with whiskers. The maximum sintering temperature was set based on the shrinkage curve. Free cooling followed. Details about the sintering runs are reported in Table 1, where T_{ON} identify the temperature at which the shrinkage started. Density was estimated by the Archimedes method.

Table 1. Composition and sintering parameters of the hot-pressed samples [1].

Label	Composition	T_{ON} (°C)	T_{MAX} (°C)	Final Density (g/cm^3)	Relative Density (%)
ZrB20f	ZrB_2 + 20% SiC Short fibers	1545	1650	4.89	94.0
ZrB20w	ZrB_2 + 20% SiC Whiskers	1560	1700	5.22	97.0

The raw materials were analyzed by Scanning Electron Microscopy (SEM) (Figure 1). The samples reinforced by the fibers show a very clear separation between the base matrix and the non-reactive additive. Fibers dispersion into the matrix was homogenous, as no agglomeration was observed in the sintered body. However, some porosity was retained in the microstructure. The fibers showed a tendency to align their long-axis perpendicular to the direction of applied pressure. It is apparent that their length was significantly reduced compared with the starting dimensions, as the maximum observed length was about 300 µm. For the sample containing whiskers reinforcement, a dense microstructure was observed and the whiskers are generally well dispersed into the matrix. The mean grain size of ZrB_2 grains was slightly lower (2.1 ± 0.2 µm) than the base material (3.0 ± 0.5 µm); furthermore, the whiskers showed a tendency to form large bundles. The addition of SiC whiskers promoted both strengthening and toughening compared with the base material ZrB_2. The maximum toughness increase (5.7 MPa·m$^{1/2}$) was of the order of 50% when whiskers were added. In the sample containing fibers, the increase in fracture toughness (5.5 MPa·m$^{1/2}$) corresponds to a strength reduction. A direct observation of the crack morphology and the comparison with theoretical models demonstrates that, besides a residual stress contribution, the toughness increase was almost entirely explained in terms of crack deflection in the whiskers and in terms of crack bowing in the fibers. No crack bridging was obtained as the reinforcement pullout was hindered by the formation of interphases or intergranular wetting phases, which promoted a strong bonding between matrix and reinforcement. The values of fracture toughness of the ZrB20f and ZrB20w were close to those of hot-pressed composites previously. This indicates that, even if a more efficient thermal treatment is used, the nature of the matrix/reinforcement interface did not change notably [2].

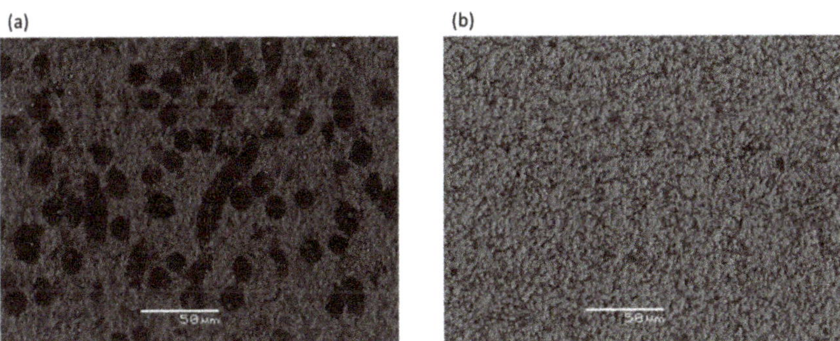

Figure 1. SEM backscattered images of typical appearance of ZrB20f (**a**) and ZrB20w (**b**).

3. Experimental Section

3.1. Experimental Set-Up

A simple circular pocket having a diameter equal to 1 mm and a depth of about 200 µm was selected as the test feature for machining experiments. These micro-features had been processed on the SARIX® SX-200 machine (Sarix, Sant'Antonino, Switzerland) by micro-EDM milling. Solid tungsten carbide electrode with a diameter equal to 300 µm was used as a tool; while the dielectric fluid was formed by hydrocarbon oil. The experiments were performed for three different process parameters settings, corresponding to different pulse shapes. It is essential to remark that the machine used for the experimental tests expresses some process parameters as indexes (e.g., peak current, width). The instantaneous values cannot be set, because the machine presents an autoregulating system. Thus, the characterization of electrical discharges population is very important not only to assess the real value of process parameters but, most of all, to evaluate the stability and repeatability of the process. Due to this characteristic of the machine, an acquisition system was developed to cover the gap. In particular, a current monitor and a voltage probe were connected to the EDM machine and to a programmable counter and a digital oscilloscope. These connections allow to acquire the current waveforms and count the discharges occurred during the process. Specifically, a current monitor with a bandwidth of 200 MHz and a Rohde & Schwarz RTO1014 oscilloscope were used. The counter has been set once the trigger value was established to avoid recording and counting the background noise.

Preliminary tests were performed to define the optimal process parameters for each combination of material and pulse type. The experimental campaign was based on a general full factorial design, featured by two factors: The additive shape, defined by two levels, and the pulse type, defined by three levels. Different levels of pulse types identify the different duration of the discharges, in particular, level A is referred to long pulses while level C identify the short pulses. Three repetitions were performed for each run.

3.2. Discharges Population Characterization

For each pulse types, combined with both materials, discharge populations have been characterized by repeated waveform samples of current and voltage signals. The current and voltage probes were connected to the digital oscilloscope having a real-time sample rate of 40 MSa/s. The trigger level of the current signal was set to 0.5 A in order to acquire all the effective discharges. The acquired waveform samples were stored in the oscilloscope buffer and then transferred to a computer to be processed by a Matlab code, written by the authors. The Matlab code was used to evaluate the numbers of electrical discharge, the current peak values, the duration (width), the voltage and to estimate the energy content in each discharge. Finally, the average value of energy per discharge (E) was estimated by integrating the instantaneous value of the power, calculated as the product of the instantaneous values of current ($i(t)$) and voltage ($v(t)$), with respect to the time (Equation (1)).

$$E = \int_0^T v(t) \cdot i(t)dt \tag{1}$$

To show the discharge population distribution for all pulse types and for both additive shapes, the peak current distribution histograms are plotted, showing the number of observed discharges, with peak current within discrete intervals, for all the (Figure 2). Histograms show good reproducibility and stability of the process, providing information regarding the frequency waveforms with different peaks of current. The discharge samples are well described by a normal distribution, which is characterized by a good reproducibility suggesting a stable process. Considering both additive shapes, the values of peak current are included in a similar range for pulse type A and B, while, for pulse type C can be identified a great difference; in fact, for whiskers, the range of variation of current peaks is smaller than for short fibers.

Thanks to the discharge characterization it was possible to define the real value of the process parameters. The pulse characteristics for the parameters setting estimated by the data elaboration with Matlab are reported in Table 2.

In terms of peak current and voltage, the differences between the two materials, machined by the same pulse type, were really tiny. This suggested that EDM machining on ZrB_2-based composites reinforced by whiskers would have been characterized by higher machining speed.

3.3. Characterization Procedure

A 3D reconstruction of micro-slots was performed by means of a confocal laser scanning microscope (Olympus LEXT, Southend, Essex, UK) with a magnification of 20×. This microscope recognizes the peaks of the reflected light intensities of multiple layers and, setting each layer as the focal point, makes it possible to analyze and measure each layer. Then, the images were analyzed with an image processor software (SPIP 6.7.3, Image Metrology, Lyngby, Denmark), firstly performing a plane correction on all the images to level the surfaces and to remove primary profiles, then the surface roughness (Sa) was assessed, on the base of the international standard UNI EN ISO 25178:2017 by the real-topography method. The process performances were evaluated through the estimation of three indicators: The material removal per discharge (MRD), the tool wear per discharge (TWD), and the tool wear ratio (TWR).

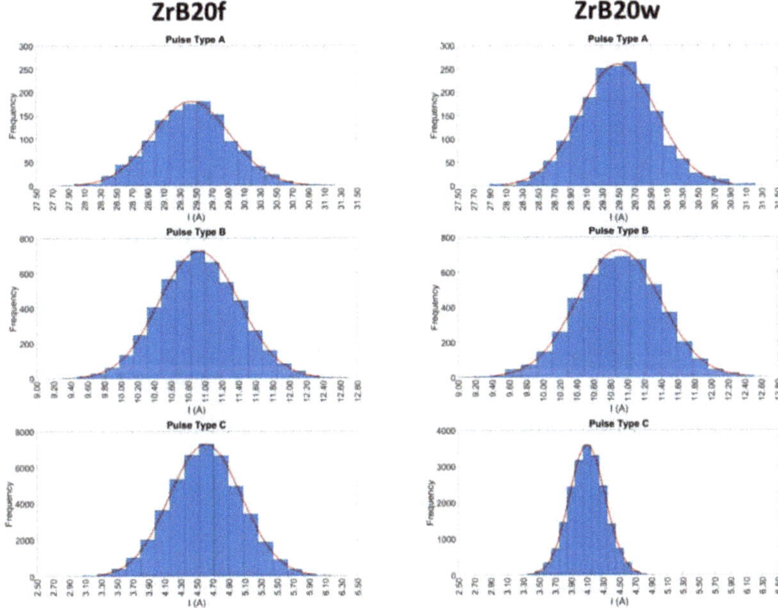

Figure 2. Examples of frequency distribution histograms for pulses occurred during ZrB20 machining.

Table 2. Mean values of pulse type characteristics.

Additive Shape	Pulse Type	Peak Current (A)	Open Circuit Voltage (V)	Width (μs)	Energy per Discharge (μJ)
Fibers (ZrB20f)	A	29.44	69.57	0.70	779.36
	B	10.93	92.89	0.33	150.30
	C	4.61	100.79	0.06	14.73
Whiskers (ZrB20w)	A	29.49	70.75	0.70	784.99
	B	10.91	91.70	0.33	161.00
	C	4.10	99.60	0.06	21.41

MRD (Equation (2)) was calculated as the ratio of material removed from the workpiece (MRW (mm^3)), estimated through SPIP, and the number of discharges (N) recorded by the programmable counter.

$$MRD = \frac{MRW}{N} \qquad (2)$$

Since this kind of materials is characterized by a high level of porosity, to get the actual values of MRD, the volume of the micro-slots was adjusted considering the relative density (δ—Table 1). for compensating the presence of porosity in the sample structure. The MRD is calculated as reported in Equation (3).

$$MRD_\delta = \frac{MRW \cdot \delta}{N} = MRD \cdot \delta \qquad (3)$$

TWD (Equation (4)) was estimated as the ratio between the material removed from the tool electrode (MRT (mm^3)) and the number of discharges [N] recorded by the programmable counter.

Tool wear was measured as the difference between the length of the electrode before and after the single milling machining. The length was measured through a touching procedure executed in a reference position. The electrode wear volume was estimated starting from the length of the tool wear and considering the tool such as a cylindrical part.

$$\text{TWD} = \frac{\text{MRT}}{\text{N}} \tag{4}$$

Tool Wear Ratio (Equation (5)) was calculated as the ratio between the previous performances' indicators considering the relative density of the workpiece material (TWR_δ).

$$\text{TWR}_\delta = \frac{\text{TWD}}{\text{MRD}} = \frac{\text{MRT}}{\text{MRW} \cdot \delta} \tag{5}$$

4. Results and Discussion

During the analysis, the energy efficiency of a single discharge was evaluated. Figure 3 shows the tool wear per discharge divided by the energy of single discharge (TWD_E), as a function of the additive fraction and the pulse type. The plot shows a lower energy efficiency for pulse type A and this is a positive aspect because it indicates less impact on the electrode wear. For both additive shapes, the medium pulses are characterized by a higher impact on the tool wear.

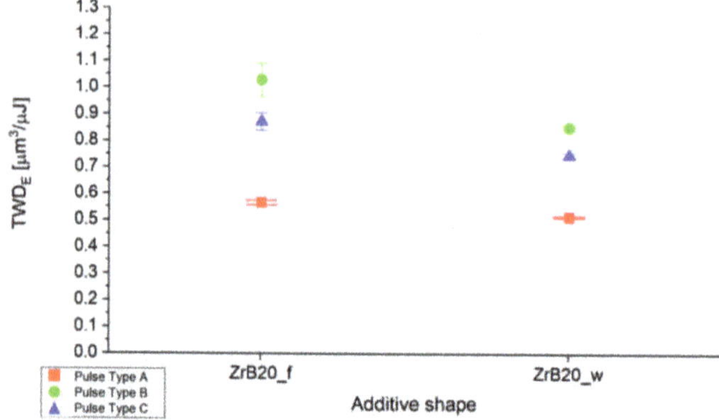

Figure 3. Average ratio between tool wear per discharge (TWD) and energy per discharge as a function of the additive shape and pulse type.

To evaluate the energy efficiency from the material removed point of view, as has already been done in the previous plot, also the material removal per single discharge was evaluate as the ratio with the energy for single discharge. In this case, the best results were obtained for pulse type C where a single discharge is characterized by lower duration and energy but and higher efficiency, such as illustrated in Figure 4.

Figure 5 shows that the TWR for all pulse type is lower for specimens containing whiskers reinforcement. In particular, the whiskers allow to reduce the tool wear and increasing the material removal rate efficiency probably thanks to the dense microstructure and the reduction of the grains size.

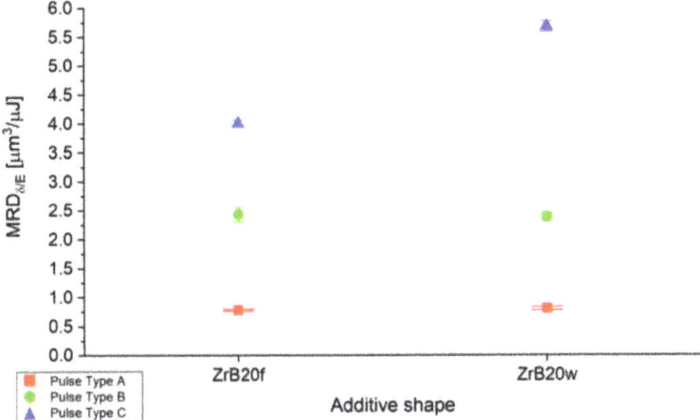

Figure 4. Average ratio between material removal per discharge (MRD) and energy per discharge estimated considering the relative density of the samples as a function of the additive shape and pulse type.

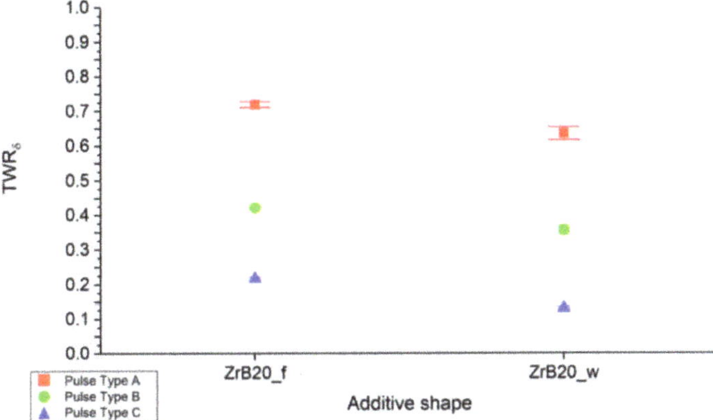

Figure 5. Average tool wear ratio (TWR) as a function of the additive shape and pulse type.

In general, for pulse type A the energy per discharge is higher in comparison to others pulse types but for both Energy efficiency of TWD (TWD_E) and Energy efficiency of MRD_δ ($MRD_{\delta/E}$), the energy efficiency is lower (positive aspect from TWD_E point of view). This can suggest a greater energy dispersion in machining performed with pulse type A than for others; in fact, energy per discharge of pulse type C is 40–50 times smaller than pulse type A but the energy efficiency is higher.

The factorial design was analyzed in order to comprehend which factors and interactions are statistically significant for the performance indicators and surface roughness. Table 3 shows the average results obtained from the experiments reporting the average and the standard deviation of MRD_δ, TWD, TWR_δ, and Sa for each level of the experimental design. A general linear model was used to perform a univariate analysis of variance, including all the main factors and their interactions.

Table 3. Results of the experimental campaign for process performances, where μ and σ identifies the average value and the standard deviation respectively.

Additive Shape	Pulse Type	μ_{MRD_δ} (μm^3)	σ_{MRD_δ} (μm^3)	μ_{TWD} (μm^3)	σ_{TWD} (μm^3)	μ_{TWR_δ}	σ_{TWR_δ}	μ_{Sa} (μm)	σ_{Sa} (μm)
Fibers	A	61.40	1.19	44.22	0.65	0.720	0.009	1.997	0.152
	B	36.63	1.93	15.48	0.95	0.422	0.004	1.086	0.036
	C	5.91	0.09	1.29	0.05	0.218	0.005	0.475	0.011
Whiskers	A	63.82	2.10	40.54	0.20	0.636	0.019	1.720	0.178
	B	38.52	1.09	13.75	0.11	0.357	0.013	0.904	0.064
	C	12.18	0.20	1.60	0.02	0.132	0.004	0.439	0.019

The Analysis of Variance (ANOVA) results of the experimental plan are reported in Table 4 omitting the values related to the pulse type, since all the *p*-values results to be very close to 0.00. The parameters are statistically significant for the process when the p-value is less than 0.05 since a confidence interval of 95% is applied. As a general remark, all the indicators resulted to be influenced by the additive shape and the pulse type. In some cases (for MRD_δ and TWD), also the interaction showed an effect in terms of ANOVA. This aspect suggested that the interaction of factors is relevant for indicators that, in some way, can be correlated to the machining duration.

Table 4. Analysis of variance *p*-values.

		FACTORS	
		Additive Shape	**Pulse Type * Additive Shape**
INDICATORS	MRD_δ (μm^3)	1.21×10^{-4}	0.03
	TWD (μm^3)	0.00	4.51×10^{-5}
	TWR_δ [-]	0.00	0.2019
	Sa (μm)	0.0045	0.156

* indicate the interaction between Pulse Type and Additive Shape

Main effects plots (Figure 6) show that indicators are mainly influenced by the pulse type that establishes the range in which process parameters can vary, and in particular, the characteristics of the pulses. For all indicators, reduction in pulse duration and in peak current intensity generate the lower value of MRD_δ, TWD, and TWR_δ. At the same time, tests with whiskers additive generate increment in MRD_δ and in surface roughness, and a reduction in TWD and TWR_δ. By increasing the pulse duration and peak current intensity from type C to type A, the machining speed (MRD_δ) is 10 times greater, but the surface quality decreases by −60%. For MRD_δ and TWD, also the interaction between pulse type and additive shape influence the results. The interaction plots (Figure 7) give more information than the *p*-value showing that the weight of the interaction is slight. What it is possible to notice is that for the samples that are machined by pulse type C, the effect on MRD_δ is more evident, while from the TWD point of view it is more evident for pulse type A). In general, tests performed on materials with whiskers additive are characterized by better results, both in terms of process performances and surface finishing. In fact, optimal performance for ED-machining are characterized by a higher level of MRD_δ, to perform a fast machining, and lower TWD, to reduce the waste of material related to the tool wear as a function of material removed from the workpiece.

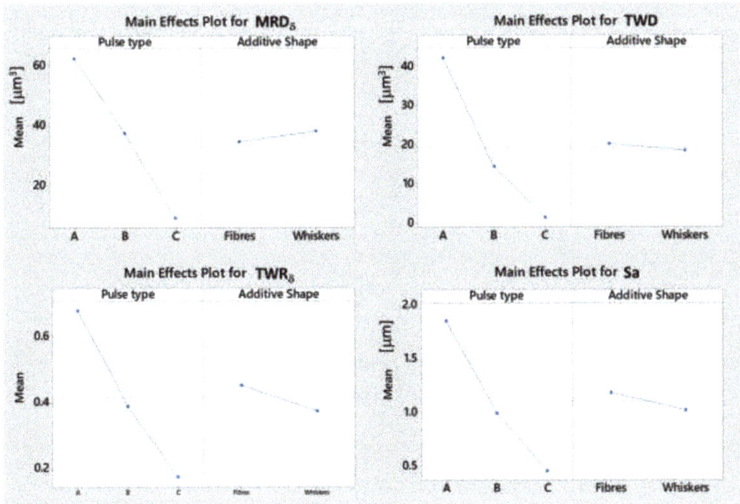

Figure 6. Main effects plot for indicators affected by pulse type and additive shape.

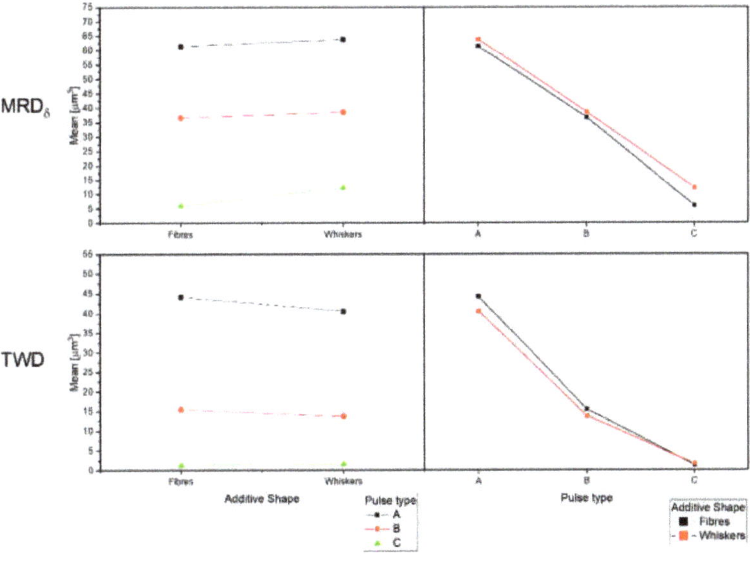

Figure 7. Interaction plot for MRD_δ and TWD.

Figures 8–10 represent each indicator as a function of the energy per discharge and, regardless of the additive shape, it is possible to observe that each indicator is well represented ($R^2 \approx 1$) by the same type of regression equation for both additive shapes. TWD shows lower values of R^2. Specifically, the MRD_δ is well described by a logarithmic regression equation (Equation (6) and Equation (7)), while the TWD and the Sa are well described by a power regression equation (Equations (8)–(11)). In all regression equation reported in the plots, y-axis is referred to the indicator, while x-axis is referred to the energy generated by a single discharge (E). For MRD_δ and TWD the differences as a function of additive shape are very small, in fact, despite they cover different ranges, the values are very close to each other. The situation is different for Sa, which is characterized by completely different ranges without overlap.

$$MRD_{\delta\_f} = 13.934 \ln(E) - 32.056 \qquad R^2 = 0.9965 \qquad (6)$$

$$MRD_{\delta\_w} = 14.28 \ln(E) - 32.328 \qquad R^2 = 0.9939 \qquad (7)$$

$$TWD_f = 0.2569 E^{0.7868} \qquad R^2 = 0.9727 \qquad (8)$$

$$TWD_w = 0.3322 E^{0.7243} \qquad R^2 = 0.9892 \qquad (9)$$

$$Sa_f = 0.1791 E^{0.3614} \qquad R^2 = 0.9999 \qquad (10)$$

$$Sa_w = 0.1359 E^{0.3784} \qquad R^2 = 0.9987 \qquad (11)$$

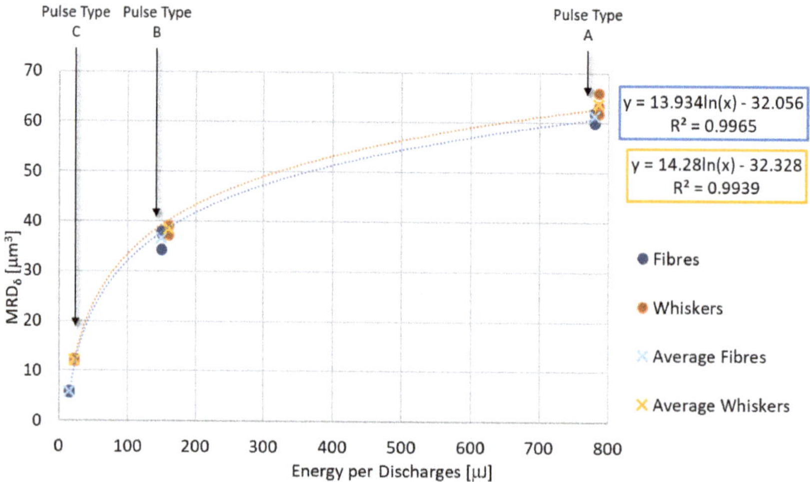

Figure 8. Material removal per discharge as a function of energy per discharge for both additive shapes and their regression equation.

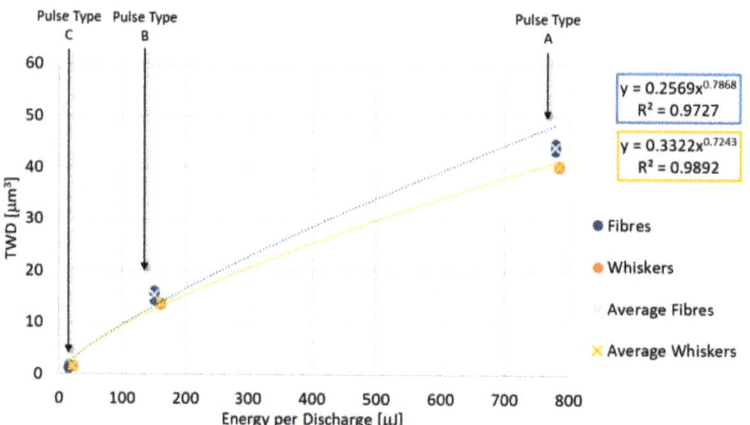

Figure 9. Tool wear per discharge as a function of energy per discharge for both additive shapes and their regression equation.

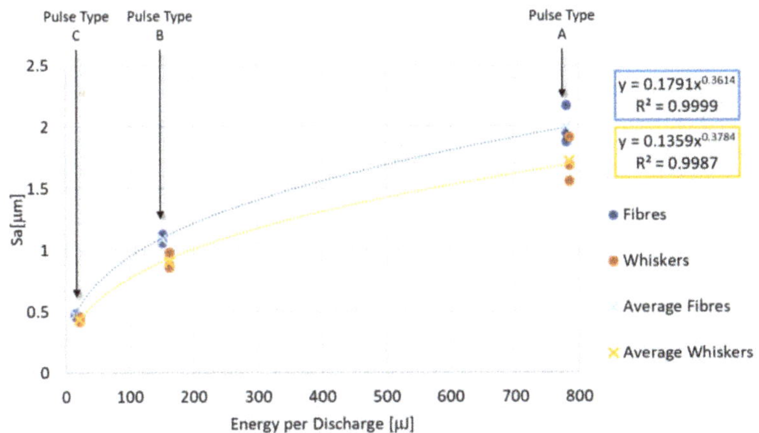

Figure 10. Surface roughness as a function of energy per discharge for both additive shapes and their regression equation.

A 3D reconstruction of an ED-machined surfaces detail scanned by the confocal laser scanning microscope with a magnification of about 100× is reported as an example in Figure 11. In particular, Figure 11a represents a portion of the machined area on ZrB20f. Here it is possible to identify very clearly, a sort of "protrusion" in correspondence of the fibers. Such as reported in a previous work [20], this aspect is probably related to incomplete machining because of SiC low electrical conductivity characteristic and the great extension of the area of the fibers. Figure 11b represents the ED-machined surface for the sample containing SiC whiskers. In this case, the surface appears uniform and homogeneous because the SiC particles, such as reported in the materials section, are better dispersed in the base matrix. These aspects justified the different results obtained in terms of surface quality and, in general, these different textures can be considered a starting point for further studies about the material removal mechanism occurred on UHTCs, in particular when there is a low-electrically conductive parts in the structure. From the 3D reconstruction it is possible to observe that the surfaces are not characterized by the typical aspect of the ED-machined surfaces which present a texture well-described by the presence of craters. The different aspect has already been presented in a previous study [20]. In his specific case, the UHTC is different, but the same considerations can be done. A sort of craters can be observed by means of a SEM (Figure 12).

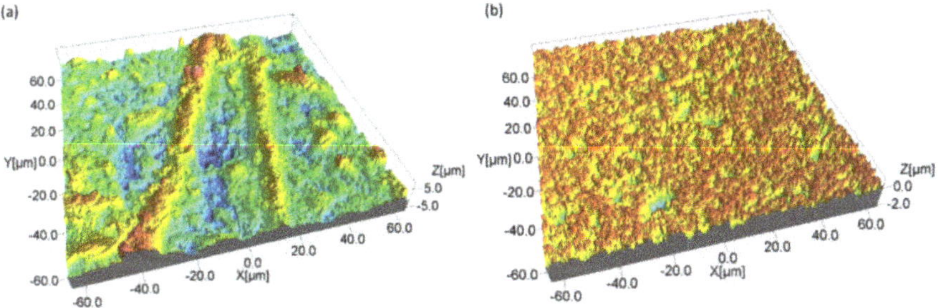

Figure 11. Details of machined surfaces for ZrB20f (**a**) and ZrB20w (**b**) machined by pulse type A.

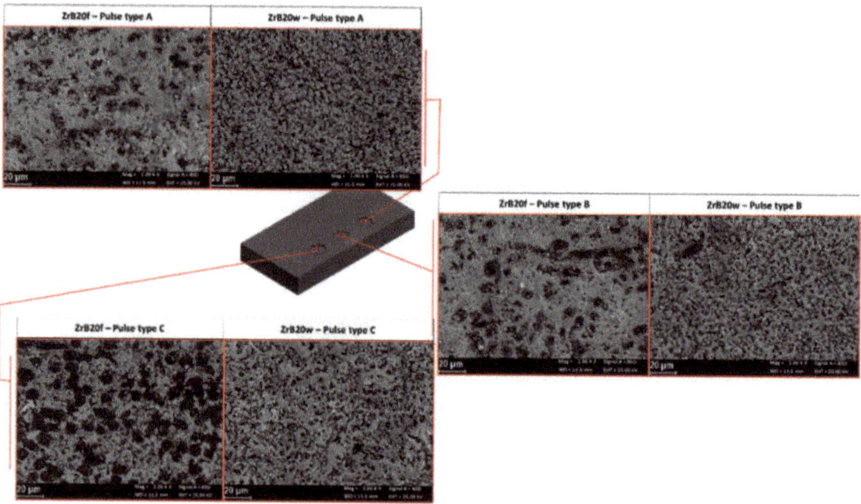

Figure 12. SEM backscatter images of the machined surface.

Machined surfaces on specimens doped with SiC whiskers are characterized by higher fragmentation of the recast layer. This aspect is particularly evident on surfaces machined by long pulses; in fact, for pulse type A and B, the surface is for the major part covered by recast materials for both additive shapes, but the specimens doped with whiskers presented smaller extensions of the single "crater" of recast materials.

Different behavior can be observed for surfaces machined by the short pulses. In this case, for specimens contained the SiC fibers can be observed a smaller area of recast material, probably because of the union of the greater dimensions of the fibers in comparison to the whiskers, the lower energy per discharge for short pulses and the low electrical conductibility of the SiC. In particular, the fibers have a bigger surface and it needs to work hard to remove the entire parts. For this reason, the surfaces contained fibers presented in their correspondence a sort of protrusion.

It is very clear that the whiskers affect in a significant way the micro-EDM process improving the energy efficiency and the machining performances increasing the machining speed and reducing the tool wear. Furthermore, the geometry and the behavior of whiskers during the samples preparation generate a homogenous surface improving the removal rate and reducing the risk of leave machining witness (as the fibers) on the surfaces which would generate better surface finishing. These considerations allow to suppose a different behavior of the discharges path as a function of the kind of material met and the distribution of the reinforcement.

5. Conclusions

An evaluation of the machinability of ZrB_2-based composites hot-pressed with different shapes non-reactive additive (SiC) was performed in this work. Stability and repeatability of the micro-EDM were evaluated to identify the effects of the additive shapes. The analysis taking into account the process performances and surface finishing.

First of all, a discharge characterization was performed to feature the different pulse type used during the machining. In general, the discharge characterization and the performances indicators allowed to identify a stable and repeatable process with a faster material removal for samples doped with whiskers.

The analysis of variance showed that both factors, pulse type, and additive shape, are statistically significant for the indicators selected in the process evaluation and in general, the use of whiskers improve the material removal rate generating lower tool wear. Furthermore, the interaction between the two factors turns out to be influential only for MRD and TWD, which are indirectly related to the machining duration, since the number of discharges occurred during the machining was considered in their estimation.

This investigation shows that the specimens having a 20 vol.% of additive in form of whiskers results to be the best solution in terms of machinability by EDM process not only for the better process performances, but also for the higher level of surface quality which is one of the essential criteria for making a proper decision for industrial application.

Author Contributions: Conceptualization, M.Q., G.B., and G.D.; methodology, M.Q., G.B. and G.D.; software, M.Q., G.B.; formal analysis, M.Q.; data curation, M.Q., G.B.; writing—original draft preparation, M.Q., writing—review and editing, G.B., G.D.; supervision, G.B., G.D. All authors have read and agreed to the published version of the manuscript.

Funding: This research received no external funding.

Acknowledgments: The authors would like to thank Diletta Sciti and Laura Silvestroni from ISTEC—CNR of Faenza (Italy) for the production and supply of the materials used in this study.

Conflicts of Interest: The authors declare no conflict of interest.

References

1. Saccone, G.; Gardi, R.; Alfano, D.; Ferrigno, A.; Del Vecchio, A. Laboratory, on-ground and in-flight investigation of ultra high temperature ceramic composite materials. *Aerosp. Sci. Technol.* **2016**, *58*, 490–497. [CrossRef]

2. Upadhya, K.; Yang, J.M.; Hoffman, W. Advanced materials for ultrahigh temperature structural applications above 2000 °C. *Am. Ceram. Soc. Bull.* **1997**, *58*, 51–56.

3. Levine, S.R.; Opila, E.J.; Halbig, M.C.; Kiser, J.D.; Singh, M.; Salem, J.A. Evaluation of ultra-high temperature ceramics for aeropropulsion use. *J. Eur. Ceram. Soc.* **2002**, *22*, 2757–2767. [CrossRef]

4. Justin, J.F.; Jankowiak, A. Ultra High Temperature Ceramics: Densification, Properties and Thermal Stability. *AerospaceLab J.* **2011**, *3*, 1–11.

5. Zhang, P.; Hu, P.; Zhang, X.; Han, J.; Meng, S. Processing and characterization of ZrB2-SiCW ultra-high temperature ceramics. *J. Alloys Compd.* **2009**, *472*, 358–362. [CrossRef]

6. Wang, H.; Wang, C.-A.; Yao, X.; Fang, D. Processing and Mechanical Properties of Zirconium Diboride-Based Ceramics Prepared by Spark Plasma Sintering. *J. Am. Ceram. Soc.* **2007**, *90*, 1992–1997. [CrossRef]

7. Zhang, X.; Xu, L.; Du, S.; Liu, C.; Han, J.; Han, W. Spark plasma sintering and hot pressing of ZrB2-SiCW ultra-high temperature ceramics. *J. Alloys Compd.* **2008**, *466*, 241–245. [CrossRef]

8. Zhang, X.; Xu, L.; Han, W.; Weng, L.; Han, J.; Du, S. Microstructure and properties of silicon carbide whisker reinforced zirconium diboride ultra-high temperature ceramics. *Solid State Sci.* **2009**, *11*, 156–161. [CrossRef]

9. Yang, F.; Zhang, X.; Han, J.; Du, S. Processing and mechanical properties of short carbon fibers toughened zirconium diboride-based ceramics. *Mater. Des.* **2008**, *29*, 1817–1820. [CrossRef]

10. Yang, F.; Zhang, X.; Han, J.; Du, S. Characterization of hot-pressed short carbon fiber reinforced ZrB2-SiC ultra-high temperature ceramic composites. *J. Alloys Compd.* **2009**, *472*, 395–399. [CrossRef]

11. Zhang, X.; Xu, L.; Du, S.; Han, J.; Hu, P.; Han, W. Fabrication and mechanical properties of ZrB2-SiCw ceramic matrix composite. *Mater. Lett.* **2008**, *62*, 1058–1060. [CrossRef]

12. Tian, W.B.; Kan, Y.M.; Zhang, G.J.; Wang, P.L. Effect of carbon nanotubes on the properties of ZrB2-SiC ceramics. *Mater. Sci. Eng. A* **2008**, *487*, 568–573. [CrossRef]

13. Guicciardi, S.; Silvestroni, L.; Nygren, M.; Sciti, D. Microstructure and Toughening Mechanisms in Spark Plasma-Sintered ZrB 2 Ceramics Reinforced by SiC Whiskers or SiC-Chopped Fibers. *J. Am. Ceram. Soc.* **2010**, *2391*, 2384–2391. [CrossRef]

14. Silvestroni, L.; Sciti, D.; Melandri, C.; Guicciardi, S. Toughened ZrB2-based ceramics through SiC whisker or SiC chopped fiber additions. *J. Eur. Ceram. Soc.* **2010**, *30*, 2155–2164. [CrossRef]

15. Rezaie, A.; Fahrenholtz, W.G.; Hilmas, G.E. The effect of a graphite addition on oxidation of ZrB2-SiC in air at 1500 °C. *J. Eur. Ceram. Soc.* **2013**, *33*, 413–421. [CrossRef]

16. Hwang, S.S.; Vasiliev, A.L.; Padture, N.P. Improved processing and oxidation-resistance of ZrB2ultra-high temperature ceramics containing SiC nanodispersoids. *Mater. Sci. Eng. A* **2007**, *464*, 216–224. [CrossRef]

17. Zhang, X.; Liu, R.; Zhang, X.; Zhu, Y.; Sun, W.; Xiong, X. Densification and ablation behavior of ZrB_2 ceramic with SiC and/or Fe additives fabricated at 1600 and 1800 °C. *Ceram. Int.* **2016**, *42*, 17074–17080. [CrossRef]
18. Zhang, L.; Kurokawa, K. Effect of SiC Addition on Oxidation Behavior of ZrB2 at 1273 K and 1473 K. *Oxid. Met.* **2016**, *85*, 311–320. [CrossRef]
19. Sciti, D.; Guicciardi, S.; Silvestroni, L. SiC chopped fibers reinforced ZrB2: Effect of the sintering aid. *Scr. Mater.* **2011**, *64*, 769–772. [CrossRef]
20. Quarto, M.; Bissacco, G.; D'Urso, G. Machinability and Energy Efficiency in Micro-EDM Milling of Zirconium Boride Reinforced with Silicon Carbide Fibers. *Materials* **2019**, *12*, 3920. [CrossRef]

 *micromachines*

Article

Investigation of a Liquid-Phase Electrode for Micro-Electro-Discharge Machining

Ruining Huang [1],* , **Ying Yi [2]** , **Erlei Zhu [1]** and **Xiaogang Xiong [1]**

1   School of Mechanical Engineering and Automation, Harbin Institute of Technology, Shenzhen 518000, China; zel15950568268@163.com (E.Z.); xiongxg@hit.edu.cn (X.X.)
2   College of Science and Engineering, Hamad Bin Khalifa University, Education City 34110, Qatar; yyi@hbku.edu.qa
*   Correspondence: hrn@hit.edu.cn; Tel./Fax: +86-755-26033774

Received: 18 September 2020; Accepted: 12 October 2020; Published: 14 October 2020

Abstract: Micro-electro-discharge machining (μEDM) plays a significant role in miniaturization. Complex electrode manufacturing and a high wear ratio are bottlenecks for μEDM and seriously restrict the manufacturing of microcomponents. To solve the electrode problems in traditional EDM, a μEDM method using liquid metal as the machining electrode was developed. Briefly, a liquid-metal tip was suspended at the end of a capillary nozzle and used as the discharge electrode for sparking the workpiece and removing workpiece material. During discharge, the liquid electrode was continuously supplied to the nozzle to eliminate the effects of liquid consumption on the erosion process. The forming process of a liquid-metal electrode tip and the influence of an applied external pressure and electric field on the electrode shape were theoretically analyzed. The effects of external pressure and electric field on the material removal rate (MRR), liquid-metal consumption rate (LMCR), and groove width were experimentally analyzed. Simulation results showed that the external pressure and electric field had a large influence on the electrode shape. Experimental results showed that the geometry and shape of the liquid-metal electrode could be controlled and constrained; furthermore, liquid consumption could be well compensated, which was very suitable for μEDM.

Keywords: Micro-electro-discharge machining (μEDM); liquid-metal electrode; Galinstan

---

1. Introduction

Micro-electro-discharge machining (μEDM) is a powerful micromachining technique with various advantages resulting from it being a noncontact and thermal process; thus, μEDM is applicable to any electrically conductive material regardless of the mechanical properties of the material. The machining process utilizes the thermal erosion of the material caused by pulses of electrical discharge generated between a microscopic electrode and the workpiece in the presence of a dielectric fluid for the removal of the workpiece material. The μEDM technique is capable of producing microholes, microchannels, and real three-dimensional (3D) microstructures. These attractive features have been leveraged for producing micromechanical components, as well as for prototyping various micro-electro-mechanical systems (MEMS) and devices. Furthermore, μEDM has been widely used in the aerospace, die, mold, and biomedical industries for machining small cavities [1–3].

It is well known that electrode miniaturization is critical for μEDM because it determines the precision of the process to manufacture smaller and more precise parts [4]. Many methods have been proposed for fabricating microelectrodes. For example, Masuzawa et al. were the first to develop wire electrode discharge grinding (WEDG) technology to make a microelectrode with a diameter of Φ2.5 μm [5]. WEDG is one of the most widely used methods to fabricate microelectrodes, and it offers the benefit of sustaining good grinding accuracy by utilizing fresh wire during the whole fabrication process. According to this method, some evolutions occurred in different ways. Egashira et al. made use of WEDG and electrochemical machining to produce a microelectrode with a diameter of Φ0.3 µm [6]. Zhang et al. proposed a tangential feed WEDG (TF-WEDG) method to improve the microelectrode accuracy [7]. Lim et al. utilized a rotating disc instead of a moving wire to grind the microelectrode [8]. Other methods reported for µEDM electrode fabrication include the single-side block electrode discharge grinding method (BEDG) [9], bilateral BEDG method [10], scanning discharge method [11], single-pulse discharge method [12], self-drilled hole reverse-EDM method [13], electroforming method [14], Lithographie, Galvanoformung and Abformung (LIGA) process method [15], and electrostatic ejection method [16]. The goal of all these methods is the same: to improve the processing efficiency and quality, reduce the processing difficulty, and make smaller and better microelectrodes. However, these methods either are complex processes or require additional devices, leading to the process being cumbersome and inefficient. It is difficult to control the size and precision of the microelectrode. Although high-quality electrodes can sometimes be obtained, the consistency is poor, and it is difficult to obtain the same size electrode again.

In addition, the mechanism of µEDM is an electrothermal physical process that removes material by repeated spark discharges, whereby the workpiece is eroded at high temperature, and the electrode itself will also wear down. Therefore, the fundamental theory determines the unavoidability of wear with the electrode tool. Due to the area effect, with the decrease in electrode diameter, the relative electrode wear rate becomes more serious [17]. Some research work has focused on electrode wear issues during the µEDM process to obtain improved machining accuracy. For example, Bissacco et al. analyzed the electrode wear rate at different energy levels in detail [18]. Wang et al. carried out quantitative research on the electrode wear amount of positive and negative pulses [19]. Tsai et al. conducted a detailed investigation on the wear rate of electrodes of different materials [20]. All these investigations indicate that electrode wear will have a serious impact on the subsequent processing performance, resulting in a decrease in machining accuracy. To address a variety of problems associated with electrode wear, many explorations on preventing and compensating electrode wear have been performed. The measures reported for preventing µEDM wear include the ultrasonic-assisted debris removal method [2], coating electrode method [21], special material electrode method [22], and discharge in gas method [23]. The methods reported for µEDM electrode compensation include the electrode uniform wear compensation method [24], electrode fixed length linear compensation method [25], effective discharge pulse monitoring compensation method [26], and prediction electrode compensation method [27]. However, the above methods are only suitable for specific occasions and have certain limitations, which lead to imperfect final results and difficulty in the promotion of these methods in industrial fields. Moreover, it is difficult to measure the actual wear of microelectrodes online, and the consistency of the online electrode is difficult to guarantee. Therefore, electrode compensation has always been one of the difficult problems in µEDM, and the various electrode compensation methods have not completely solved the problem caused by electrode wear. Thus, it is urgent to find a new processing method to solve the issues of electrode wear.

To address the wear-related problems associated with µEDM, Huang et al. proposed a novel µEDM method that uses a liquid alloy as the machining electrode instead of traditional electrodes [17,28], in which the liquid metal consumed in the process can be compensated over time, and the capillary containing the liquid metal does not participate in the discharge; thus, its shape remains unchanged and solves the problem of electrode wear. Nevertheless, these references reported the experimental characterizations and demonstration of the process on the basis of a preliminary study and lacked an analysis of the liquid-metal electrode morphology, which will affect the machining accuracy. Therefore, this study specifically focuses on analyzing the influence of the external pressure and electric field on the shape of liquid electrodes, as well as the influence of different electrode shapes on machining characteristics. The characterization of the process with varying discharge parameters is discussed, along with the arbitrary patterning of silicon substrates using the developed method.

2. Method

The principle of conducting μEDM with a liquid-metal electrode is illustrated in Figure 1. A liquid metal is held through a metallic capillary nozzle coated with a dielectric film so that the liquid protrudes on the nozzle tip and forms a droplet. Under the action of an appropriate pressure and electric field, the droplet suspended at the tip will countervail part of the surface tension and become a conical tip that serves as a microdischarge electrode. A pulse generator is applied between the liquid metal and the workpiece. Moving the workpiece toward the liquid electrode initiates a spark discharge when the gap meets the breakdown condition and produces a high temperature to melt and erode the workpiece material. Similar to conventional μEDM, the liquid-metal electrode material will also be consumed, but liquid metal is continuously supplied to the nozzle tip to compensate for the consumption at the right pressure, thereby eliminating its impact on the removal process. In the current method, the system is configured so that the discharge pulses are generated only between the liquid electrode and workpiece through the controlled supply of liquid, and the capillary itself does not participate in the discharge; therefore, it remains intact. Thus, this method uses a liquid metal tip as a microdischarge electrode, which can be automatically compensated. Furthermore, its geometric shape can be controlled and constrained, which is highly desirable for microsized thin-walled flexible devices, thin-film sensors, or workpiece surface etching.

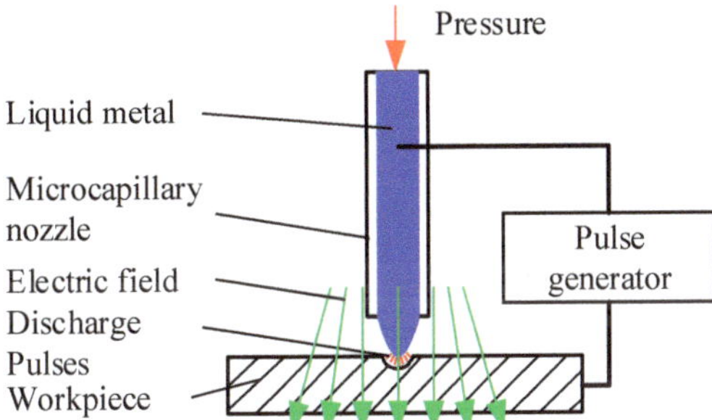

Figure 1. Principle of micro-electro-discharge machining (μEDM) using a liquid-metal electrode.

3. Shape of the Liquid-Phase Electrode

3.1. Mathematical Model

To obtain a liquid-metal electrode that meets the discharge demand, the formation conditions and morphology of the electrode are extremely critical.

We know that the surface tension at any point on the plane of the static liquid surface is the same in all directions, counteracting each other. In other words, the pressure just outside the surface P_o and just inside the surface P_i is equal. Therefore, there is no additional pressure on the plane of the static liquid surface. However, if the pressure inside the surface P_i is greater than the pressure outside the surface P_o, an additional pressure P_f will be generated and induce a curved liquid surface (if the drop is small, the effect of gravity may be neglected and the shape may be assumed to be spherical), and vice versa. Figure 2 shows that the tip of the liquid-metal electrode appears to be a convex sphere projected over the flat bottom of the coated needle (the liquid metal does not wet the coated needle).

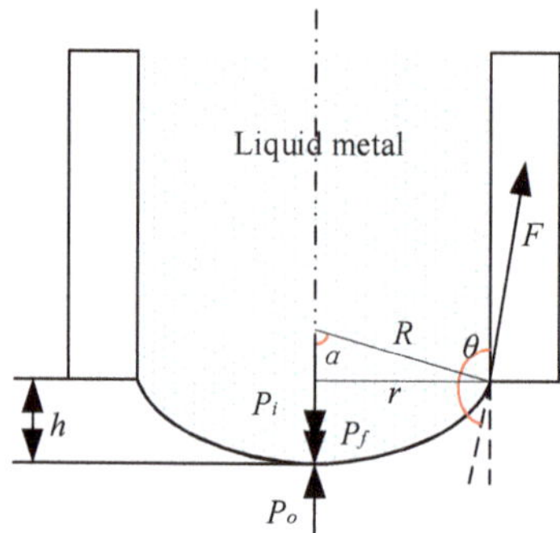

Figure 2. Tip shape of the liquid-metal electrode.

From the geometric relationship of the solid–liquid contact surface shown in Figure 2, it can be found that the relationship between θ and α is

$$\theta = \frac{\pi}{2} + \alpha, \tag{1}$$

where θ is the liquid–solid contact angle (°), and α is the angle between the boundary radius of the curved liquid surface and the axis of the capillary (°).

The additional pressure at the tip of the liquid electrode is as follows [29]:

$$P_f = \frac{2\pi r^2 \delta}{R} / \pi r^2 = \frac{2\delta}{R} = \frac{2\delta \sin\left(\theta - \frac{\pi}{2}\right)}{r}, \tag{2}$$

where δ is the coefficient of surface tension (N/m), r is the inner radius of the capillary (m), and R is the radius of the curved liquid surface (m).

There is a trigonometric relationship between the contact angle θ and the additional pressure P_f produced by the curved liquid surface. When the contact angle is $\theta = \pi/2$, the additional pressure is $P_f = 0$ Pa. When the contact angle is $\theta < \pi/2$, the additional pressure is $P_f < 0$ Pa. This result indicates a negative pressure. The liquid surface at the tip of the liquid metal is concave, and the direction of additional pressure is downward, which will not happen in our research case. When the contact angle is $\theta = \pi$, the additional pressure P_f is at the maximum, and this state usually leads to a liquid flow or the ejection of liquid.

However, under natural conditions, the P_f is too small, and the radius R of the convex sphere suspended at the end of the coated needle tip is too large. This result leads to the sagittal h (the height of a segment) being too small; thus, the protruding convex sphere used as an electrode cannot meet the requirement for the μEDM process. Therefore, extra pressure is applied to the needle. The pressure on the tip of the liquid-metal electrode is shown in Figure 3.

From Figure 3, the liquid metal remains stationary inside the needle at equilibrium, the same as the tip of the liquid metal. The additional pressure P_f generated on the convex liquid surface of the liquid metal is

$$P_f = P_1 + P_{l1} - P_{l2}, \tag{3}$$

where P_1 is the extra applied pressure (Pa), P_{l1} is the liquid-metal gravity (Pa), and P_{l2} is the dielectric-liquid gravity (Pa).

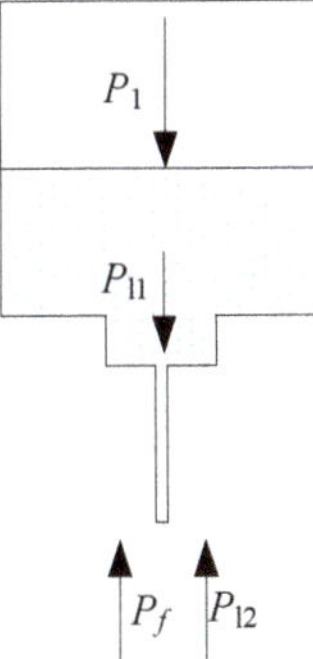

Figure 3. Pressures of the liquid-metal electrode on the nozzle tip.

Hence, the tip of the liquid-metal electrode will maintain its shape in the equilibrium state. Then, the contact angle θ of the liquid-metal electrode tip is

$$\theta = \frac{\pi}{2} + \arcsin\left[\frac{(P_1 + P_{l1} - P_{l2}) \cdot r}{2\delta}\right]. \tag{4}$$

During the EDM process, a pulse generator is applied to the electrode and workpiece. This applied voltage will generate an electric field between the liquid-metal electrode and the workpiece. The electric field will exert an electric field force at the tip of the liquid-metal electrode, which will affect or even change the tip shape of the liquid-metal electrode. At the same time, the change in the tip shape of the liquid-metal electrode will change the electric field distribution between the electrode and the workpiece, and the electric field force on the tip of the electrode will also change.

In the case without an electric field force, as shown in Figure 4, the tip shape of the liquid-metal electrode will form a spherical shape due to surface tension, as discussed before.

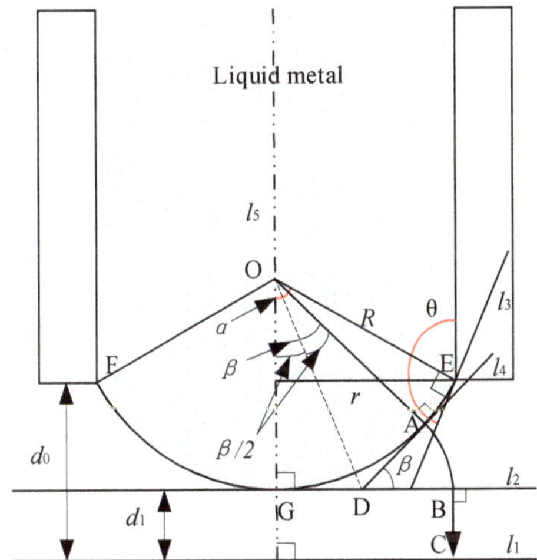

Figure 4. Electric field force of the liquid-metal electrode tip.

Suppose that the liquid metal is the ideal conductor (its resistivity is assumed to be 0 (as conductivity approaches infinity), which makes calculations easy to perform), the surface of the liquid metal is smooth, and the electric field is an irrotational field; then, the electric field line at point A follows arc

$\overset{\frown}{AB}$ and line BC. Suppose line segment DA is equal to r_0; then, according to the geometric relationship in Figure 4, the length of line segment BC can be obtained as follows:

$$d_1 = d_0 - R(1 - \cos\alpha). \tag{5}$$

The length of arc $\overset{\frown}{AB}$ can be calculated as follows:

$$r_0\beta = R\tan(\frac{\beta}{2})\beta = \beta R\frac{1 - \cos\beta}{\sin\beta}. \tag{6}$$

The relationship between the electric potential U at point A and the electric field intensity E can be expressed as

$$U = \int_{\overset{\frown}{AB}+BC} \vec{E}d\vec{l} = E(r_0\beta + d_1) = E(\frac{\beta R(1 - \cos\beta) + d_1\sin\beta}{\sin\beta}). \tag{7}$$

Then, the average electric field intensity E on line ABC can be obtained as follows:

$$E = \frac{U\sin\beta}{\beta R(1 - \cos\beta) + d_1\sin\beta}. \tag{8}$$

In the spherical coordinate system with point O as the center of the ball and R as the radius, by taking an area differential element ds near point A, the charge of point A in this differential element can be calculated as

$$dq = \varepsilon Eds = \frac{\varepsilon U\sin\beta}{\beta R(1 - \cos\beta) + d_1\sin\beta}R^2\sin\beta d\beta d\gamma, \tag{9}$$

where γ is the azimuth in the coordinate system with a range of 0–2π, and ε is the dielectric constant.

Integrating the entire convex spherical surface, the charge of the entire spherical surface can be obtained as follows:

$$\begin{aligned} q &= \int_0^{2\pi}\int_0^\alpha \frac{\varepsilon U\sin\beta}{\beta R(1-\cos\beta)+d_1\sin\beta}R^2\sin\beta d\beta d\gamma \\ &= 2\pi\varepsilon UR^2\int_0^\alpha \frac{\sin^2\beta}{\beta R(1-\cos\beta)+d_1\sin\beta}d\beta. \end{aligned} \tag{10}$$

According to the principle of virtual displacement, the electric field force received by the convex spherical surface is given as

$$\begin{aligned} Fe &= \frac{1}{2}\frac{\partial(q/U)}{\partial d_1}U^2 \\ &= \pi\varepsilon U^2R^2\int_0^\alpha \frac{\sin^3\beta}{[\beta R(1-\cos\beta)+d_1\sin\beta]^2}d\beta. \end{aligned} \tag{11}$$

Since $\beta < \pi/2$, β can be approximately replaced by $2\sin(\beta/2)$, substituting this into Equation (11) provides the following:

$$\begin{aligned} Fe = {} & \frac{\pi\varepsilon U^2}{2}\left[\frac{2(d_1+R)}{d_1} + \frac{2R(d_1-4R)}{16R^2+d_1^2} + \ln\frac{d_1}{2R\sin^2(\alpha/2)+d_1\cos(\alpha/2)}\right. \\ & + \frac{4R(8R^2+d_1^2)+2d_1(12R^2+d_1^2)\cos(\alpha/2)}{[R(\cos\alpha-1)-d_1\cos(\alpha/2)](16R^2+d_1^2)} \\ & \left. - \frac{d_1(24R^2+d_1^2)}{(16R^2+d_1^2)^{3/2}}\ln\frac{\left[\left(\sqrt{16R^2+d_1^2}+4R\right)[1-\cos(\alpha/2)]+d_1[1+\cos(\alpha/2)]\right]^2}{2R\sin^2(\alpha/2)+d_1\cos(\alpha/2)}\right]. \end{aligned} \tag{12}$$

Equation (12) shows the electric field force of the entire liquid-metal convex sphere. This is to analyze the entire spherical surface as a whole. However, to analyze the influence of the electric field force on the tip shape of the liquid-metal electrode, it is necessary to divide the spherical surface of the liquid-metal tip into an infinite number of microelements to analyze the force situation of each microelement. The electric field force near any point A on the spherical surface is

$$dF_e \;=\; \frac{1}{2}\,\frac{\varepsilon U^2 R^2 \sin^3\beta}{[\beta R(1-\cos\beta)+d_1\sin\beta]^2}\,d\beta d\gamma. \tag{13}$$

If $r = 10$ µm, $d_0 = 100$ µm, and $U = 100$ V, the value range of β is $[-\pi/2{-}\pi/2]$, and $d\beta$ and $d\gamma$ are constant at 1; then, the curve of the electric field force dFe around β at any point A on the spherical surface is as shown in Figure 5.

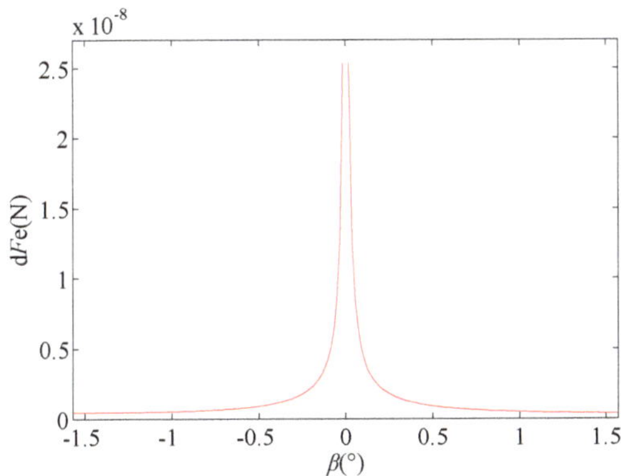

Figure 5. Relationship between the electric field force dFe and β at any point A on the spherical surface.

As shown in Figure 5, the electric field force on the spherical surface is not evenly distributed. The electric field force is the largest at the lowest point G of the spherical surface, and, with point G as the center, the electric field force gradually decreases toward point E, which is similar to point F. This nonuniform distribution of the electric field force may cause the tip shape of the liquid-metal electrode to change from a spherical shape to a Taylor cone.

3.2. Simulation

3.2.1. Geometric Model and Boundary Condition

Figure 6 shows the geometric size and boundary conditions of the model used for simulation. The size of the simulation area is 0.4×0.6 mm, the length of the capillary nozzle is 0.5 mm, and the distance between the nozzle tip and the workpiece is 0.1 mm. Zone A is the liquid metal inside the electrode, zone B is stainless steel, zone C is the EDM oil, and zone D is a parylene C coating with a thickness of 20 µm, as shown in Table 1, where E the electrostatic field on the boundary when a voltage is applied between liquid-metal electrode and workpiece.

At present, most metals or alloys are in the solid state at room temperature. Exceptions include francium, cesium, rubidium, mercury, sodium–potassium alloys, and gallium-based alloys, which can be defined as liquid metals. Their melting points are either lower than or close to room temperature, which enable them to remain in the liquid state at room temperature. Unfortunately, the intrinsic radioactivity of cesium, extreme instability of francium and rubidium, flammability and corrosivity of sodium–potassium alloys, and toxicity of mercury limit their applications to certain specific areas. On the other hand, gallium-based alloys, such as Galinstan (68.5 wt.% gallium, 21.5 wt.% indium, and 10 wt.% tin), a commercially available eutectic liquid alloy, is a low-activity and nontoxic liquid with a low melting point ($-19\,^{\circ}$C) and low viscosity (0.0024 Pa·s at $20\,^{\circ}$C) that allows it to be easily transferred through microscale nozzles. Its high electrical conductivity ($\sim$3.5 $\times$ 10^6 S/m at $20\,^{\circ}$C), high thermal conductivity (16.5 W·M^{-1}·K^{-1}), and high boiling point (>1300 °C) are desirable features for a µEDM electrode application. Therefore, Galinstan is used as the liquid electrode in this study.

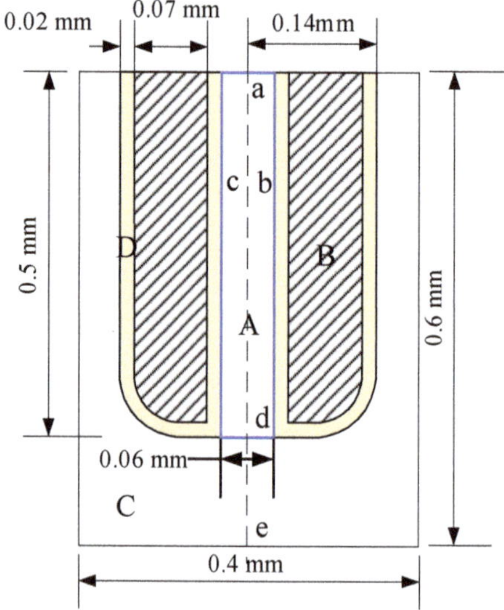

Figure 6. Geometric size and the boundary conditions of the model.

Table 1. Boundary conditions of the model.

Boundary	Fluid Field	Electrostatic Field
a	Entrance, extra pressure P_1	E
b,c	No slip wall	E
d	Initial interface	E
e	Export	0

3.2.2. Tip Shape of the Liquid Electrode at Different Extra Pressures

From the theoretical analysis in Section 2, it is known that the extra pressure is an important factor affecting the contact angle θ of the liquid-metal electrode. To evaluate this effect in detail, the influence of extra pressure on the tip shape of the liquid-metal electrode was simulated by varying extra pressures. The simulation results are shown in Figure 7. When the extra pressure P_1 is 0.1 atm, the liquid metal cannot be extruded from the needle (Figure 7a). The reason may be that the extra pressure is too small to overcome the internal frictional resistance caused by Galinstan's viscosity. As the extra pressure increases, the liquid-metal electrode tip gradually becomes a cone (Figure 7b,c). A larger extra pressure results in a faster change in the tip shape of the liquid-metal electrode. The tip shape of the liquid-metal electrode gradually changes into a spherical shape under the action of pressure. The liquid metal gradually extends to the parylene coating at the bottom of the needle, making the radius of the liquid-metal electrode tip larger than the inner diameter of the needle. This result implies that the width of the groove patterning by liquid-metal electrode µEDM may be larger than the inner diameter of the needle. Moreover, as the extra pressure increases, the sagitta of the tip of the liquid-metal electrode grows longer (Figure 7d). This result verifies the analysis results of Section 2. It is worth noting that, for an extra pressure over 2 atm, the liquid metal will promptly eject and touch the workpiece (Figure 7e). Therefore, it is necessary to set an appropriate pressure to obtain an electrode that meets processing needs.

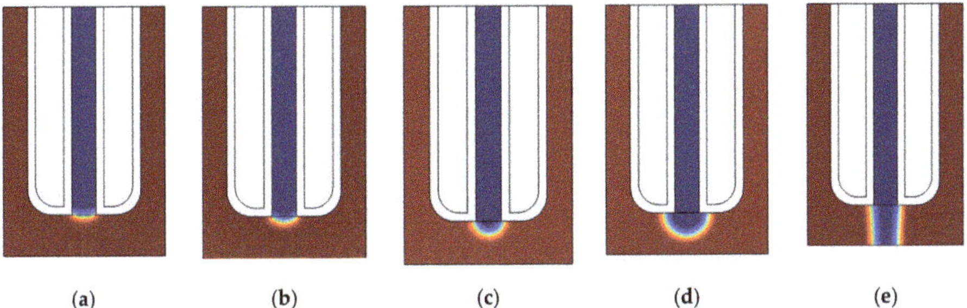

Figure 7. Simulation results at different extra pressures. (**a**) 0.1 atm. (**b**) 0.8 atm. (**c**) 1.2 atm. (**d**) 1.4 atm. (**e**) 2 atm.

3.2.3. Tip Shape of the Liquid Electrode at Different Voltages

Figure 8 shows the simulation result of the tip shape of the liquid-metal electrode at different voltages (tested up to 2 kV, with an extra pressure of $P_1 = 1.2$ atm). As shown in Figure 8a,b, it can be seen that the tips of the liquid-metal electrode are almost the same in the case of low voltage. This outcome is reasonable because the electric force at low voltage is not high enough to affect the tip of the liquid-metal electrode.

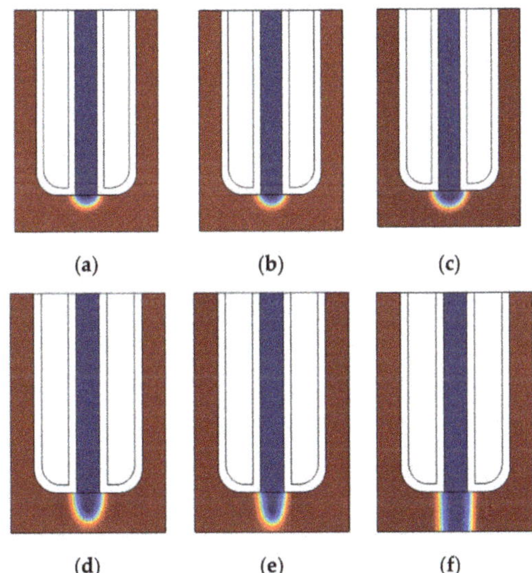

Figure 8. Simulation results at different voltages. (**a**) 50 V. (**b**) 100 V. (**c**) 0.5 kV. (**d**) 1 kV. (**e**) 1.5 kV. (**f**) 2 kV.

When the voltage is 0.5 kV, as shown in Figure 8c, the tip of the liquid-metal electrode is gradually stretched into an ellipsoidal shape under the combined action of the electric field force and the extra pressure. Compared with Figure 8a or 8b, the sagitta of the tip shape of the liquid-metal electrode is even longer and changes more rapidly over time. This may be attributed to the tensile effect of the electric field force generated by the high voltage on the liquid metal. As the voltage continues to increase, the tip of the liquid-metal electrode will clearly change. As seen from Figure 8d,e, the tip of the liquid-metal electrode changes over time and is gradually stretched into an inverted cone. Compared with Figure 8c, the tip of the liquid-metal electrode is now more of a conical shape, with a smaller diameter. The electric field force generated by the voltage constrains the tip shape of the liquid-metal electrode, and the constraint becomes stronger with increasing voltage. When the electric field force generated by the additional high voltage is sufficiently large, the liquid-metal electrode terminal can be stretched into a reverse cone.

It is also worth noting that for a voltage greater than 2 kV, the tip of the liquid-metal electrode is gradually stretched into an inverted cone and then ejected to the workpiece (as shown in Figure 8f). This result can be explained as the stretching effect of the liquid metal under the extra pressure and electric field force. This explanation is reasonable because the electric field force is too strong.

From the simulation, it can be seen that the effect of a low voltage on the electrode tip is not obvious, but the effect of a high electric field force is very obvious. Therefore, the external electric field may be an appealing way to restrain the tip of the liquid-metal electrode in future research.

4. Experimental Set-Up and Procedure

The schematic in Figure 9 illustrates the μEDM process experiments. The set-up has a servo-controlled three-axis stage with a 100 nm positioning resolution, comprising an *XY* stage with the holder on which the work tank is held and a *Z* stage to vertically position the syringe containing the needle with a liquid-metal electrode. The work tank is configured to have a sample holder made of plastic, in which the workpieces are immersed in a dielectric EDM fluid, fixed and electrically coupled with the discharge circuit using conductive adhesive tape. The nozzle is connected to the syringe that stores Galinstan, and a pressurizing unit is used to apply pressure to the syringe to feed Galinstan to the needle. To prevent the oxidation of Galinstan, a low-concentration (2%) sulfuric acid solution is added to the syringe and floats on top of Galinstan. Furthermore, the H_2SO_4 solution keeps the liquid metal clean to avoid clogging the needle. Galinstan is electrically coupled with the discharge circuit using a conductive copper wire immersed in it and sealed with glue on the syringe wall.

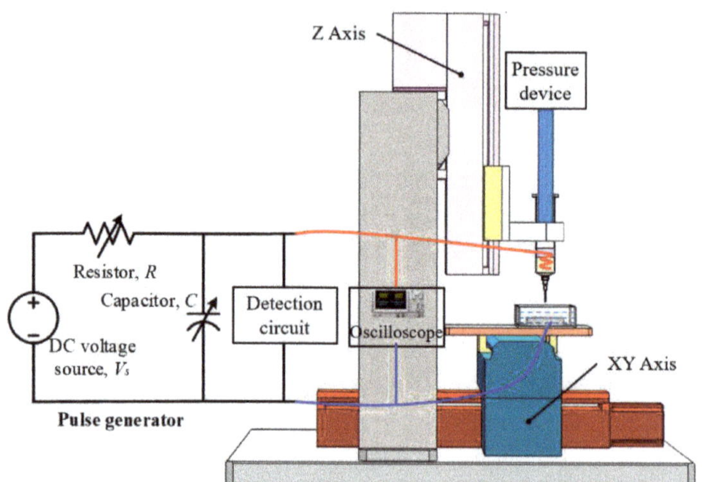

Figure 9. Experimental set-up for liquid-metal electrode μEDM.

The system employs a relaxation-type (resistor–capacitor, or RC) pulse generator with a variable direct current (DC) voltage source. In this type of pulse generation circuit, the loop-voltage equation can be expressed as

$$U_c = E(1 - e^{-\frac{t}{T}}),\tag{14}$$

where U_c is the voltage on the capacitor C (V), E is the voltage of the DC source S (V), and $T = RC$. In this case, the discharge energy is stored in the capacitor. When the voltage of the capacitor meets the condition of $U_c \geq U_d$ (U_d is the breakdown voltage of the discharge gap), the gap discharges and instantly releases the energy forming a pulse current. The generated discharge pulses are monitored using a detection circuit, and the readouts of the signals are obtained via a General Purpose Interface Bus (GPIB) interface connected to a computer; moreover, a probe is coupled with the discharge circuit and connected to an oscilloscope for visual observations.

A stainless-steel needle (outer diameter D_O = 240 µm, inner diameter D_I = 100 µm, length L = 13 mm) is used as the needle in the set-up. The needle is coated with a dielectric film, parylene C (thickness, 20 µm), to eliminate the effects of discharge on the needle surface. The coating consequently modifies the D_O and D_I of the needle to 280 and 60 µm, respectively. Although the distance, D, between the workpiece surface and the needle tip cannot be directly determined via electrical surface detection, it can make use of a nozzle with the presence of Galinstan at its tip. Briefly, a low pressure (less than 0.5 atm) is applied via a precision pressure device to ensure that the liquid metal in the needle is just flush with the tip of the needle. Then, using a small voltage (10 V) between them electrical surface detection is carried out using the Z-axis of the stage at a low speed, and retracting the needle (50 µm) to determine the gap, as shown in Figure 10a–c. To decrease the measuring error, electrical surface detection both toward and away from the gap is repeated continuously three times, and the average is taken as the gap position. The µEDM process with this setting is performed in the die-sinking mode, and the workpiece is electrically coupled with the discharge circuit cathode while the electrode is connected to the anode. After applying a normal pressure at the nozzle and applying a machining voltage between the nozzle and workpiece, the gap d between the liquid metal and the workpiece is adaptively determined during discharging, with horizontal scanning of the workpieces using the XY stage with respect to the liquid electrode having a constant gap distance D for the lateral removal of the material using the liquid electrode, as shown in Figure 10d. The experimental conditions are outlined in Table 2.

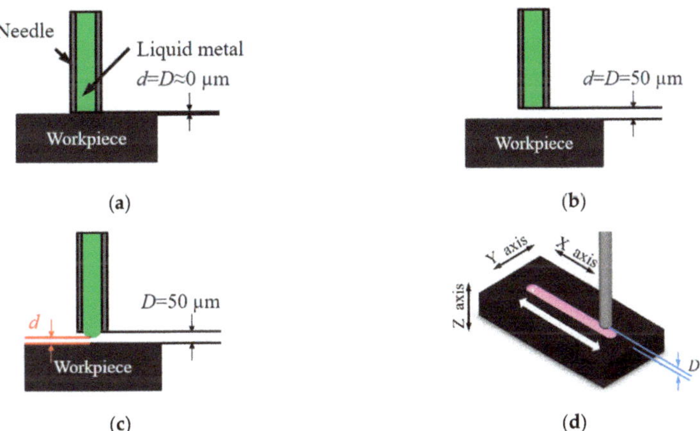

Figure 10. Electrical surface detection and scanning erosion. A low pressure (0.1 atm) is applied via a precision pressure device to ensure that the liquid metal in the needle is just flush with the tip of the needle; then, a small voltage (10 V) is used between them to carry out electrical surface detection using the Z-axis of the stage at a low speed from (**a**) D = 0 µm (**b**) to D = 50 µm (retracting the needle) to determine the gap. (**c**) The above electrical surface detection is repeated three times, and the average is taken as the gap position. Additionally, the gap d between the liquid metal and the workpiece is adaptively determined during discharging. (**d**) Horizontal scanning of the workpieces using the XY stage with respect to the liquid electrode at a constant gap distance D for the lateral removal of the material using the liquid electrode.

Table 2. Experimental conditions.

Parameters	Conditions
Workpiece	Doped silicon (*p* type)
Voltage (V_s)	60–220 V
Resistance (R)	0.2–10 kΩ
Capacitance (C)	10–470 nF
Pressure (P_1)	1.1–1.5 atm
Working medium	Kerosene
Lateral scan speed	0.5 mm/s

Machining characteristics such as the material removal rate (MRR) (mm^3/min) and liquid-metal consumption rate (LMCR) (mm^3/min) are adopted to evaluate the effects of the machining parameters on the liquid-metal electrode μEDM processes. The MRR is computed as the ratio of the material removed from the workpiece (approximated as the volume of the frustum of the cone) to the recorded machining time of the EDM system. The LMCR is calculated as the ratio of the liquid metal consumed to the machining time.

5. Experimental Results and Discussion

5.1. Microgroove and Arbitrary Patterning

Microgroove erosion was performed through programming the motions of the three-axis stage. An example of a microgroove pattern using the coated needle is shown in Figure 11. The electrical conditions were $C = 27$ nF, $R = 1$ kΩ, $V_s = 100$ V, and $P_1 = 1.2$ atm. The tool scanning path in the XY plane was repeated for 200 scans at scanning speeds of up to 1.0 mm/s with a constant gap distance of $D = 100$ μm. A pressure of 1.2 atm was applied to the syringe. The profile of the produced groove was analyzed using a laser confocal microscope, and the sample results are included in Figure 12. The linewidth and depth of the produced groove were measured to be approximately 280 and 90 μm, respectively. The cross-sectional profile of the groove was similar to the shape of the liquid electrode tip. This result is reasonable because EDM is a copying process mode, and the groove will match the electrode. In addition, randomly distributed pits appeared at the edge of the processed groove in Figure 11a. The reason for the micropits near the edges of the produced groove is that some pulses were likely generated between the suspended liquid microelectrode and debris comprising the removed material of the workpiece and the liquid metal that had already dripped on the workpiece surface. By programming the contour of the nozzle motion with respect to the XY plane (with the same D value), arbitrary patterning was successfully demonstrated on the silicon sample (Figure 12).

(a)

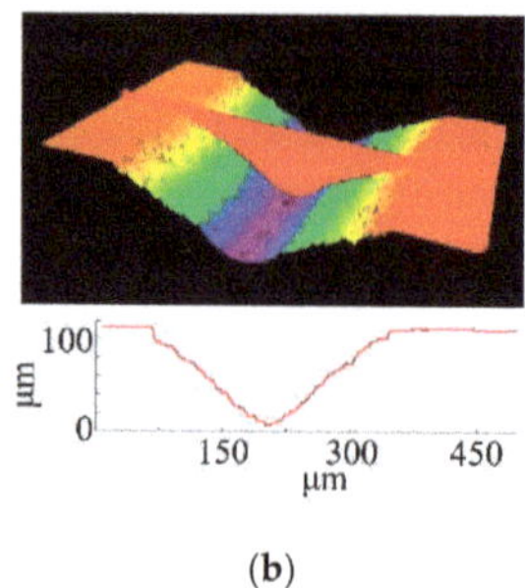

(b)

Figure 11. Linear groove pattern. (**a**) Sample of the microgroove and (**b**) profile of the microgroove.

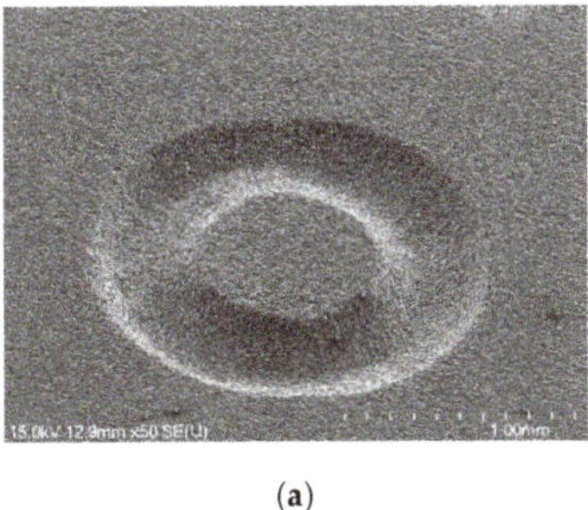
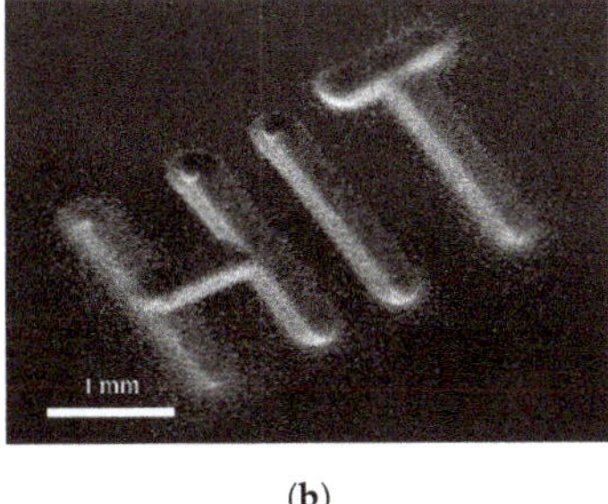

(**a**)   (**b**)

Figure 12. Sample result of arbitrary scanning-mode patterning performed using the developed process: (**a**) convex circular pillar and (**b**) "HIT" characters.

5.2. Process Characterizations

5.2.1. Dependence of Machining Characteristics on the Pressure

The widths of the linear patterns produced by lateral scanning of the Galinstan electrode were characterized while varying the pressure. The line widths of the produced patterns were measured using an optical measuring microscope. Figure 13 shows the relationship between the measured line widths and pressure (with V_s = 100 V, C = 100 nF, R = 0.5 kΩ, as shown in Table 2). As shown in the graph, an optimal pressure value for the minimum line width was found under the same discharge conditions.

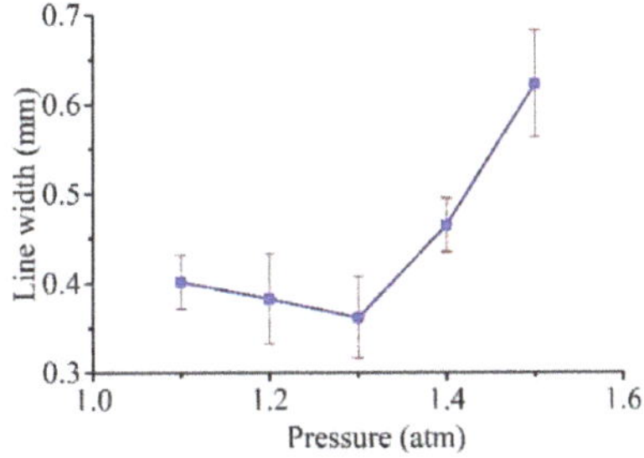

Figure 13. Effect of pressure P_1 on the line widths.

When the applied pressure was small, the liquid-solid contact angle θ was also small; hence, the tip of the liquid electrode was large. With an increase in applied pressure, the contact angle θ gradually increased, and the radius r of the curved liquid surface gradually decreased. That is, the end of the electrode correspondingly decreased; thus, the width of the machined groove decreased directly. When the applied pressure continued to increase, although the contact angle θ gradually increased, the radius r of the curved liquid surface gradually decreased, forming an electrode with a finer end; however, the width of the groove gradually increased. The reason for this result may be that when the applied pressure is greater, the liquid metal at the tip is more likely to fall onto the workpiece, which will affect the processing and cause discharge instability. Moreover, we also found that, when the external pressure was greater than 2 atm, the liquid metal ejected like a continuous stream of water; thus, it was impossible to carry out EDM.

In this connection, it is clear that there was a similar trend in the relationship between the MRR and pressure P_1, as shown in Figure 14. During the discharge process, when the pressure was small, the consumed Galinstan could not be compensated in time, leading to an open-circuit state most of the time; this resulted in the small value of the MRR. The MRR increases with increasing pressure because the frequency of discharge increases with an increasing Galinstan compensation.

After the optimal pressure value, as the pressure increases, excess Galinstan is suspended on the needle tip, and this droplet easily falls due to the discharge and electrostatic forces and causes unstable discharge, thus resulting in the MRR decreasing.

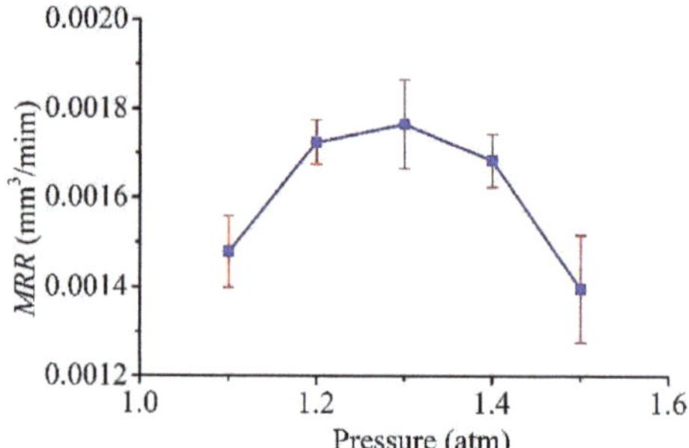

Figure 14. Effect of pressure P_1 on the material removal rate (MRR).

Figure 15 shows the relationship between the LMCR and pressure P_1. As shown in the graph, the LMCR increased with increasing pressure in a consistent manner. This outcome is reasonable because the flow rate, which determines the consumption of Galinstan based on the velocity of the flow, increased with increasing pressure. It is worth noting that, for pressures over 1.3 atm, the amount of fed Galinstan was much greater than the amount of Galinstan consumed by discharge at an excessive flow velocity. This result is disadvantageous for both the MRR and LMCR. Therefore, the feeding of consumed Galinstan is very important for high efficiency and stable processing, which largely depends on the applied pressure.

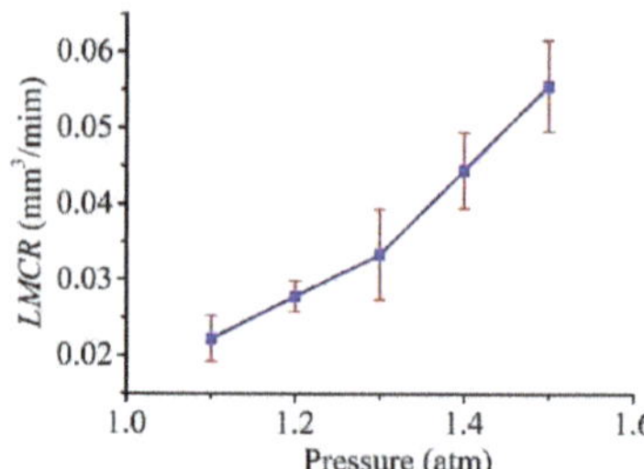

Figure 15. Effect of pressure P_1 on the liquid-metal consumption rate (LMCR).

5.2.2. Dependence of Machining Characteristics on the Voltage

Figure 16 plots the measured line widths as a function of the voltage (tested up to 220 V, with constant $C = 100$ nF, $R = 0.5$ kΩ, and $P_1 = 1.3$ atm, as shown in Table 2), and the figure clearly shows that the line width increased with the applied voltage in a consistent manner. This result is reasonable as the discharge energy, which determines the amount of material removed by a single pulse, increased with the voltage. In this regard, as shown in Figures 17 and 18, a similar trend was obtained for both the MRR and LMCR by varying the voltage V_s; the trends were consistent given the dependence of the discharge energy on the voltage. Increasing the discharge energy caused a greater spark intensity, which generated more melted material in the spark region. Thus, the workpiece and the electrode material were both subjected to an increase in sparks, which increased both the MRR and the LMCR. As a result, the MRR and LMCR substantially increased with increasing discharge energy.

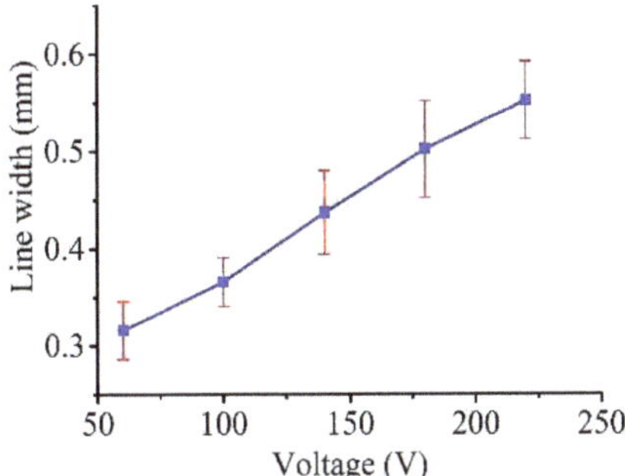

Figure 16. Effect of voltage V_s on the line widths.

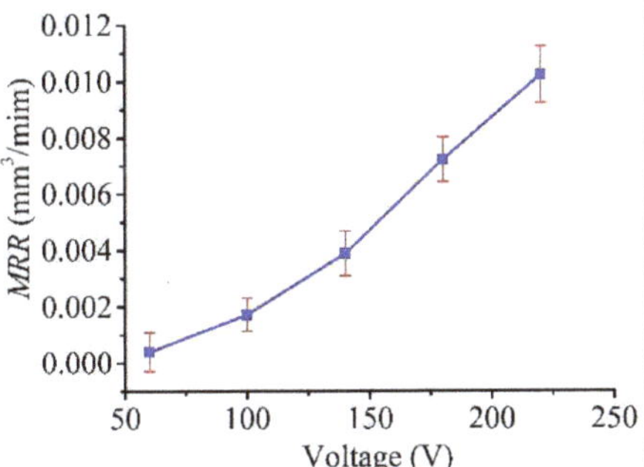

Figure 17. Effect of open voltage V_s on the MRR.

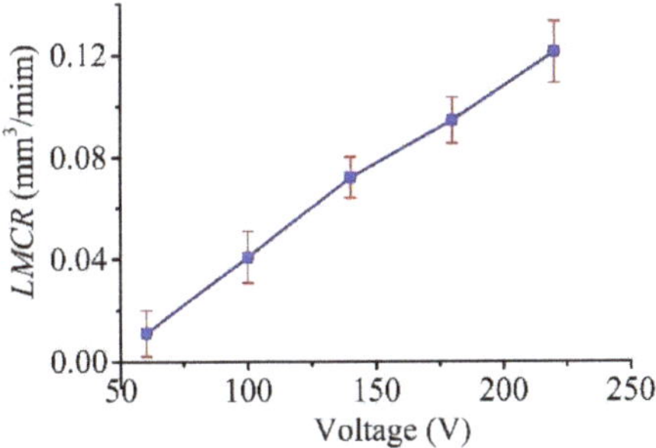

Figure 18. Effect of voltage V_s on the LMCR.

It can be seen from the simulation that a higher voltage resulted in a greater electric field, whereas, a larger contact angle θ resulted in a smaller radius r of the curved liquid surface. That is, the electrode tip was correspondingly smaller, and the widths of the linear patterns produced by the lateral scanning of the Galinstan electrode should be smaller as the voltage increases. However, the line width showed the opposite trend in this experiment, in that the line width increased with the voltage. There may be two reasons. First, the material removal depends on the energy of a single electric discharge, E_{SED}, which is expressed as

$$E_{SED} = \frac{1}{2}CU^2,\tag{15}$$

where C is the capacitance of the RC circuit and U is the machining voltage. From Equation (15), it can be seen that, as the voltage increases, the discharge energy of a single pulse increases squarely with the voltage; thus, the amount of removed material increases. The effect of increasing the voltage on the energy for material removal is much greater than that on the size of the electrode.

Second, the applied voltage for μEDM is relatively small, and the voltage changes within a small range of 50 to 220 V. The size of the electrode tip does not change significantly at low voltages. Only when the voltage is over 1 kV will a change in the liquid electrode tip be apparent. However, it is impossible to use such a high voltage for μEDM.

Therefore, changing the size of the electrode by increasing the voltage shows a much smaller effect on groove width than the effect of increased energy by increasing the voltage. Notably, from an energy point of view, an increase in voltage is beneficial to the MRR.

6. Conclusions

In this paper, the shape of the electrode, which is used in liquid-electrode μEDM processing for the purpose of resolving the problems related to electrode wear in traditional μEDM, was analyzed and simulated. Moreover, experiments were performed for verification. Arbitrary patterns on silicon samples using a liquid-metal electrode were well presented. The influence of extra pressure and open voltage V_s on the line widths, MRR, and LMCR were also investigated. The external pressure had a significant impact on the tip shape of the liquid electrode and its compensation. There was an optimum pressure value for the minimum line widths and maximum MRR. The LMCR increased with pressure; hence, an appropriate pressure was also beneficial to the LMCR. Although the voltage also influenced the tip shape of the liquid electrode, it was negligible at low voltage. However, at low voltages, the line widths, MRR, and LMCR consistently exhibited the same trend as the voltage. That is, they increased with voltage. Thus, the single-pulse discharge energy, which determines the amount of material removed by a single pulse, increased with the voltage.

Author Contributions: Conceptualization, R.H. and Y.Y.; methodology, R.H.; software, E.Z.; validation, R.H., Y.Y., E.Z. and X.X.; formal analysis, E.Z. and R.H; investigation, R.H. and E.Z.; resources, R.H.; data curation, E.Z.; writing—original draft preparation, R.H.; writing—review and editing, R.H. and X.X.; visualization, X.X.; supervision, R.H.; project administration, R.H.; funding acquisition, R.H. All authors have read and agreed to the published version of the manuscript.

Funding: This work was funded by the National Natural Science Foundation of China (No. 51975155, No. 51475107) and the Shenzhen Basic Research Program (No. JCYJ20170811160440239, No. JCYJ202008240825330001).

Acknowledgments: The author would like to thank Professor K. Takahata from the University of British Columbia for providing technical support in this research.

Conflicts of Interest: The authors declare no conflict of interest.

References

1. Bilal, A.; Jahan, M.P.; Talamona, D.; Perveen, A. Electro-Discharge Machining of Ceramics: A Review. *Micromachines* **2019**, *10*, 10. [CrossRef] [PubMed]
2. Chavoshi, S.Z.; Luo, X.C. Hybrid Micro-machining Processes: A Review. *Precis. Eng.* **2015**, *41*, 1–23. [CrossRef]
3. Raju, L.; Hiremath, S.S. A State-of-the-art Review on Micro Electro-Discharge Machining. *Procedia Technol.* **2016**, *25*, 1281–1288. [CrossRef]
4. Oliaei, S.N.B.; Karpat, Y.; Paulo Davim, J.; Preveen, A. Micro Tool Design and Fabrication: A review. *J. Manuf. Process.* **2018**, *36*, 496–519. [CrossRef]
5. Masuzawa, T.; Fujino, M.; Kobayashi, K.; Suzuki, T.; Kinoshita, N. Wire Electro-Discharge Grinding for Micro-Machining. *CIRP Ann. Manuf. Technol.* **1985**, *34*, 431–434. [CrossRef]
6. Egashira, K.; Morita, Y.; Hattori, Y. Electrical Discharge Machining of Submicron Holes Using Ultrasmall-diameter Electrodes. *Precs. Eng.* **2010**, *34*, 139–144. [CrossRef]
7. Zhang, L.; Tong, H.; Li, Y. Precision Machining of Micro Tool Electrodes in Micro EDM for Drilling Array Micro Holes. *Precis. Eng.* **2015**, *39*, 100–106. [CrossRef]
8. Lim, H.S.; Wong, Y.S.; Rahman, M.; Lee, M.K.E. A Study on the Machining of High Aspect Ratio Micro-structures Using micro-EDM. *J. Mater. Process. Technol.* **2003**, *140*, 318–325. [CrossRef]
9. Zhao, W.S.; Jia, B.X.; Wang, Z.L.; Hu, F.Q. Study on Block Electrode Discharge Grinding of Micro Rods. *Key Eng. Mater.* **2006**, *304–305*, 21. [CrossRef]

10. Yin, Q.F.; Wang, X.Q.; Wang, P.; Qian, Z.Q.; Zhou, L.; Zhang, Y.B. Fabrication of Micro Rod Electrode by Electrical Discharge Grinding Using Two Block Electrodes. *J. Mater. Process. Technol.* **2016**, *234*, 143–149. [CrossRef]

11. Mohri, N.; Tani, T. Micro-pin Electrodes Formation by Micro-Scanning EDM Process. *CIRP Ann. Manuf. Technol.* **2006**, *55*, 175–178. [CrossRef]

12. Liu, Y.; Cai, H.T.; Li, H.S. Fabrication of Micro Spherical Electrode by One Pulse EDM and Their Application in Electrochemical Micromachining. *J. Manuf. Process.* **2015**, *17*, 162–170. [CrossRef]

13. Singh, A.K.; Patowari, P.K.; Deshpande, N.V. Analysis of Micro-rods Machined Using Reverse Micro-EDM. *J. Braz. Soc. Mech. Sci.* **2019**, *41*, 15. [CrossRef]

14. Ming, P.M.; Zhu, D.; Zeng, Y.B.; Hu, Y.Y. Wear Resistance of Copper EDM Tool Electrode Electroformed from Copper Sulfate Baths and Pyrophosphate Baths. *Int. J. Adv. Manuf. Technol.* **2010**, *50*, 635–641. [CrossRef]

15. Hu, Y.Y.; Zhu, D.; Qu, N.S.; Zeng, Y.B.; Ming, P.M. Fabrication of High-aspect-ratio Electrode Array by Combining UV-LIGA with Micro Electro Discharge Machining. *Microsyst. Technol.* **2009**, *15*, 519–525. [CrossRef]

16. Zhang, Y.O.; Zhao, W.S.; Kang, X.M. Intensive Study on the Electrostatic Field Induced Electrolyte Jet Micro Electrical Discharge Machining. *Procedia CIRP* **2018**, *68*, 599–603. [CrossRef]

17. Huang, R.N.; Yi, Y.; Yu, W.B.; Takahata, K. Liquid-phase Alloy as a Microfluidic Electrode for Micro-electro-discharge patterning. *J. Mater. Process. Technol.* **2018**, *258*, 1–8. [CrossRef]

18. Bissacco, G.; Valentincic, J.; Hansen, H.N. Towards the Effective Tool Wear Control in Micro-EDM Milling. *Int. J. Adv. Manuf. Technol.* **2010**, *47*, 3–9. [CrossRef]

19. Wang, J.; Ferraris, E.; Galbiati, M.; Qian, J.; Reynaerts, D. Simultaneously Counting of Positive and Negative Pulse Parts to Predict Tool Wear in Micro-EDM Milling. In Proceedings of the 9th International Conference on Micromanufacturing, Singapore, 25–28 March 2014; Available online: https://lirias.kuleuven.be/handle/123456789/445665 (accessed on 25 March 2014).

20. Tsai, Y.Y.; Masuzawa, T. An Index to Evaluate the Wear Resistance of the Electrode in Micro-EDM. *J. Mater. Process. Technol.* **2004**, *149*, 304–309. [CrossRef]

21. Chiou, H.; Tsao, C.C.; Hsu, C.Y. A Study of the Machining Characteristics of Micro EDM Milling and its Improvement by Electrode Coating. *Int. J. Adv. Manuf. Technol.* **2015**, *78*, 1857–1864. [CrossRef]

22. Tsai, H.C.; Yan, B.H.; Huang, F.Y. EDM Performance of Cr/Cu-based Composite Electrodes. *Int. J. Mach. Tool. Manuf.* **2003**, *43*, 245–252. [CrossRef]

23. Zou, R.M.; Yu, Z.Y.; Yan, C.Y.; Li, J.Z.; Liu, X.; Xu, W.J. Micro Electrical Discharge Machining in Nitrogen Plasma Jet. *Precs. Eng.* **2018**, *51*, 198–207. [CrossRef]

24. Wang, J.; Qian, J.; Ferraris, E.; Reynaerts, D. In-situ Process Monitoring and Adaptive Control for Precision Micro-EDM Cavity Milling. *Precs. Eng.* **2017**, *47*, 261–275. [CrossRef]

25. Zhang, L.N.; Du, J.Y.; Zhuang, X.S.; Wang, Z.L.; Pei, J.Y. Geometric Prediction of Conic Tool in Micro-EDM Milling with Fix-length Compensation Using Simulation. *Int. J. Mach. Tool. Manuf.* **2015**, *89*, 86–94. [CrossRef]

26. Nirala, C.K.; Saha, P. Precise µEDM-drilling Using Real-time Indirect Tool Wear Compensation. *J. Mater. Process. Technol.* **2017**, *240*, 176–189. [CrossRef]

27. Lee, C.S.; Heo, E.Y.; Kim, J.M.; Choi, I.H.; Kim, D.W. Electrode Wear Estimation Model for EDM drilling. *Robot. Cim-Int. Manuf.* **2015**, *36*, 70–75. [CrossRef]

28. Huang, R.N.; Xiong, X.G.; Yi, Y.; Zhu, E.L. Liquid alloy electrode for no-wear micro electrical discharge machining. *Int. J. Adv. Manuf. Technol.* **2020**, *106*, 1281–1290. [CrossRef]

29. Liu, Y.C.; Wang, Y. Analysis and Discussion on Related Problems of Surface Tension. *Chin. J. Chem. Educ.* **2018**, *39*, 12–15. [CrossRef]

Publisher's Note: MDPI stays neutral with regard to jurisdictional claims in published maps and institutional affiliations.

 *micromachines*

Article

Pulse-Type Influence on the Micro-EDM Milling Machinability of Si₃N₄–TiN Workpieces

Valeria Marrocco [1],* , Francesco Modica [1] , Vincenzo Bellantone [1] , Valentina Medri [2] and Irene Fassi [3]

1 STIIMA CNR, Institute of Intelligent Industrial Technologies and Systems for Advanced Manufacturing, National Research Council, Via P. Lembo 38/F, 70124 Bari, Italy; francesco.modica@stiima.cnr.it (F.M.); vincenzo.bellantone@stiima.cnr.it (V.B.)

2 ISTEC CNR, Institute of Science and Technologies for Ceramics, National Research Council, Via Granarolo, 64-48018 Faenza (RA), Italy; valentina.medri@istec.cnr.it

3 STIIMA CNR, Institute of Intelligent Industrial Technologies and Systems for Advanced Manufacturing, National Research Council, Via A. Corti 12, 20133 Milan, Italy; irene.fassi@stiima.cnr.it

* Correspondence: valeria.marrocco@stiima.cnr.it; Tel.: +39-080-548-1265-1; Fax: +39-080-548-2533

Received: 15 September 2020; Accepted: 12 October 2020; Published: 13 October 2020

Abstract: In this paper, the effect of the micro-electro discharge machining (EDM) milling machinability of Si₃N₄–TiN workpieces was investigated. The material removal rate (MRR) and tool wear rate (TWR) were analyzed in relation to discharge pulse types in order to evaluate how the different pulse shapes impact on such micro-EDM performance indicators. Voltage and current pulse waveforms were acquired during micro-EDM trials, scheduled according to a Design of Experiment (DOE); then, a pulse discrimination algorithm was used to post-process the data off-line and discriminate the pulse types as short, arc, delayed, or normal. The analysis showed that, for the considered process parameter combinations, MRR was sensitive only to normal pulses, while the other pulse types had no remarkable effect on it. On the contrary, TWR was affected by normal pulses, but the occurrence of arcs and delayed pulses induced unexpected improvements in tool wear. Those results suggest that micro-EDM manufacturing of Si₃N₄–TiN workpiece is relevantly different from the micro-EDM process performed on metal workpieces such as steel. Additionally, the inspection of the Si₃N₄–TiN micro-EDM surface, performed by SEM and EDS analyses, showed the presence of re-solidified droplets and micro-cracks, which modified the chemical composition and the consequent surface quality of the machined micro-features.

Keywords: ceramic composite; micro-EDM milling; pulse discrimination

---

1. Introduction

Silicon nitride-based ceramics constitute a class of structural materials characterized by high strength, fracture toughness, thermal shock resistance, wear resistance, low coefficient of friction, and hardness. These ceramics can often be sintered by adding electrically conductive reinforcements, such WC, MoSi₂, TiN, TiC, TiCN, TiB₂ and ZrN, with variable content, thus obtaining electro-conductive ceramic composites [1–3]. This feature allows for the manufacture of these materials via non-contact technologies such as electrical discharge machining (EDM). In particular, EDM has shown feasibility for the realization of free form features and shapes required in high-temperature and aggressive environments for the production of various heating elements (e.g., ceramic glow plugs, igniters, ceramic heaters [4]. Moreover, due to the fact of their biocompatibility, some of these ceramic composites were investigated for the realization of load-bearing prostheses and bone mini-fixation devices [3–6]. In this field, micro-EDM displayed a high potential for the fabrication of micro-free-form features in Si₃N₄–TiN ceramic composite workpieces needed to realize such medical components [7].

Issues related to micro-EDM manufacturing of Si_3N_4–TiN ceramic composite have been already explored by the research community. Indeed, machining performance and surface quality were analyzed by Liu et al. [8] in relation to different machining regimes and to the manufacturing of a mesoscopic gas turbine. From the process viewpoint, the main results demonstrated that when a relaxation-type pulse generator was chosen for the micro-EDM process, faster machining and optimized material removal rate (MRR) and low tool wear rate (TWR) could be accomplished at the expense of surface quality and roughness. In particular, surface quality inspection showed that chemical reactions, triggered by the spark occurrence, induced a modification of the composite structure and also took part in the material removal mechanism. Subsequently, Liu et al. [9] presented an analysis on the micro-EDM capability of manufacturing Si_3N_4–TiN ceramic composites, considering different machining regimes. In that work, the authors presented evidence of the variability ascribed to the pulse shape and duration of the discharge waveforms, not entirely due to the process setting. Therefore, also in this case, Si_3N_4–TiN material seemed to have played an active role in the material removal process, almost independently of the machining regimes considered for the analyses. It was evident that the optimization of micro-EDM related to ceramic composites deserved further investigation, since the relatively low value of the electrical conductivity of such material, the chemical composition of the matrix, and the consequent intrinsic inhomogeneity of the composite introduced several challenges. Very recently, a work by Selvarajan et al. [10] pointed out the effect of micro-EDM process parameters (voltage, current, pulse on time, and pulse off time) on MRR and surface quality related to two Si_3N_4-based composites. The results showed once more the challenge of assessing optimized process parameters and accomplishing the required feature accuracy.

In the last decade, several authors have proposed alternative approaches to investigate micro-EDM process more thoroughly. In particular, the monitoring of voltage (gap and/or open) and discharge current waveforms became established to improve the general understanding of the micro-EDM process. However, the majority of the papers dealing with the micro-EDM monitoring approach considered metals as workpieces, and their main goal was the assessment of tool wear. For instance, the authors in Reference [11] exploited discharge monitoring during micro-EDM milling to measure tool wear per discharge and material removal per discharge. A method based on pulse counting was proposed in Reference [12] to estimate the total energy associated to discharge pulses and investigate material removal and tool wear characteristic for different micro-EDM machining types (shape-up and flat-head) and conditions (spindle rotation and tool electrode vibration). The discrimination of positive and negative parts of voltage and current waveforms was proposed in Reference [13] with the goal of estimating tool wear and tool wear error. However, the reported results held validity for relatively stable micro-manufacturing and features having small depths.

A different use of micro-EDM monitoring was proposed in Reference [14]. In this work, a first classification of pulses in relation to micro-EDM milling and wire processes was presented, and four main pulse categories were identified: normal, effective arc, transient short circuit, and complex pulses. A pulse-type discrimination strategy was also implemented and applied to micro-EDM milling of hardened steel in order to investigate the influence of process parameters on discharge shapes and process performance [15]; the study emphasized the importance of sparking gap and feed rate for the process stability and identified that tool wear increase could be associated to the increase of arc number. Recently, numerical simulations and experiments were used to evaluate the occurrence of pulse types in dependence of debris accumulation [16,17]. The studies were carried out considering reverse micro-EDM (RMEDM) and confirmed that the increase of debris quantity induced a modification of pulse shapes. In particular, the authors found different pulse orders, where the first one was identified as a normal pulse, and the other two pulse types (second and higher orders, characterized by lower voltage values) occurred more frequently as the concentration of debris increased within the sparking gap. In order to evaluate the differences among pulse types and shapes, a different approach based on the analysis of power spectral density (PSD) applied to the micro-EDM milling process was proposed considering two different workpiece materials: hardened steel [18] and Si_3N_4–TiN [19]. The results underlined that especially for Si_3N_4–TiN, the number of discharges per acquisition windows underwent a huge variability, and the discharge probability was very low compared to steel workpiece manufacturing. Moreover, different energetic contributions to the material removal process from normal pulses was observed for the micro-EDM milling of Si3N4–TiN workpieces independent of the energy settings.

Although relevant investigations of the manufacturing of Si_3N_4–TiN ceramic composites are currently available in the literature, the influence of discharge pulse shapes on micro-EDM performance indicators has not been fully assessed. Therefore, in the present paper, micro-EDM milling of Si_3N_4–TiN ceramic composite was analyzed by means of pulse-type characterization and MRR and TWR evaluation. The experiments were performed by implementing a design of experiment (DOE): the finishing regime was applied, thus implying the use of relaxation-type generator producing short pulses. The voltage and current waveforms were acquired during 54 trials resulting from the DOE, where process parameters, frequency (F) and gap and pulse width (W), were varied. The pulse classification was obtained by an off-line pulse discrimination strategy. The post-processed data related to pulse types and distribution were then put in relation to MRR and TWR. Finally, chemical and surface characterization of the Si_3N_4–TiN ceramic composite workpiece, before and after the micro-EDM process, is also reported and discussed.

2. Material, Micro-EDM Settings, and Design of Experiment (DOE)

The workpiece considered in this study was a sintered billet (25 mm in diameter) of Si_3N_4–TiN composed by a mixture of commercial raw powders. A TiN (grade C, HC Starck Ltd., Munich, Germany) 35% in volume was used as secondary electro-conductive phase and Y_2O_3 (grade C, HC Starck Ltd., Munich, Germany) and Al_2O_3 (Ultra-High Purity, Baikowski Chemie SA, Poisy, France) were used as sintering aids. The electrical resistivity was equal to 5.88×10^{-4} Ω cm. Figure 1 reports the SEM micrograph of the mirror polished and plasma-etched surface of the Si_3N_4–TiN-35 vol% composite: the typical microstructure of silicon nitride, composed of elongated silicon nitride grains in a matrix of nearly equiaxed sub-micrometric grains, was highlighted.

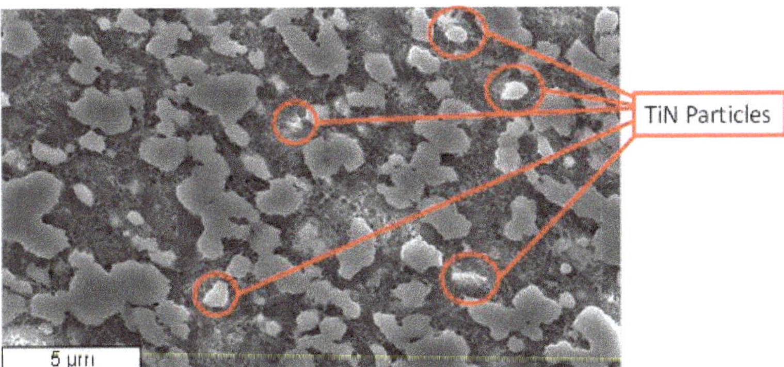

Figure 1. SEM micrograph of a mirror polished and plasma-etched surface of the Si_3N_4-35 vol% composite: dark grains are Si_3N_4 particles, while coarser, light grey grains are TiN particles.

As highlighted in Figure 1, TiN particles are coarser (from 1 to 5 µm) than the silicon nitride grains. The grain boundary phase, consisting of silicates and oxynitrides of the cations from sintering aids, locates preferentially at triple points, and a continuous film of amorphous phase is present at the interfaces between Si_3N_4 grains. Furthermore, after cutting the sintered billet by diamond tool machining (DTM), two different Ra were measured: a longitudinal surface roughness equal to Ra = 0.16 µm and a transversal surface roughness equal to Ra = 0.30 µm.

The basic feature used to implement the experimental plan was a micro-channel 5 mm long, 50 μm deep, and 0.42 mm wide, manufactured onto the sintered Si_3N_4–TiN workpiece. The micro-EDM machine was the Sarix SX 200. The tool was a tungsten carbide (WC) cylindrical rod having a nominal diameter of 0.4 mm; hydrocarbon oil was used as dielectric fluid in the experiments. All technological parameters adopted for the trials are reported in Table 1.

Table 1. The micro-electro discharge machining (EDM) milling process parameter settings.

Micro-EDM Process Parameter	Unit	Values
Energy (E)	Index	110 (finishing, RC generator)
Open Circuit Voltage (OCV)	V	100
Layer Thickness (LT)	μm	1
Frequency (F)	kHz	120–160
Pulse Width (W)	μs	1–3–5
Gap	Index	40–60–80

For Sarix SX 200, E is an index identifying the pulse generator type selected for the machining. In this case, E110 indicates finishing regime actuated by RC relaxation-type generator capable of producing short pulses. The OCV is the open voltage value, i.e., the maximum voltage achievable before discharge. Since the micro-EDM milling approach is implemented via a layer-by-layer strategy, a layer thickness (LT) of 1 μm was set. Frequency (F) indicates the frequency of the micro-EDM pulse generator, while pulse width (W) indicates the time interval in which the micro-EDM generator is disconnected from the tool and the workpiece and the discharge occurs within the sparking gap. Finally, Gap is the index related to the servo control loop responsible for the sparking gap stability (i.e., the distance between the tool and the workpiece). In order to estimate the statistical impact of the parameters' variability on machining performance and pulse distribution, a general full factorial Design of Experiment (DOE) was implemented: the varied process parameters were frequency (F) and pulse width (W) and Gap, which were grouped in 18 sets of parameters (SoP); three replicas of the experimental plan were performed, resulting in a total of 54 trials.

3. Micro-EDM Monitoring Setup and Pulse Discrimination Strategy

The sketch of the micro-EDM monitoring setup, along with the actual machining setup, is reported in Figure 2.

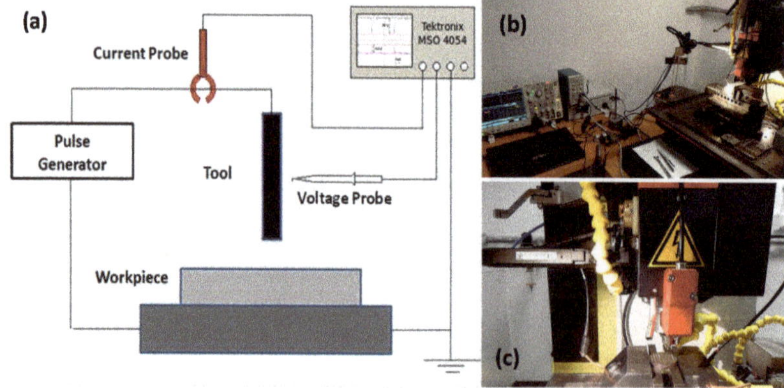

Figure 2. (**a**) Sketch of the micro-EDM monitoring setup; (**b**) actual monitoring setup connected to the Sarix SX 200 via voltage and current probes; (**c**) voltage and current probes mounted on the Sarix SX 200.

The acquisition of the voltage and current waveforms was done by means of two probes—a standard voltage probe and a Tektronix TCP312 current probe with a bandwidth of 100 MHz. The voltage probe was placed close to the tool tip in order to record voltage values (OCV, gap voltage). As the current probe works through Hall effect, it was hooked to the cable supplying the tool. In order to acquire and record the waveforms, both probes were connected to the oscilloscope—Tektronix MSO4054. The pulse generator was not connected to the oscilloscope.

The oscilloscope sampling frequency (Fs = 50 MHz) and the number of samples for each waveform ($Ns = 10^6$) were set considering the bandwidth of signals to be acquired and the duration of the phenomenon. With this setting, the duration of each acquisition window was equal to $T_{acq} = 20$ ms. The acquired data were delivered to a PC through the oscilloscope USB port: due to the USB port speed, an elapsed time of 0.8 s between two subsequent windows in each trial must be considered. The acquired waveforms were then supplied as input to the off-line classification algorithm which provided the number of each pulse type as an output. The pulse types were gathered in four classes: short, arc, delayed, and normal [14,15,18,19]. The discharge events occur within the sparking gap during the time interval equal to T–W, being T = 1/F, but only if the spark condition is fulfilled. When the discharge happen, a drop of OCV to its minimum value is observed. In this case, a normal pulse is detected. If the spark condition is not favorable, OCV keeps its constant value, as no discharge occurs. Nonetheless, local machining conditions, such as the presence of unremoved debris in the sparking gap, can promote discharge events independently of the pulse generator; in this case, two types of pulses can be observed: arcs and delayed. The distinction between these two categories can be made based on the maximum voltage value. Generally, arcs have lower Vmax than delayed. Short pulses happen when both tool and workpiece electrodes come into contact due to the instantaneous local debris concentration, and they are characterized by low voltage values approaching the ground level. Short pulses do not provide any material removal process and generally produce a waste of time.

The flow chart of the pulse discrimination strategy is reported in Figure 3; all routines were implemented in MATLAB®. The beginning of the algorithm foresees an initialization phase, where all pulse type counters—N_p (total), N_s (shorts), N_a (arcs), N_d (delayed), N_n (normal)—are reset. Subsequently, the acquisition windows of one trial are analyzed one by one via the algorithm, to assess the presence and type of pulses. The first iteration comprises the computation of the time derivative of the V waveform (dV); then, this value is compared to a set threshold (dV_{th}). If dV < dV_{th}, one pulse is identified, and the total counter N_p is incremented by one (N_p + 1). If this condition is not satisfied, the program starts searching again for a pulse. If a pulse is found, the algorithm proceeds to its identification, starting from shorts: in this case, a comparison among mean values of V and I and defined thresholds $V_{thrShort}$ and $I_{thrShort}$ is performed. If this check succeeds, the counter N_s is incremented by one (N_s + 1). If this check fails, then the algorithm searches for delayed pulse. A delayed pulse is identified by comparing the maximum I recorded during the discharge with the defined threshold I_{thrDel}. If this check succeeds the counter N_d is incremented by one (N_d + 1), otherwise the algorithm searches for arcs. The criterion for this comparison is the same as before, along with the comparison of the mean V with a programmable threshold V_{thrArc}. If this check succeeds the counter N_a is incremented by one (N_a + 1), otherwise the pulse is classified as normal and the counter N_n is incremented by one (N_n + 1). At the end of the classification, the index of the derivative of the voltage array is incremented of a programmable amount (about a half of the pulse width) in order to speed up the next pulse search, and the loop starts over again to analyze the subsequent acquired observation window.

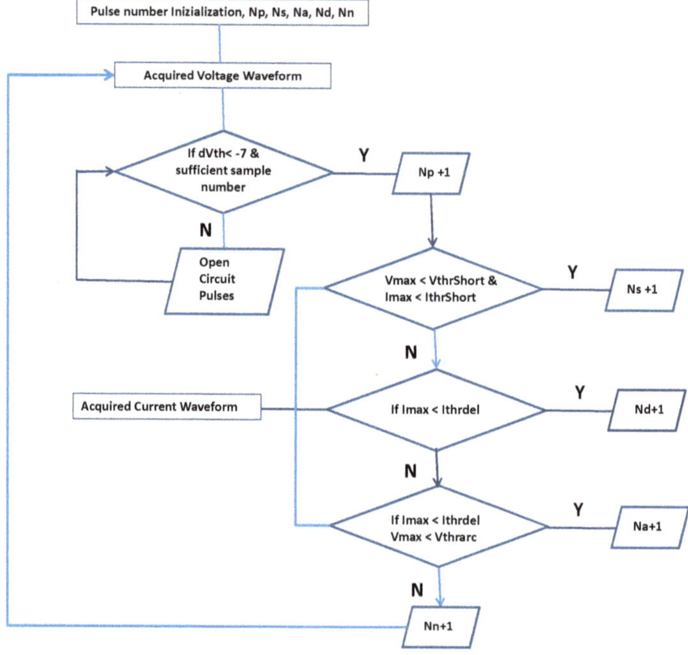

Figure 3. Flow chart of the discrimination strategy algorithm applied to all acquisition windows at each trial.

The defined current and voltage threshold values shown in Table 2.

Table 2. I and V threshold settings.

Threshold Value	(% of OCV and Imax)
I_{thr}	50% Imax (A)
I_{thr_short}	10% Imax (A)
V_{thr_short}	30% OCV (V)
V_{thr_arc}	95% OCV (V)

4. Results

In this section, pulse-type distribution, statistical analysis, and the results regarding the influence of pulse-type number and occurrence on MRR and TWR are presented and discussed.

4.1. Pulse-Type Distribution

The monitoring of each trial required a certain number of acquisition windows, approximately 250 for each process. The monitored trials related to all micro-EDM processes displayed relevant variability of discharge occurrence. In order to explain this fact, we reported the count of all pulse types recorded in the last 40 acquisition windows (i.e., the last part of the machining) and pertaining two distinct trials: F = 120 kHz, G = 60, W = 3 μs and F = 120 kHz, G = 60, W = 1 μs. As shown in Figure 4, the total number of all discharges occurring in each acquisition window was very variable. This fact confirms issues about the repeatability of the micro-EDM process on this kind of ceramic composite, mainly ascribable to the workpiece itself rather than to the parameter settings. Nonetheless, it is evident that in all acquired subsequent widows, the number of normal pulses always exceeds the number of arcs, delayed, and shorts.

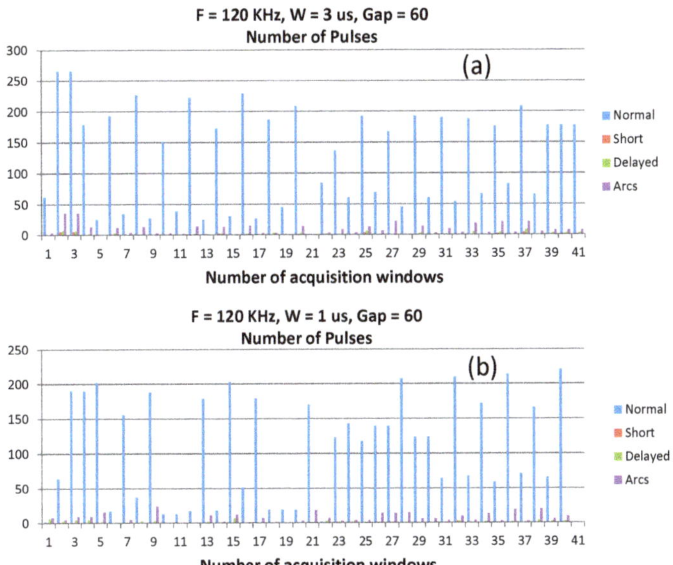

Figure 4. Number of normal pulses counted in the latest 40 acquisition windows for two different trials: (**a**) F = 120 kHz, W = 3 µs, G = 60; (**b**) F = 120 kHz, W = 1 µs, G = 60.

Figure 5 shows the average distribution in percentage of all pulse types for all trials calculated during the last part of the micro-EDM manufacturing. The histograms highlight that normal pulses are the most numerous, indicating the general stability experienced during all micro-EDM trials. This fact is also confirmed by the negligible number of shorts, arcs, and delayed pulses. It is worth stressing that an anomalous behavior was found in trial 45, which was repeated at the end of all trials.

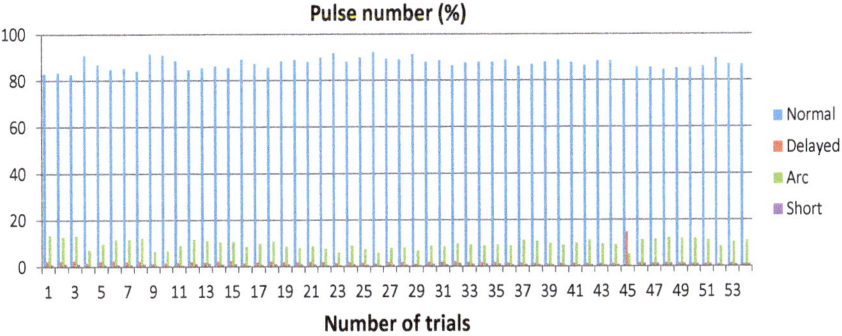

Figure 5. Number of normal, arc, short, and delayed pulses reported in percentage.

4.2. Analysis of Variance: Pulse Type and Process Parameters

Before proceeding with the statistical analysis of pulse types, the analysis of variance (ANOVA) was carried out to assess the relation between the selected process parameters (F, W and Gap) and the performance indicators, MRR, and TWR. The coefficients of determination R^2, which defines how much variation in the response is explained by the model, were 86.7% for MRR and 37.4% for TWR. These values suggested that the used model described quite properly the relation among F, W and Gap, and MRR, but it was not accurate in relation to TWR. It is worth stressing that the tool wear lengths measured by control touch procedure during micro-EDM milling process were generally very low (ranging between 3.5–6 µm) and this could have prevented the model from describing the statistical relation properly.

The analysis of variance was applied to the pulse types in order to identify the statistical relation with the process parameters. The coefficients of determination R^2 for normal, arc, delayed, and short pulses were 81.5%, 85.7%, 85.2%, and 68.8%, respectively. In the following analysis, short pulses were not considered; since their number was low, they were not statistically relevant, as indicated by the lowest coefficient of determination, and they did not take part as a material removal mechanism.

The first step of the analysis comprised the examination of residuals for normal, arc, and delayed pulses to assess the validity of the ANOVA method. The Pareto charts are reported in Figure 6. For all cases, the Pareto showed normal distribution and homogeneity of variances, thus confirming the validity of the assumptions for the ANOVA method application.

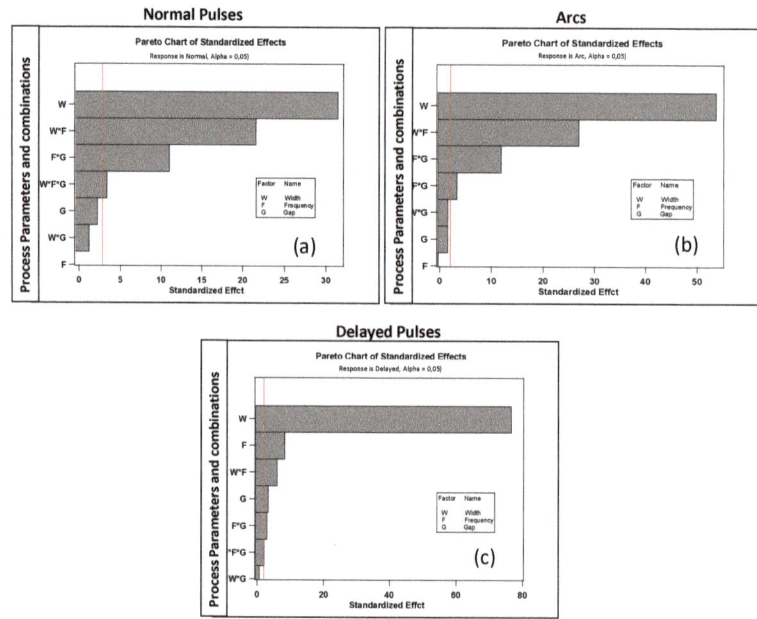

Figure 6. Pareto charts for (**a**) normal, (**b**) arcs, and (**c**) delayed pulses.

As evident from the charts, all pulses were influenced by combinations of the chosen process parameters. Only delayed pulses displayed less sensitivity to gap, while it was mainly influenced by W and F. All regression equations obtained from the statistical analysis and related to normal, arc, and delayed pulses are reported in Appendix A.

Replica diagrams and main effects plots were computed and reported in Figures 7–9. Figure 7a,b refer to normal pulses; the graphs shows that their number increased when W was equal to 3 µs. In this case, the generator was completely recharged within W, so that the rest of the period t–W was reserved for discharge events. As one can notice, in the considered range of values, parameter F had almost no effects on pulse occurrence, whereas higher gap values (farther distances between tool and workpiece) favored normal pulses.

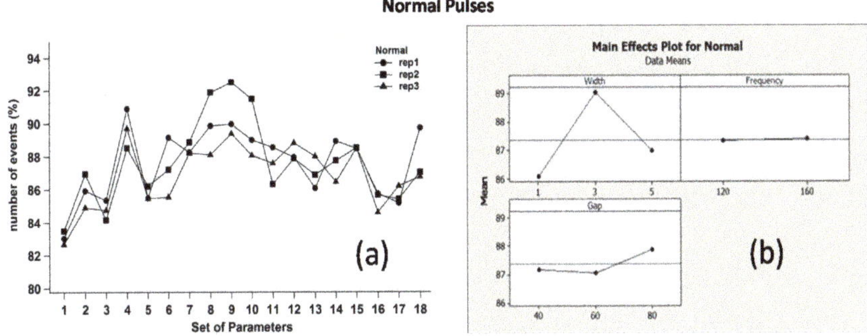

Figure 7. (**a**) Replica diagrams and (**b**) main effects plot related to normal pulses.

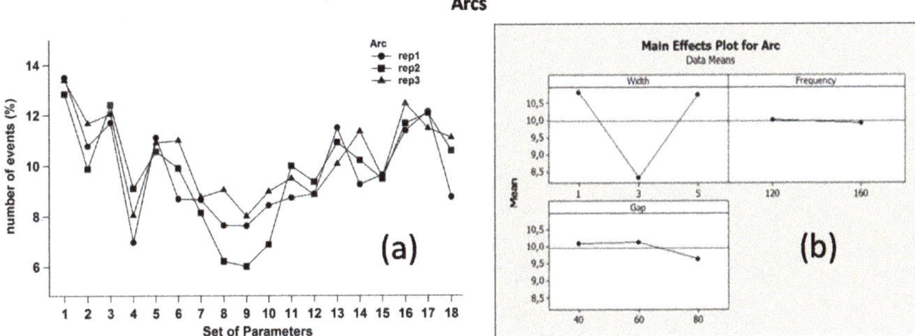

Figure 8. (**a**) Replica diagrams and (**b**) main effects plot related to arc pulses.

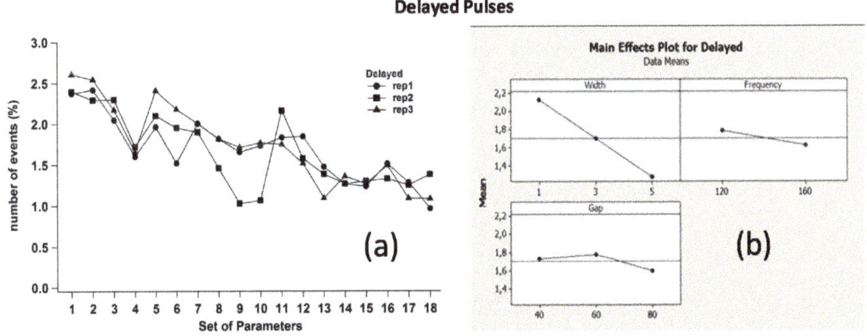

Figure 9. (**a**) Replica diagrams and (**b**) main effects plot related to delayed pulses.

Figure 8a,b refer to arcs, where the minimum number of such a pulse type was detected in correspondence of W = 3 μs and Gap = 80, thus confirming that these parameter values could actually provide a stable process. Nonetheless, also in this case, F seemed to have negligible impact on arcs. It is worth noting that normal and arc pulses exhibited a dual response in relation to the process parameter effects, i.e., an increase in normal pulses should correspond a decrease in arcs.

The plots related to delayed pulses are reported in Figure 10a,b.

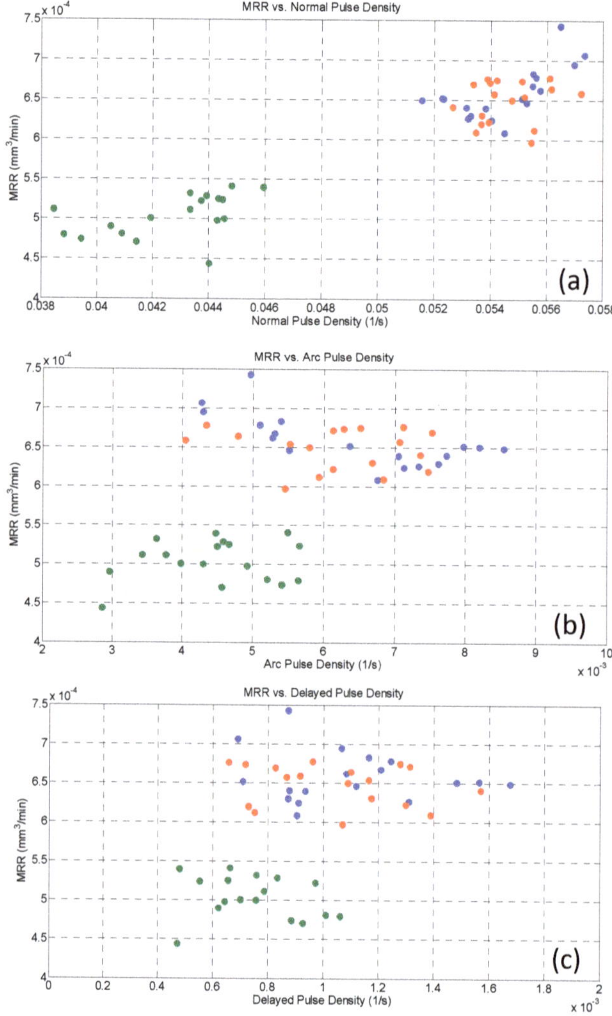

Figure 10. MRR versus (**a**) normal, (**b**) arc, and (**c**) delayed pulse density: blue dots refer to gap = 40, red dots to gap = 60, and green dots to gap = 80.

Delayed pulses have different dependences on W, F and Gap compared to the previous pulse types. First, they exhibit a strong linear dependence on W. Moreover, if F is reduced, the discharge probability time t = T–W increases slightly, thus promoting the probability for a delayed to happen. Furthermore, for a higher gap value, the delayed pulse occurrence is certainly reduced. In other words, when the tool and the workpiece are farther from each other, the required energy to allow spark ignition is the highest possible, i.e., set OCV. However, when maximum voltage is applied to the load, it is more likely to trigger normal pulses rather than delayed, which are characterized by lower OCV values. The result is that delayed number decreases. Furthermore, the plots show quite clearly that the delayed–Gap relation is similar to the arcs.

4.3. Relation between Pulse Types and Machining Performance

In previous works [15,18], the monitoring, pulse discrimination, and statistical analysis were performed on hardened steel workpieces with similar process parameter settings. In particular, those previous results suggested that a higher number of normal pulses had a positive effect on process stability and machining speed, so that MRR results would be good, while TWR increased its value, thus following a dual trend with respect to MRR. It was also found that as arcs increased in number and occurrence, MRR had a beneficial turn, and TWR worsened. Considering this roadmap, similar results were partly expected in the present analysis, as the role of the ceramic workpiece composition would certainly affect the final results. In order to represent MRR and TWR as a function of pulse type number and occurrence, a normalized pulse density (NPD) was defined according to the following equation:

$$\text{NPD} = \text{average } (N_{\text{PulseType}}/T_{\text{acq}}) \tag{1}$$

where $N_{\text{PulseType}}$ is the number of each pulse type (normal, arc, short or delayed) recorded in the last 25 µm of micro-channel manufacturing, and T_{acq} is the duration of the acquisition window. Figures 10 and 11 depict the corresponding average values of MRR and TWR. The diversity in the dots color indicates the process parameter chosen to discriminate the behavior, in this case, Gap.

Figure 10a shows that MRR increases as the occurrence of normal pulses increases as well. When Gap is higher, a lower number of overall pulses is expected as the tool and workpiece are farther from each other. Moreover, in this condition, the erosion process was slower thus decreasing MRR. On the contrary, arcs (Figure 10b) did not induce any variability on MRR, while a slight decrease of it was observed when a delayed occurrence was higher (Figure 10c). Reasonably, by considering the nature of delayed pulses, the increase in the machining times could explain this trend. Nonetheless, these results reveal that arcs and delayed did not affect MRR significantly; so, even though these pulse types are capable of removing material, their occurrence was neither problematic nor positive to the process, as long as their percentage was kept smaller than normal pulses.

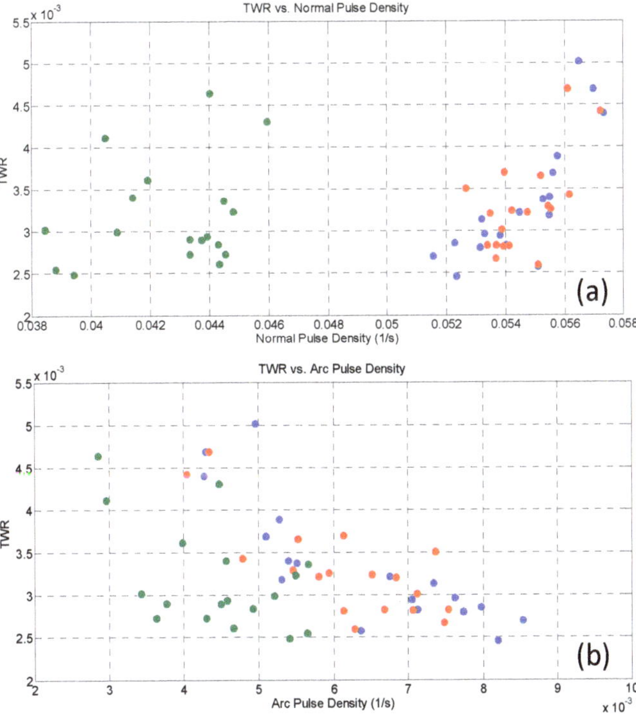

Figure 11. *Cont.*

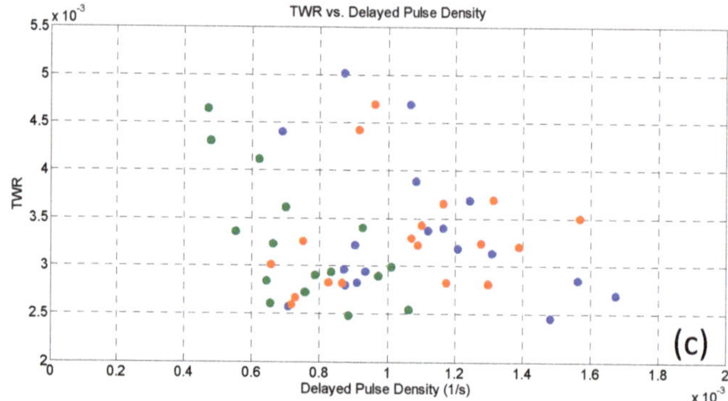

Figure 11. TWR versus (**a**) normal, (**b**) arc, and (**c**) delayed pulse density: blue dots refer to gap = 40, red dots to gap = 60, and green dots to gap = 80.

Figure 11 reports TWR behavior in the function of normal, arc, and delayed normalized pulse density.

First, it must be pointed out that the overall recorded TWR values are very small, in particular if these results are compared to those obtained with hardened steel (TWR is one order of magnitude higher for steel, and it ranged between 0.02–0.04 [15,18]). However, the relation between TWR and normal pulses confirms the expected trend, i.e., higher pulse occurrence leads to higher tool consumption. Conversely, a peculiar trend can be observed in Figure 11b: when arcs increase, TWR decrease, and this is in contrast with the expected results. Indeed, arcs are capable of removing material from both workpieces and tools, and an increase of TWR should have actually been measured. It was also verified that the tool wear volume was actually lower. Two reasons may explain these results: the first relies on the energy density of arcs, which was lower than normal pulses, and so less material could have been removed from both electrodes. Furthermore, a protective layer, generated by the melting, decomposition, and evaporation of Si_3N_4 particles during the micro-EDM process, was deposited on the tool surface and prevented the tool from more severe wearing. Finally, Figure 11c displays the trend for delayed pulses which is akin to that reported for arcs, thus suggesting a similarity of the nature for such pulses.

4.4. Surface Evaluation of Si_3N_4–TiN Workpiece before and after the Micro-EDM Process

The SEM and EDS analyses were also performed to evaluate the surface quality after micro-EDM milling of the Si_3N_4–TiN ceramic composite. Figure 12 evidences a foamy and porous structure, as also observed in Reference [8] with many spark-induced craters and melt-formation droplets appearing on the composite surface after micro-EDM.

The super-positioning of craters with varying diameters, which is not clearly distinguishable, derives from the material removal mechanism, linked to the interaction between the materials characteristics and chemical reactions triggered by the erosion process. The quantity of material that solidified and adhered to the surface is a function of the composition and microstructure of the starting ceramic materials, whereas the re-solidified droplets are almost entirely due to the TiN particles; conversely, Si_3N_4 particles are removed by evaporation. We also observed that, for a length of about 1 mm, the tool tip was covered by a protective layer, typically composed of Si_3N_4 particles. The presence of micro-cracks was due to the thermal expansion mismatch between silicon nitride, titanium nitride ($3.0 \cdot 10$-6 °C-1 and $9.4 \cdot 10$-6 °C-1, respectively) and re-solidified particles induced by the plasma breakdown. It can be also noticed that, since the material exhibited a degree of porosity, this could have affected the discharge probability, thus reducing the number of total discharges. For sake of completeness, surface roughness Ra of the machined micro-channels was also measured after micro-EDM processes: the Ra values ranged between 0.77–0.98 µm.

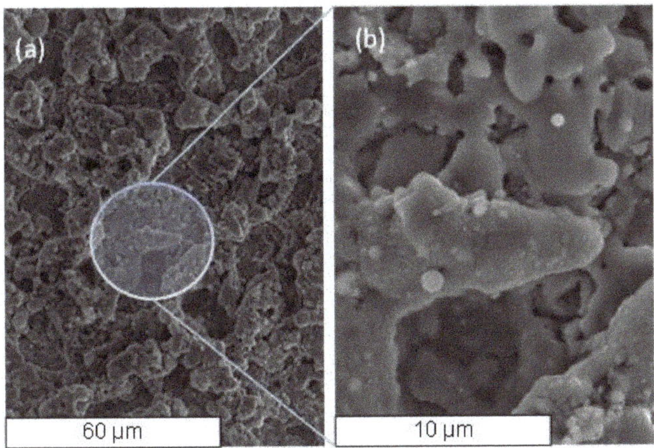

Figure 12. SEM micrograph of the EDM surface (**a**) and magnified (**b**).

Elemental composition was checked by EDS and reported in Figure 13, where DTM and micro-EDM surfaces are compared. The element concentration is explicated in Table 3.

From the analysis of the micro-EDM surfaces, the presence of C and W belonging to the tool was detected. Moreover, a marked decrease of N due to the fact of its evaporation during the erosion process was evident with respect to the DTM surface (Table 3). The elemental composition was checked on the micro-EDM surfaces also by point EDS analysis on a merging particle. Besides the presence of C and W, Co was also detected due to the contribution of the tool (Co is commonly used in WC composition), while N was absent. It can be noticed that the Si/Ti atomic ratio was not constant, as it depends on the dimension of the checked area and on the surface modification due to the preferential ablation/oxidation of Si_3N_4 or TiN during the micro-EDM process.

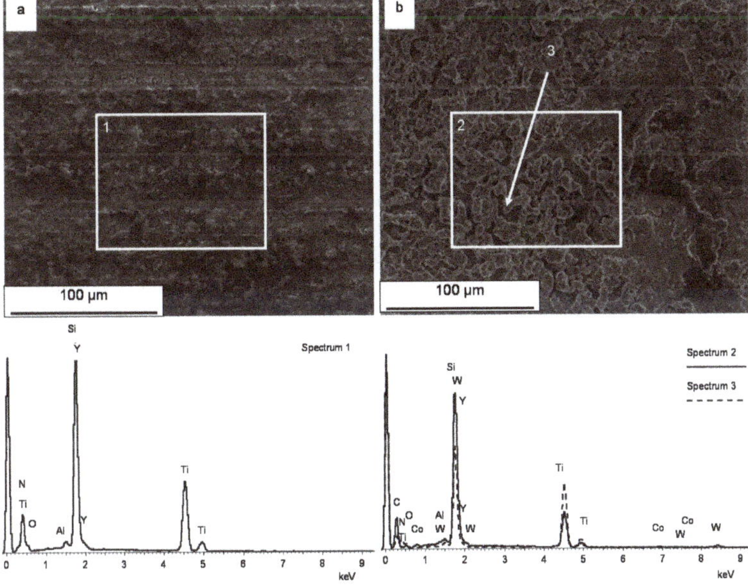

Figure 13. SEM-EDS characterization of Diamond Tool Machining (DTM) (**a**) and micro-EDM (**b**) surfaces.

Table 3. EDS element concentration.

Element	Spectrum 1: DTM Window (Atomic %)	Spectrum 2: EDM Window (Atomic %)	Spectrum 3: EDM Point (Atomic %)
C K	-	35.21	66.68
N K	45.01	10.91	-
O K	9.36	10.50	3.23
Al K	0.77	0.59	0.59
Si K	26.34	18.42	16.22
Ti K	17.79	23.41	9.45
Co K	-	-	0.51
Y L	0.75	0.46	0.60
W M	-	0.50	2.71

5. Discussion

The experimental results highlighted that the micro-EDM machinability of Si_3N_4–TiN ceramic composite workpieces differ drastically from experienced practice on metal workpieces. In particular, the statistical analysis of the pulse types in relation to process parameters showed that the triggering of normal, arc, and delayed pulses cannot be simply controlled through the setting of a single process parameter, but rather through a combination of them, such as pulse width W, frequency F and Gap. Moreover, when the pulse type occurrence is put in relation to micro-EDM performance indicators, it is clear that the MRR results were reasonably influenced by normal pulses, while arcs and delayed had a negligible effect on it. On the contrary, TWR exhibited very low values and did not display the common response expected from the different pulse types; in particular, arcs and delayed pulses did not affect TWR negatively.

6. Conclusions

The micro-EDM machinability of Si_3N_4–TiN ceramic composite workpieces was investigated by putting in relation pulse-type distribution with process parameters and performance indicators. To this aim, a Design of Experiment was implemented by considering the variation of pulse width W, frequency F and Gap. During the experiments, voltage and discharge current waveforms were acquired: the data were then post-processed to discern pulse type (i.e., normal, arcs, delayed, and shorts) via the pulse discrimination strategy. First, the pulse distribution showed that normal pulses were the most numerous, whereas other pulse types occurred very sporadically. The collected data were then statistically evaluated to elicit the relation existing among process parameter combinations and pulse-type distribution. The statistical analysis proved statistical reliability only for normal, arc, and delayed pulses, whereas short pulses were neglected in the discussion due to the numerical and statistical irrelevance. Pareto charts, regression equations, replica diagrams, and main effects plots revealed that all pulses were influenced by combinations of all selected process parameters.

The final analysis of the influence between pulse type and process performance indicators (MRR and TWR) showed that:

- MRR values increase as the number of normal pulses grows. Arcs and delayed pulses do not produce any positive or negative effects on MRR. A reasonable explanation is found in their low number and in their effective contribution to the material removal process.
- TWR displays very low values in all trials. Despite this, it was observed that the increase in normal pulses leads to the increase in TWR, as expected, whereas the increase in arcs and delayed pulses induces a peculiar decrease in TWR.

Finally, it was shown that the Si_3N_4–TiN workpiece features intrinsic degrees of electrical-conductivity inhomogeneity before machining as assessed by SEM and EDS evaluation. Additionally, same analyses performed on the micro-EDM surface displayed re-solidified droplets and micro-cracks, induced by chemical mechanism, which witnessed a surface roughness increase with respect to un-machined surface. Indeed, the surface roughness Ra measured after micro-EDM ranges between 0.77–0.98 μm, whereas the starting Ra of the workpiece ranges between 0.16 and 0.30 μm.

Author Contributions: Conceptualization, V.M. (Valeria Marrocco) and F.M.; methodology, V.M. (Valeria Marrocco), F.M. and V.B.; software, V.M. (Valeria Marrocco); validation, V.M. (Valeria Marrocco), F.M., V.B. and V.M. (Valentina Medri); formal analysis, V.M. (Valeria Marrocco) and V.B.; investigation, V.M. (Valeria Marrocco) and V.M. (Valentina Medri); data curation, V.M. (Valeria Marrocco), V.B. and V.M. (Valentina Medri); writing—original draft preparation, V.M. (Valeria Marrocco); writing—review and editing, V.M. (Valeria Marrocco) and V.M. (Valentina Medri); supervision, I.F. All authors have read and agreed to the published version of the manuscript.

Funding: This research received no external funding.

Conflicts of Interest: The authors declare no conflict of interest.

Appendix A

Since two of the considered factors (W and Gap) had three levels, a general full factorial design was planned for the experiments. The corresponding regression equations obtained by the statistical analyses for normal, arcs, and delayed pulses are reported below:

W = Width, G = Gap, F = Frequency

$$
\begin{aligned}
\text{Normal} =\ & 87.379 - 1292\,W\_1 + 1667\,W\_3 - 0.375\,W\_5 - 0.042\,F\_120 + 0.042\,F\_160 - 0.198 \\
& G\_40 - 0.305\,G\_60 + 0.503\,G\_80 - 1.455\,W \times F\_1\,120 + 1.455\,W \times F\_1\,160 + 0.677\,W \times F\_3 \\
& 120 - 0.677\,W \times F\_3\,160 + 0.779\,W \times F\_5\,120 - 0.779\,W \times F\_5\,160 + 0.495\,W \times G\_1\,40 + \\
& 0.055\,W \times G\_1\,60 - 0.550\,W \times G\_1\,80 + 0.141\,W \times G\_3\,40 - 0.015\,W \times G\_3\,60 - 0.126\,W \times \\
& G\_3\,80 - 0.636\,W \times G\_5\,40 - 0.039\,W \times G\_5\,60 + 0.676\,W \times G\_5\,80 - 0.968\,F \times G\_120\,40 + \\
& 0.840\,F \times G\_120\,60 + 0.128\,F \times G\_120\,80 + 0.968\,F \times G\_160\,40 - 0.840\,F \times G\_160\,60 - 0.128 \\
& F \times G\_160\,80 - 0.858\,W \times F \times G\_1\,120\,40 + 0.761\,W \times F \times G\_1\,120\,60 + 0.097\,W \times F \times G\_1 \\
& 120\,80 + 0.858\,W \times F \times G\_1\,160\,40 - 0.761\,W \times F \times G\_1\,160\,60 - 0.097\,W \times F \times G\_1\,160 \\
& 80 - 0.203\,W \times F \times G\_3\,120\,40 - 0.238\,W \times F \times G\_3\,120\,60 + 0.441\,W \times F \times G\_3\,120\,80 + \\
& 0.203\,W \times F \times G\_3\,160\,40 + 0.238\,W \times F \times G\_3\,160\,60 - 0.441\,W \times F \times G\_3\,160\,80 + 1.061\,W \times \\
& F \times G\_5\,120\,40 - 0.523\,W \times F \times G\_5\,120\,60 - 0.538\,W \times F \times G\_5\,120\,80 - 1.061\,W \times F \times G\_5\,160 \\
& 40 + 0.523\,W \times F \times G\_5\,160\,60 + 0.538\,W \times F \times G\_5\,160\,80
\end{aligned}
\tag{A1}
$$

$$
\begin{aligned}
\text{Arc} =\ & 9.963 + 0.848\,W\_1 - 1.643\,W\_3 + 0.795\,W\_5 + 0.051\,F\_120 - 0.051\,F\,160 + 0.136\,G\_40 + \\
& 0.181\,G\_60 - 0.318\,G\_80 + 1.169\,W \times F\_1\,120 - 1.169\,W \times F\_1\,160 - 0.580\,W \times F\_3\,120 + \\
& 0.580\,W \times F\_3\,160 - 0.589\,W \times F\_5\,120 + 0.589\,W \times F\_5\,160 - 0.298\,W \times G\_1\,40 - 0.170\,W \times \\
& G\_1\,60 + 0.468\,W \times G\_1\,80 - 0.141\,W \times G\_3\,40 + 0.022\,W \times G\_3\,60 + 0.120\,W \times G\_3\,80 + \\
& 0.439\,W \times G\_5\,40 + 0.148\,W \times G\_5\,60 - 0.587\,W \times G\_5\,80 + 0.721\,F \times G\_120\,40 - 0.632\,F \times \\
& G\_120\,60 - 0.089\,F \times G\_120\,80 - 0.721\,F \times G\_160\,40 + 0.632\,F \times G\_160\,60 + 0.089\,F \times G\_160 \\
& 80 + 0.668\,W \times F \times G\_1\,120\,40 - 0.632\,W \times F \times G\_1\,120\,60 - 0.036\,W \times F \times G\_1\,120\,80 - 0.668 \\
& W \times F \times G\_1\,160\,40 + 0.632\,W \times F \times G\_1\,160\,60 + 0.036\,W \times F \times G\_1\,160\,80 + 0.016\,W \times F \times \\
& G\_3\,120\,40 + 0.274\,W \times F \times G\_3\,120\,60 - 0.290\,W \times F \times G\_3\,120\,80 - 0.016\,W \times F \times G\_3\,160\,40 - \\
& 0.274\,W \times F \times G\_3\,160\,60 + 0.290\,W \times F \times G\_3\,160\,80 - 0.684\,W \times F \times G\_5\,120\,40 + 0.358\,W \times \\
& F \times G\_5\,120\,60 + 0.326\,W \times F \times G\_5\,120\,80 + 0.684\,W \times F \times G\_5\,160\,40 - 0.358\,W \times F \times G\_5 \\
& 160\,60 - 0.326\,W \times F \times G\_5\,160\,80
\end{aligned}
\tag{A2}
$$

$$
\begin{aligned}
\text{Delayed} =\ & 1.7033 + 0.4235\,W\_1 - 0.0001\,W\_3 - 0.4234\,W\_5 + 0.0827\,F\_120 - 0.0827\,F\_160 + \\
& 0.0260\,G\_40 + 0.0799\,G\_60 - 0.1059\,G\_80 + 0.1428\,W \times F\_1\,120 - 0.1428\,W \times F\_1\,160 - 0.0747 \\
& W \times F\_3\,120 + 0.0747\,W \times F\_3\,160 - 0.0681\,W \times F\_5\,120 + 0.0681\,W \times F\_5\,160 - 0.0924\,W \times \\
& G\_1\,40 + 0.0834\,W \times G\_1\,60 + 0.0090\,W \times G\_1\,80 + 0.0165\,W \times G\_3\,40 + 0.0237\,W \times G\_3\,60 - \\
& 0.0403\,W \times G\_3\,80 + 0.0759\,W \times G\_5\,40 - 0.1071\,W \times G\_5\,60 + 0.0312\,W \times G\_5\,80 + 0.1045\,F \times \\
& G\_120\,40 - 0.0603\,F \times G\_120\,60 - 0.0441\,F \times G\_120\,80 - 0.1045\,F \times G\_160\,40 + 0.0603\,F \times G\_160 \\
& 60 + 0.0441\,F \times G\_160\,80 + 0.0708\,W \times F \times G\_1\,120\,40 - 0.0340\,W \times F \times G\_1\,120\,60 - 0.0368 \\
& W \times F \times G\_1\,120\,80 - 0.0708\,W \times F \times G\_1\,160\,40 + 0.0340\,W \times F \times G\_1\,160\,60 + 0.0368\,W \times F \times \\
& G\_1\,160\,80 + 0.1106\,W \times F \times G\_3\,120\,40 - 0.0560\,W \times F \times G\_3\,120\,60 - 0.0545\,W \times F \times G\_3\,120 \\
& 80 - 0.1106\,W \times F \times G\_3\,160\,40 + 0.0560\,W \times F \times G\_3\,160\,60 + 0.0545\,W \times F \times G\_3\,160\,80 - \\
& 0.1814\,W \times F \times G\_5\,120\,40 + 0.0900\,W \times F \times G\_5\,120\,60 + 0.0913\,W \times F \times G\_5\,120\,80 + \\
& 0.1814\,W \times F \times G\_5\,160\,40 - 0.0900\,W \times F \times G\_5\,160\,60 - 0.0913\,W \times F \times G\_5\,160\,80.
\end{aligned}
\tag{A3}
$$

Micromachines **2020**, *11*, 932

References

1. Martin, C.; Cales, B.; Vivier, P.; Mathieu, P. Electrical discharge machinable ceramic composites. *Mater. Sci. Eng. A* **1989**, *109*, 351–356. [CrossRef]
2. Liu, C.-C. Microstructure and tool electrode erosion in EDMed of TiN/ Si$_3$N$_4$ composites. *Mater. Sci. Eng. A* **2003**, *363*, 221–227. [CrossRef]
3. Bucciotti, F.; Mazzocchi, M.; Bellosi, A. Perspectives of the Si3N4–TiN ceramic composite as a biomaterial and manufacturing of complex-shaped implantable devices by electrical discharge machining (EDM). *J. Appl. Biomater. Biomech.* **2010**, *8*, 28–32.
4. Bracisiewicz, M.; Medri, V.; Bellosi, A. Factors inducing degradation of properties after long term oxidation of Si3N4–TiN electroconductive composites. *Appl. Surf. Sci.* **2002**, *202*, 139–149. [CrossRef]
5. Mazzocchi, M.; Bellosi, A. On the possibility of silicon nitride as a ceramic for structural orthopaedic implants. Part I: Processing, microstructure, mechanical properties, cytotoxicity. *J. Mater. Sci. Mater. Med.* **2008**, *19*, 2881–2887. [CrossRef] [PubMed]
6. Mazzocchi, M.; Gardini, D.; Traverso, P.L.; Faga, M.G.; Bellosi, A. On the possibility of silicon nitride as a ceramic for structural orthopaedic implants. Part II: Chemical stability and wear resistance in body environment. *J. Mater. Sci. Mater. Med.* **2008**, *19*, 2889–2901. [CrossRef] [PubMed]
7. Modica, F.; Pagano, C.; Marrocco, V.; Fassi, I. Micro-EDM studies of the fabrication of customized internal fixation devices for orthopedic surgery. In Proceedings of the ASME 2015 IDETC/CIE, Boston, MA, USA, 2–5 August 2015.
8. Liu, K.; Reynaerts, D.; Lauwers, B. Influence of the pulse shape on the EDM performance of Si 3 N 4 –TiN ceramic composite. *CIRP Ann.* **2009**, *58*, 217–220. [CrossRef]
9. Liu, K.; Lauwers, B.; Reynaerts, D. Process capabilities of Micro-EDM and its applications. *Int. J. Adv. Manuf. Technol.* **2009**, *47*, 11–19. [CrossRef]
10. Selvarajan, L.; Rajavel, R.; Prakash, B.; Mohan, D.G.; Gopi, S. Investigation on spark electrical discharge machining of Si3N4 based advanced conductive ceramic composites. *Mater. Today Proc.* **2020**, *27*, 2174–2178. [CrossRef]
11. Bissacco, G.; Valentincic, J.; Hansen, H.N.; Wiwe, B.D. Towards the effective tool wear control in micro-EDM milling. *Int. J. Adv. Manuf. Technol.* **2009**, *47*, 3–9. [CrossRef]
12. Mahardika, M.; Mitsui, K. A new method for monitoring micro-electric discharge machining processes. *Int. J. Mach. Tools Manuf.* **2008**, *48*, 446–458. [CrossRef]
13. Wang, J.; Ferraris, E.; Galbiati, M.; Qian, J.; Reynaerts, D. Simultaneously counting of positive and negative pulse parts to predict tool wear in micro-EDM milling. In Proceedings of the ICOMM 2014, Singapore, 25–28 March 2014.
14. Liao, Y.S.; Chang, T.Y.; Chuang, T.J. An on-line monitoring system for a micro electrical discharge machining (micro-EDM) process. *J. Micromech. Microeng.* **2008**, *18*, 35009. [CrossRef]
15. Modica, F.; Guadagno, G.; Marrocco, V.; Fassi, I. Evaluation of micro-EDM milling performance using pulse discrimination. In Proceedings of the ASME 2014, IDETC/CIE, Buffalo, NY, USA, 17–20 August 2014.
16. Roy, T.; Datta, D.; Balasubramaniam, R. Debris based discharge segregation in reverse micro EDM. *Measurement* **2020**, *153*, 107433. [CrossRef]
17. Roy, T.; Datta, D.; Balasubramaniam, R. Numerical modelling and simulation of surface roughness of 3-D hemispherical convex micro-feature generated by reverse micro-EDM. *Int. J. Adv. Manuf. Technol.* **2018**, *97*, 979–992. [CrossRef]
18. Marrocco, V.; Modica, F.; Guadagno, G.; Fassi, I. Estimate of power spectral density of discharge pulses in micro-EDM milling. In Proceedings of the 4M-IWMF, Copenhagen, Denmark, 12–15 September 2016.
19. Marrocco, V.; Modica, F.; Fassi, I. Analysis of discharge pulses in micro-EDM milling of Si3N4–TiN composite workpiece by means of power spectral density (PSD). *J. Manuf. Process.* **2019**, *43*, 112–118. [CrossRef]

 *micromachines*

Article

Surface Characterization and Tribological Performance Analysis of Electric Discharge Machined Duplex Stainless Steel

Timur Rizovich Ablyaz [1,*], Evgeny Sergeevich Shlykov [1], Karim Ravilevich Muratov [1], Amit Mahajan [2], Gurpreet Singh [3], Sandeep Devgan [2] and Sarabjeet Singh Sidhu [3]

[1] Mechanical Engineering Faculty, Perm National Research Polytechnic University, 614000 Perm, Russia; kruspert@mail.ru (E.S.S.); karimur_80@mail.ru (K.R.M.)

[2] Mechanical Engineering Department, Khalsa College of Engineering and Technology, Amritsar 143001, India; amitmahajan291@gmail.com (A.M.); devgan.sandeep186@gmail.com (S.D.)

[3] Mechanical Engineering Department, Beant College of Engineering and Technology, Gurdaspur 143521, India; singh.gurpreet191@gmail.com (G.S.); sarabjeetsidhu@yahoo.com (S.S.S.)

* Correspondence: lowrider11-13-11@mail.ru

Received: 22 September 2020; Accepted: 5 October 2020; Published: 7 October 2020

Abstract: The present article focused on the surface characterization of electric discharge machined duplex stainless steel (DSS-2205) alloy with three variants of electrode material (Graphite, Copper-Tungsten and Tungsten electrodes). Experimentation was executed as per Taguchi L18 orthogonal array to inspect the influence of electric discharge machining (EDM) parameters on the material removal rate and surface roughness. The results revealed that the discharge current (contribution: 45.10%), dielectric medium (contribution: 18.24%) majorly affects the material removal rate, whereas electrode material (contribution: 38.72%), pulse-on-time (contribution: 26.11%) were the significant parameters affecting the surface roughness. The machined surface at high spark energy in EDM oil portrayed porosity, oxides formation, and intermetallic compounds. Moreover, a pin-on-disc wear analysis was executed and the machined surface exhibits 70% superior wear resistance compared to the un-machined sample. The surface thus produced also exhibited improved surface wettability responses. The outcomes depict that EDMed DSS alloy can be considered in the different biomedical and industrial applications.

Keywords: material processing; DSS-2205 alloy; electric-discharge machining; surface integrity; wear resistance; surface wettability

---

1. Introduction

Today, electric discharge machining notably established itself for the processing of hard and complicated geometrical contours, which are difficult to fabricate by traditional machining techniques [1,2]. This non-traditional machining technique showed its proficiency for the applications in the manufacturing of aerospace products, moulds, dies, etc [3,4]. The process is also recommended to fabricate the bio-implants owing to its favorable results in orthopedic fields [5–7]. The input process parameters namely pulse-on-time, pulse-off-time, current, dielectric medium, spark gap voltage, type of electrode and polarity (negative or positive) play a momentous role in the machining of diverse materials [8,9]. The optimum set of these machining performance parameters has promisingly enhanced the material removal rate and efficiently improves the surface properties [10,11].

Razavykia et al. [12] reported that the discharge current, pulse-on-time, electrode material, and voltage significantly influence the MRR and surface quality of Co-Cr-Mo alloy. Similarly, Mahajan and Sidhu [13] concluded that pulse-on-time, discharge current and electrode material were the dominant parameters for the improvement of corrosion resistance, wear characteristics and biocompatibility of Co-Cr samples. Furthermore, Philip et al. [14] employed the EDM for Ti6Al4V alloy and compared the tribological characteristics of a machined and unmachined specimen. The results of their study exhibited the improved specific wear rate and coefficient of friction due to the formation of oxides and carbide layers on the machined surface. Besides, Devgan and Sidhu [15] executed electro discharge treatment for investigating the surface wettability and corrosion resistance responses of β-titanium alloy. Simao et al. [16] scrutinized the impact of EDM process parameters on various responses such as material removal rate (MRR), tool wear rate (TWR) and surface hardness of AISI H13 tool steel. They concluded that the machining performance and surface properties of materials could be enhanced by an appropriate combination of EDM operation parameters. For instance, Singh et al. [17] reported dielectric medium and discharge current as eminent parameters for improved microhardness and wear resistance of stainless steel 316L using electro-discharge treatment. The processing of duplex stainless steel (DSS-2205) was carried out by Pramanic et al. [18]. They reported that input parameters i.e., pulse-on-time, pulse-off-time, and wire tension significantly influence the MRR, surface properties and kerf width. Alshemary et al. [19] reported that pulse-on and pulse-off time significantly influenced the wire-EDMing of DSS alloy. However, Rajmohan et al. [20] investigated the impact of wire EDM process parameters on the DSS-2205 alloy. It was observed that the current and pulse-on-time parameters had a considerable effect on MRR and SR. Rajaram et al. [21] utilized EDM for drilling of small holes (3 mm dia.) on DSS 2205. The experimental outcomes revealed that the input parameter such as current significantly contributed to the MRR of DSS alloy. Recently, Mahajan et al. [22] reported the excellent hemocompatibility and corrosion resistance outcomes of ED machined DSS-2205 alloy.

Along with the conventional tool electrodes, researchers also explored the performance of composite electrodes for the machining of hard to machine materials. Khanra et al. [23] considered a ZrB_2-Cu composite electrode for machining of mild steel workpiece and reported the improved MRR and diminished TWR as compared to the Cu tool. Similar findings were observed by Tsai et al. [24] who utilized Cr/Cu composite electrode and confirmed the formation of a recast layer on the surface that improved the corrosion resistance. The results also demonstrated higher MRR and lower TWR as compared to other metal electrodes. Grisharin et al. [25] observed the improved wear resistance and machining efficiency when a copper-colloidal graphite composite electrode was used to machine different alloys. However, Teng et al. [26] employed a Cu-Ni composite tool for the processing of polycrystalline diamond specimens and suggested better MRR as well as surface roughness responses as compared to the Cu electrode.

According to the literature survey, as briefly discussed above, it was observed that various materials had been machined using EDM. However, this technique is not considerably reported yet for the machining of duplex stainless steel (DSS-2205) alloy. DSS-2205 can be used as an alternative for austenitic stainless steel (316L) owing to its enormous applications in industry as well as in the biomedical field. This paper reported the effect of different types of tool materials and dielectric medium on material removal rate, surface roughness, morphology, phase transformation, tribological performance, and surface wettability of the EDMed surface. The first step examined the effect of chosen process parameters on the MRR and SR of machined samples, and statically scrutinizes the significant factors. The next step studied the surface morphology and phase analysis of samples depicting superior results using field emission scanning electron microscopy (FE-SEM), x-ray diffractometer (XRD) and energy dispersive x-ray analysis (EDX) techniques. The DSS alloy is commonly utilized in mining industries, heat exchangers and oil or gas processing industries, where wear characteristics and surface wettability play an important role for long term usage of alloy. Therefore, contact angle measurement and pin-on-disc wear tests were performed and compared with the results with an un-machined sample to investigate the tribological and wettability behavior of the EDMed sample.

2. Material and Methods

2.1. Tool and Workpiece

In this research, duplex stainless steel (DSS-2205) in the form of a square plate of 90 mm with a thickness of 20 mm was procured from Solitaire Impex, Mumbai, India. The workpiece chemical composition of Fe: 69.93%; Cr: 22.81%; Ni: 5.2%; Mo: 3.05%; Mn: 1.43%; Si: 0.5%; C: 0.028%; P: 0.03%; S: 0.02%, and density 7.8 g/cm^3; melting point 1350 °C, thermal conductivity 19.4 at 100 °C·W/mK, and electrical resistivity 0.085×10^{-6} Ω cm. Duplex stainless steel consists of chromium as its main content after iron and molybdenum which make greater utility of DSS-2205 alloy in the biomedical domain. The traditional machining processes are inappropriate to handle such hard materials. Therefore, under such circumstances, spark erosion commonly known as EDM employed as an emerging technique for treating such hard materials [27,28].

In this study, three different tool electrodes viz. graphite (C), copper-tungsten (25-Cu/75-W) and tungsten (W) were chosen for treating the DSS substrates in die-sinking EDM. Table 1 listed the specifications of all three electrodes. Initially, the emery paper (material: silicon carbide (SiC), grit-800) was employed for the surface finishing of the alloy plate. Further, the plate surface was cleaned with ethanol solution (C_2H_5OH) before ED machining.

Table 1. Properties of Graphite, Copper-Tungsten and Tungsten electrodes.

Property	Graphite	Copper-Tungsten	Tungsten
Diameter (mm)	10	10	10
Density (g/cm^3)	2.26	14.5	18.8
Melting point (°C)	3650	3410	3400
Electrical resistivity (Ω cm)	6.0×10^{-3}	4.5	5.6×10^{-3}
Thermal conductivity (W/mK)	24	189	163.3
Thermal expansion coefficient (μm/mK)	6	11.7	4.5
Specific heat capacity (J/Kg °C)	720	214	133
Hardness (HB)	10	195	2570

2.2. Design of Experiment

The Taguchi methodology was used to design the experimental array. In this investigation, an orthogonal array of L18 mixed-level design matrix was used to scrutinize the effects of five controllable parameters on two responses i.e., MRR, and SR. The chosen process parameters and their corresponding levels are tabulated in Table 2. The Minitab-17 statistical software was used to prepare the experimental design matrix. Further, analysis of variance (ANOVA) was utilized to analyze the dominance of process parameters on the MRR and SR.

Table 2. Parameters descriptions and values.

Parameter	Units	Levels		
		Level 1	Level 2	Level 3
Current (I)	ampere	5	10	16
Pulse-on-time (P-on)	μ-seconds	60	150	200
Pulse-off-time (P-off)	μ-seconds	60	150	200
Electrode	–	Gr	W	W-Cu
Dielectric medium	–	EDM oil	Deionized water	–

2.3. Experimental Procedure

All the experimental trials were performed on a die-sinker EDM (Electronica, India: Smart ZNC S50) with constant gap voltage (140 V) and machining depth (0.5 mm) for each run. Also, negative polarity (tool (+) workpiece (−)) has opted throughout the experimentation. The material removal process during the EDM technique depends upon the generation of heat on the substrate due to the abundance of electric sparks between the electrode [29,30]. Both tool electrode as well as alloy substrate immersed in the dielectric fluid tank that provides proper stability during this thermo-electric process.

An in-house fabricated tank (18″ × 18″ × 24″) was used of capacity 10 liters, containing a stirrer and circulation pump for appropriate flushing and avoiding debris within the working area. Figure 1 represented the schematic arrangement of EDM, experimental set up of machining and FE-SEM image of un-machined DSS-2205 substrate.

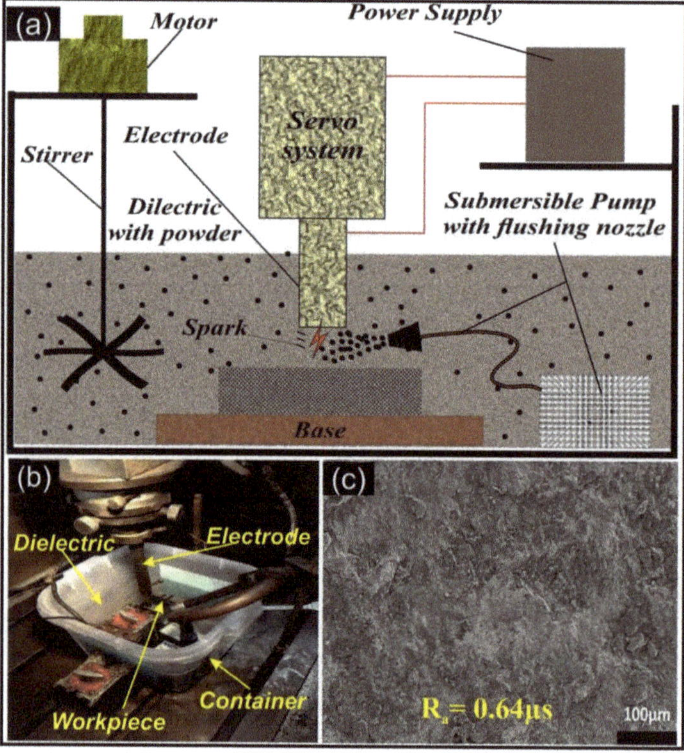

Figure 1. (**a**) Schematic arrangement of EDM; (**b**) Pictorial view of experimental set up of machining; (**c**) FE-SEM image of un-machined DSS-2205 substrate (Ra = 0.64 µs).

2.4. Calculations of Material Removal Rate (MRR) and Surface Roughness (SR)

The weight of the workpiece was measured before and after each trial using a precise weighing balance (Citizen CY220) for calculating the MRR using Equation (1).

$$MRR(\text{mm}^3/\text{min.}) = \frac{(w_2 - w_1) \times 1000}{\rho \times t} \tag{1}$$

where;

"w_2 and w_1" correspond to the workpiece weight (g) before and after each trial,
"ρ" is the density of workpiece, and
"t" (minutes) is the machining time.

The other output response i.e., surface roughness of the machined DSS substrates were measured using the Mitutoyo SJ-201 surface profilometer. The roughness of each machined sample was measured diametrically at three different points, and an average value was considered (Ra) for further investigation.

2.5. FE-SEM, EDX and XRD Analysis

After machining, the morphology of the surface was examined using FE-SEM (Hitachi SU-8810, Japan) at 11.0 kV of accelerating voltage. FE-SEM was also utilized to examine the surface after the wear test and also for the recast layer thickness. The phase transformation analysis (XRD; PANalytical X'Pert Pro MPD, The Netherlands) was performed using Cu-Kα X-ray radiation, and with generator settings of 40 mA and 45 kV. The elemental composition of the machined DSS-2205 specimen was analyzed via energy-dispersive X-ray spectroscopy (EDX; incorporated with FE-SEM) to observe the EDMed samples.

2.6. Investigation of Tribological Characteristics

Additionally, the sample exhibiting superior output responses i.e., high material removal and roughness were investigated for their tribological performance using a pin-on-disc type tribometer (DUCOM Instrument, Bangalore, India). ASTM G99-17, a standard for pin-on-disc wear analysis was followed and the EN31 steel disc of diameter 120 mm and thickness 20 mm was used to test the wear of the specimens at 100 rpm rotation speed. Test lubricant (ringer solution), track diameter (80 mm), steady load (70 N) and running time (3600 s) remains constant for each experimental run. The working operation of the tribometer, and calculations of wear, friction values obtained via associated software (TR-20LE) built-in with the attached computer system.

2.7. Microhardness and Recast Layer Thickness Measurement

German made; Mitutoyo microhardness tester was used under low-force hardness scale (HV 0.2) with a test force-load of 1.96 N for a dwell time of 10 s. The microhardness was figured thrice at distinct points, and an average value was noted for the calculation. The EDMed sample with superior output responses and correlated to tribological performance was cut cross-sectionally for measuring the recast layer thickness. The diamond paste was utilized for the mirror-polished of the substrate. The surface morphological investigation of a cross-section of the EDMed substrate depicted the recast layer thickness that was measured at five different positions at the transverse section and its average value was recorded.

2.8. Surface Wettability (Contact Angle Measurement)

The surface wettability of the alloy is the crucial property that impacts the other significant characteristics and also influences the enduring usage of the alloy substrate [31]. The hydrophobic or hydrophilic nature of the surface represented the wettability which was measured by the water contact angle (WCA). If WCA is greater than 90°, the surface is represented as hydrophobic, whereas, the angle lesser than 90° with the surface is considered as hydrophilic [32]. The wettability investigation was executed by utilizing a contact angle goniometer (Model 790; make: Rame–Hart instrument, USA) where the contact angle was computed in an environmental chamber through the sessile drop technique at 28 °C. The surface was cleaned with acetone solution before the experimentation. The contact angle of the substrate was measured at five different positions and an average value considered and reported as WCA. The digital camera captured a 20 μL distilled water profile of the droplet set on top of the substrate surface by a Gilmont microsyringe.

3. Results and Discussion

The present study predicts and optimizes the ED machining performance parameters for duplex stainless steel (DSS-2205). All the experimental trials were carried out thrice (i.e., 18 × 3 = 54 runs) to minimize the error and for precise outcomes. The respective results in Table 3 signified the average material removal rate and surface roughness values attained from each experimental run, followed by the standard deviation (i.e., Avg. ± S.D). Further, these results were investigated statistically via ANOVA and the most significant parameters that influence these outcomes were investigated.

Table 3. Experimental L18 array design matrix and output response observations.

Exp. Trial	Levels of Controllable Parameters					Output Responses			
	I (A)	P-on (µs)	P-off (µs)	Electrode	Dielectric Medium	MRR (mm³/min) Avg. ± SD	S/N Ratio (MRR)	SR (µm) Avg. ± SD	S/N Ratio (SR)
1.	1	1	1	1	1	4.06 ± 0.23	12.1585	0.14 ± 0.05	−17.68
2.	1	2	2	2	1	3.85 ± 0.29	11.6704	0.56 ± 0.07	−5.17
3.	1	3	3	3	1	3.98 ± 0.14	11.9995	0.71 ± 0.07	−3.00
4.	2	1	1	2	1	13.41 ± 0.96	22.5151	0.26 ± 0.02	−11.75
5.	2	2	2	3	1	18.38 ± 0.49	25.2822	1.0 ± 0.09	−0.36
6.	2	3	3	1	1	10.31 ± 0.13	20.2695	0.2 ± 0.04	−17.25
7.	3	1	2	1	1	14.49 ± 0.21	23.2229	0.13 ± 0.04	−19.02
8.	3	2	3	2	1	27.82 ± 0.04	28.8871	1.4 ± 0.08	2.79
9.	3	3	1	3	1	39.4 ± 0.98	31.9058	1.21 ± 0.06	1.65
10.	1	1	3	3	2	3.1 ± 0.05	9.8251	0.23 ± 0.04	−13.14
11.	1	2	1	1	2	4.36 ± 0.67	12.6451	0.40 ± 0.01	−7.82
12.	1	3	2	2	2	0.14 ± 0.04	−17.2078	0.30 ± 0.06	−11.44
13.	2	1	2	3	2	4.29 ± 0.6	12.5315	0.19 ± 0.05	−14.93
14.	2	2	3	1	2	5.38 ± 0.37	14.5921	0.08 ± 0.02	−22.50
15.	2	3	1	2	2	1.01 ± 0.04	0.1421	0.18 ± 0.03	−15.14
16.	3	1	3	2	2	7.45 ± 0.45	17.4242	0.28 ± 0.05	−11.27
17.	3	2	1	3	2	28.18 ± 0.66	28.9952	1.3 ± 0.08	2.24
18.	3	3	2	1	2	12.84 ± 0.30	22.1710	0.26 ± 0.02	−11.78

3.1. MRR Results Investigation by ANOVA

Table 4 detailed the ANOVA results for the material removal rate of DSS-2205 alloy. The F-values, with a confidence level of 95%, acquainted the influential factors that extremely affect the substrate surface responses after machining. The parameters with higher F-value reveal its superior impact on the output machining responses. Likewise, the *p*-value indicated the significance level of the input controllable factor. The signal-to-noise ratios (S/N ratios) results of MRR represented current as of the most significant factor that majorly contributes in removing the material from DSS-2205 alloy, followed by the dielectric medium and electrode material.

Table 4. ANOVA for Material Removal Rate.

Source	DF	Sum of Squares	Mean Squares	F-Value	*p*-Value	% Contribution
Current	2	1036.56	518.28	18.34	0.001 *	45.18
Pulse-on-time	2	232.70	116.35	4.12	0.059	10.14
Pulse-off-time	2	89.57	44.78	1.58	0.263	3.90
Electrode	2	290.77	145.39	5.15	0.037 *	12.67
Dielectric medium	1	418.50	418.50	14.81	0.005 *	18.24
Error	8	226.05	28.26			
Total	17	2294.14				

* Significant at 95% confidence level

However, the *p*-values (>0.05) for pulse-on-time and pulse-off-time are not considerable; consequently, both are in-significant factors for machining of DSS-2205 under the selected range of parametric settings. The machined alloy sample as per the parametric settings of trial 9, demonstrates the highest material removal rate (39.4 ± 0.98 mm³/min). These results are in accordance with previously reported studies, where researchers reported discharge current, pulse-on duration, dielectric type and electrode material as the most significant factors for the EDM performance affecting the output responses [33–35]. The optimum parameters for maximum material removal rate of DSS substrates were examined from the S/N ratios plot (Figure 2) as 16 A of current, P-on = 150 µs, P-off = 60 µs, and use of W-Cu electrode in EDM oil.

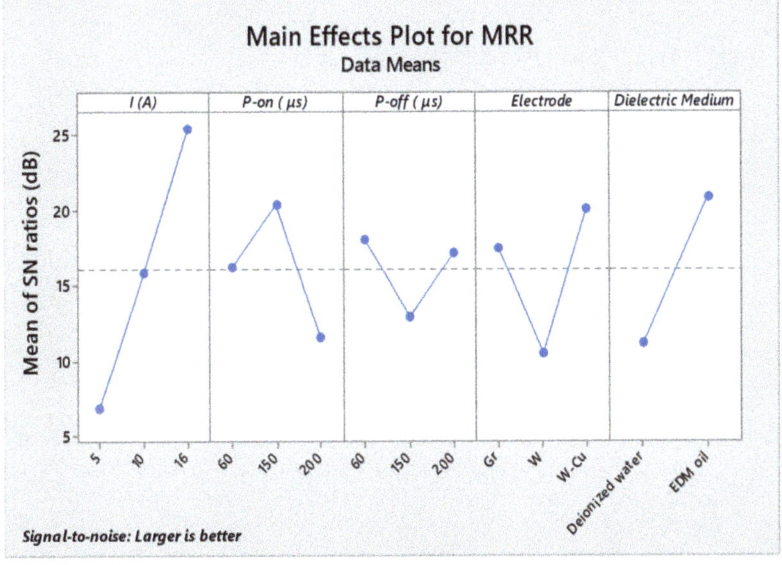

Figure 2. Main effect plot for S/N ratios of Material Removal Rate.

3.2. SR Results Investigation by ANOVA

Table 5 depicts the analysis of variance results for the input factors in order to observe their dominance affecting the S/N ratios outcome of the surface roughness. From ANOVA results, electrode material (p-value: 0.001) was the most significant factor with a confidence level of 95% that influences the surface roughness of the machined substrates. The other factors such as, pulse-on-time (p-value: 0.003), current (p-value: 0.011) and dielectric medium (p-value: 0.033) also play a momentous role in producing the rough surfaces. The S/N ratios plot (Figure 3) disclosed that the DSS alloy samples machined in EDM oil with copper- tungsten (Cu/W) electrode at 16 A current, 150 µs pulse-on-time with a 60 µs pulse-off-time provided the more substantial surface roughness responses.

Table 5. ANOVA for Surface Roughness.

Source	DF	Sum of Squares	Mean Squares	F-Value	p-Value	% Contribution
Current	2	180.62	90.31	8.31	0.011 *	17.39
Pulse-on-time	2	271.23	135.62	12.48	0.003 *	26.11
Pulse-off-time	2	25.38	12.69	1.17	0.359	2.50
Electrode	2	402.08	201.04	18.50	0.001 *	38.73
Dielectric medium	1	72.03	72.03	6.63	0.033 *	6.94
Error	8	86.94	10.87			
Total	17	1038.28				

* Significant at 95% confidence level

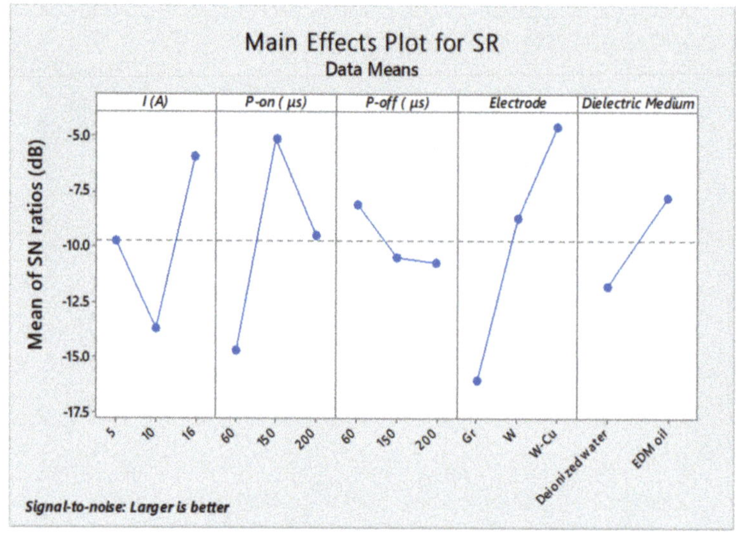

Figure 3. Main effect plot for S/N ratios of Surface Roughness.

Among all the trials, specimen machined according to the parameter set of trial 8 exhibits a highly rough surface (Ra = 1.4 ± 0.08 µm). The results showed that some other machined substrates also have improved surface roughness as compared to the un-machined surface (Ra = 0.64 µs). These outcomes also endorsed the prominence of EDM in the biomedical field, where surface roughness plays a crucial role in the adequate engagement of human tissues and bones with implant surface [36–38].

3.3. Surface Morphology and Compositional Analysis of Machined Surface

Figure 4 illustrates the surface morphology of the EDMed substrates (trial 8 and trial 9) exhibiting excellent results of material removal rate and surface roughness. Both the machined substrates showed micro and macropores and re-solidified metallic droplets on the surface. It is observable from the images (Figure 4a,b) that higher spark energy (Spark Energy = Current × P-on × Voltage) generate pores along with small peaks and valleys on the EDT surface [39,40]. The presence of the pores and molten metal droplets on the surface promotes the biological performance of the substrates [41].

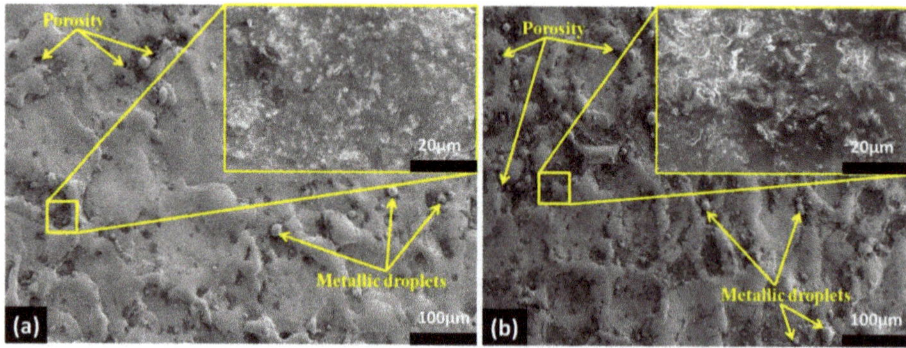

Figure 4. FE-SEM images illustrate surface roughness (**a**) Sample 8 (Ra = 1.4 µs), and (**b**) Sample 9 (Ra = 1.21 µs).

The EDX spectrum of the EDMed sample (trial 9) exhibiting higher material removal and surface roughness represented in Figure 5a. The presence of basic elements of DSS alloys viz. Fe, Ni, Cr,

Mo, Mn and C was observed on the surface. Moreover, apart from base elements of alloy, the high percentage of oxygen element was also observed on the machined substrate. The EDX outcome of phase transformation of the machined substrate is also affirmed by the XRD pattern (Figure 5b). The formation of compounds viz. rhombohedral structured iron oxide (Fe_2O_3), major phase of $CrMn_{1.5}O_4$, tetragonal structured iron-chromium (Fe-Cr), hexagonal structured chromium oxide (Cr_2O_3) and tungsten carbide (WC) on the machined surface improves the wear resistance of the alloy. The presence of these oxides and carbides on the surface also improves the biocompatibility of alloy substrate due to which DSS alloy can also utilize in the biomedical domain.

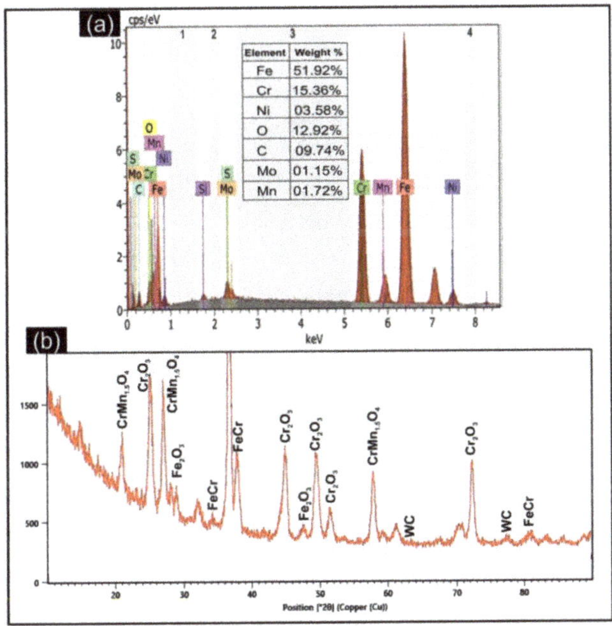

Figure 5. (**a**) EDS elemental spectrum and (**b**) XRD pattern of EDMed substrate.

3.4. Tribological Performance Analysis of Machined Surface

The machined sample with superior outcomes i.e., trial 9 was further assessed for their tribological performance. Moreover, the wear rate and coefficient of friction of the EDMed substrates were compared with the untreated substrate of DSS alloy. Table 6 demonstrates the wear characteristics of both treated and untreated substrates. It has been noticed that the wear rate of an untreated substrate ($3.52 \pm 0.15 \times 10^{-5}$ mm^3/Nm) was higher than the EDMed substrate ($1.23 \pm 0.11 \times 10^{-5}$ mm^3/Nm). Figure 6a represented the wear rate comparison of both specimens, whereas, the coefficient of friction with time for both pin substrates is showed in Figure 6b. It has been portrayed that the co-efficient of friction (μ) value of the un-machined specimen ($\mu_{average}$ = 0.32) is greater than the EDMed substrate ($\mu_{average}$ = 0.23).

Table 6. Wear rate and co-efficient of friction (COF) of samples.

Sr. No.	Sample	N = 1		N = 2		N = 3		Avg. ± SD	
		COF	Wear Rate (mm^3/Nm)	COF	Wear Rate (mm^3/Nm)	COF	Wear Rate (mm^3/Nm)	COF	Wear Rate (mm^3/Nm)
1.	Untreated	0.317	3.52×10^{-5}	0.322	3.68×10^{-5}	0.312	3.37×10^{-5}	0.317 ± 0.005	$3.52 \times 10^{-5} \pm 0.15$
2.	Treated	0.237	1.05×10^{-5}	0.246	1.13×10^{-5}	0.231	0.97×10^{-5}	0.238 ± 0.007	$1.05 \times 10^{-5} \pm 0.08$

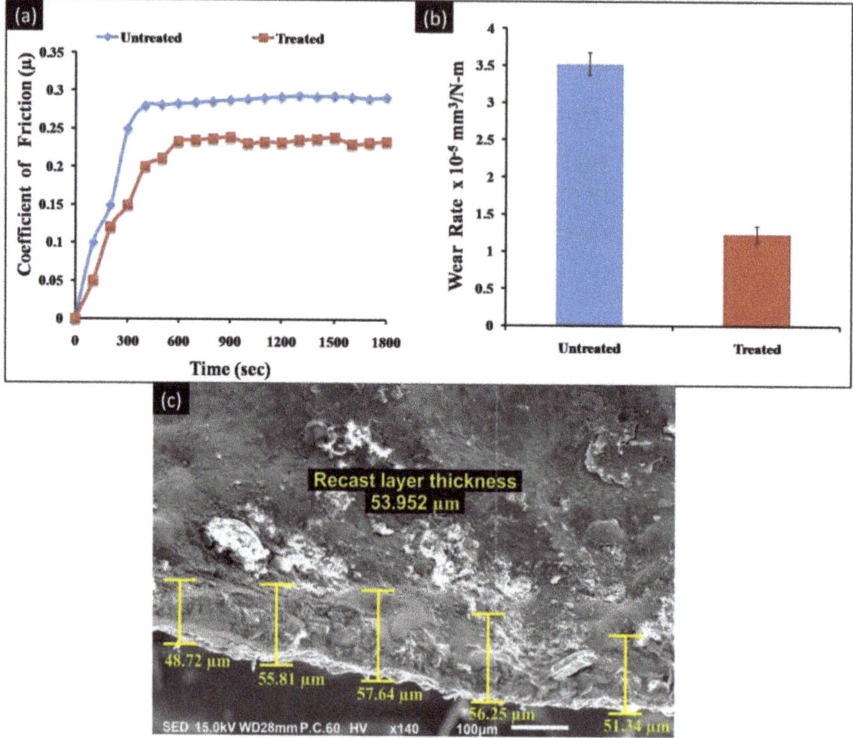

Figure 6. (**a**) Wear rate comparison of untreated and treated samples; (**b**) Variation of the co-efficient of friction of samples; (**c**) Cross-section of the EDMed substrate.

Wear appearances for the EDMed and un-machined specimens were depicted by FE-SEM images. The un-machined surface was witnessed with flakes, pits and deep grooves (Figure 7a). However, the machined specimen was found with black patches that symbolized the tribochemical reaction (specimen surface wear in the high oxide atmosphere) (Figure 7b). Moreover, the machined surface was noticed with light scratches and no delamination that reveal the high wear resistance of the substrate [42–44]. Evidently, electric discharge machining at elevated temperature results in the chemical reaction between the dielectric fluid and the workpiece material elements. It results in the formation of carbide and oxide layers on the substrate that improves the wear resistance of the surface.

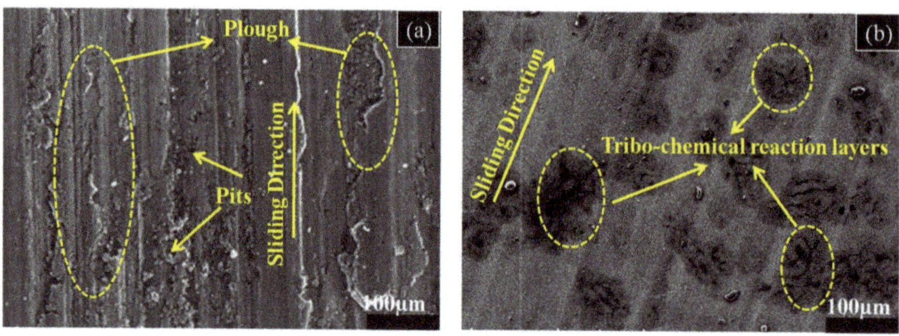

Figure 7. FE-SEM images represent the wear appearances of (**a**) Untreated substrate; (**b**) EDMed substrate.

3.5. Analysis of Microhardness and Recast Layer Thickness

Micro-hardness results are also in-line with wear resistance outcomes. EDMed DSS substrate (trial 9) showed 487 $HV_{0.2}$ micro-hardness, which was 1.46 (approx.) times better than the un-machined surface (334 $HV_{0.2}$) of DSS alloy. These results were further affirmed by the investigation of the recast layer thickness of the machined surface. The cross-section image of the treated substrate was witnessed with the thick recast layer (53.952 µm) on the surface (Figure 6c). The re-deposition of melting material droplets from a working specimen as well as a tool electrode could be the possible reason for the recast layer on the machined substrate [45,46]. The outcomes portrayed improved resistance towards wear and micro-hardness of the machined substrate in comparison with an untreated substrate that admires the EDMed substrate in various industrial applications.

3.6. Surface Wettability Analysis

The WCA scrutinization demonstrated the hydrophobic or hydrophilic nature of tested substrates. The contact angles procured on untreated and treated substrates after 10s are represented in Figure 8. The results clearly showed that the EDM considerably improved the wettability of the surface. The WCA of the machined surface was 78.27 ± 0.41° respectively signifying the hydrophilic surface. The un-machined surface was hydrophobic with WCA of 105.96 ± 0.52° (above 90°). These results were in accordance with surface roughness outcomes. However, the surface roughness of the substrate has a huge impact on the wettability of the surface. Some researchers reported the direct relation between surface roughness and wettability [47]. The increased surface roughness leads to enhance the surface free energy by offering the expanded surface area to water droplet [48,49]. Therefore, surface alteration enhances the wetting responses of surface that endorse the machined substrate applications in the biomedical domain, lubrication and for different coatings [50,51].

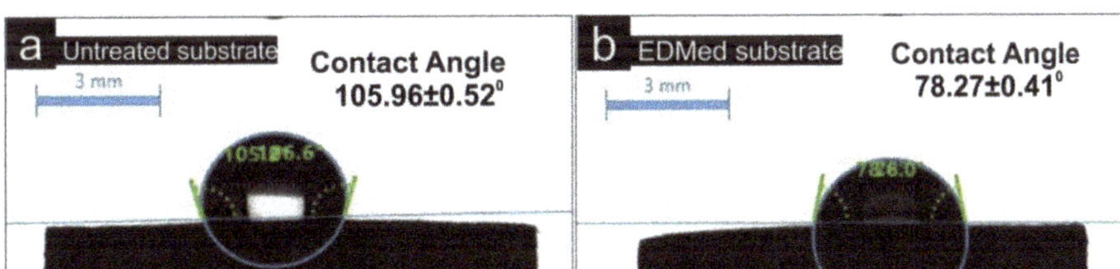

Figure 8. Contact angle illustration of (**a**) Untreated surface; (**b**) EDMed surface.

4. Conclusions

The present study described the processing of duplex stainless steel (DSS-2205) alloy by EDM using graphite, copper-tungsten and tungsten as electrodes in two different dielectric mediums, namely EDM oil and deionized water. Based on the result, following conclusions have been drawn.

The dominating factors depicting maximum material removal rate (39.4 ± 0.98 mm^3/min) were current (contribution: 45.10%), dielectric medium (contribution: 18.24%), and electrode (contribution: 12.67%).

The higher surface roughness accelerates the osseointegration process of bioimplant. The most significant parameters for higher surface roughness (Ra = 1.4 ± 0.08 µm) were electrode (contribution: 38.72%), pulse-on-time (contribution: 26.11%), and current (contribution: 17.39%).

From the S/N ratios plot, the optimum parametric combinations for favorable MRR and SR values by ED machining of DSS substrate are, EDM oil as dielectric medium, W/Cu electrode and current at 16 A, pulse-on-time at 150 µs coupled with the lowest pulse-off-time (i.e., 60 µs).

The FE-SEM and XRD examination confirmed the evenly distributed porous surface, and formation of oxides and other intermetallic compounds on the DSS-2205 surface machined at higher spark energy in the presence of EDM oil as a dielectric medium.

Moreover, EDMed substrate with higher MRR and surface roughness also exhibited enhanced wear resistance and surface wettability responses as compared to untreated DSS alloy substrate. The results also illustrated that EDMed DSS-2205 could be employed in the biomedical field.

Author Contributions: Conceptualization, T.R.A., S.S.S. and A.M.; methodology, S.S.S. and T.R.A.; software, G.S. and S.D.; validation, A.M., G.S., K.R.M. and E.S.S.; formal analysis, T.R.A., E.S.S. and S.S.S.; investigation, A.M., G.S., S.D.; resources, S.S.S.; data curation, T.R.A., E.S.S. and S.S.S.; writing—original draft preparation, A.M. and G.S.; writing—review and editing, A.M., G.S., S.S.S. and T.R.A.; visualization, K.R.M., E.S.S., G.S., and A.M.; supervision, S.S.S. and T.R.A.; project administration, T.R.A., A.M. and S.S.S.; funding acquisition, T.R.A., E.S.S. and K.R.M. All authors have read and agreed to the published version of the manuscript.

Funding: This work was supported by Russian Science Foundation, grant number 20-79-00048.

Conflicts of Interest: The authors declare no conflict of interest. The funders had no role in the design of the study; in the collection, analyses, or interpretation of data; in the writing of the manuscript, or in the decision to publish the results.

References

1. Talla, G.; Gangopadhyay, S.; Biswas, C.K. Influence of graphite powder mixed EDM on the surface integrity characteristics of Inconel 625. *Part. Sci. Technol.* **2017**, *35*, 219–226. [CrossRef]

2. Singh, G.; Ablyaz, T.R.; Shlykov, E.S.; Muratov, K.R.; Bhui, A.S.; Sidhu, S.S. Enhancing corrosion and wear resistance of Ti6Al4V alloy using CNTs mixed electro-discharge process. *Micromachines* **2020**, *11*, 850. [CrossRef] [PubMed]

3. Rajeswari, R.; Shunmugam, M.S. Finishing performance of die-sinking EDM with ultrasonic vibration and powder addition through pulse train studies. *Mach. Sci. Technol.* **2019**, *24*, 245–273. [CrossRef]

4. Kumar, V.; Diyaley, S.; Chakraborty, S. Teaching-learning-based parametric optimization of an electrical discharge machining process. *FU Ser. Mech. Eng.* **2020**, *18*, 281–300. [CrossRef]

5. Mahajan, A.; Sidhu, S.S. Enhancing biocompatibility of Co-Cr alloy implants via electrical discharge process. *Mater. Technol.* **2018**, *33*, 524–531. [CrossRef]

6. Singh, G.; Lamichhane, Y.; Bhui, A.S.; Sidhu, S.S.; Bains, P.S.; Mukhiya, P. Surface morphology and microhardness behavior of 316L in HAp-PMEDM. *FU Ser. Mech. Eng.* **2019**, *17*, 445–454. [CrossRef]

7. Umar-Farooq, M.; Pervez-Mughal, M.; Ahmed, N.; Ahmad-Mufti, N.; Al-Ahmari, A.M.; He, Y. On the investigation of surface integrity of Ti6Al4V ELI using Si-mixed electric discharge machining. *Materials* **2020**, *13*, 1549. [CrossRef]

8. Al-Amin, M.; Abdul-Rani, A.M.; Abdu-Aliyu, A.A.; Abdul-Razak, M.A.; Hastuty, S.; Bryant, M.G. Powder mixed-EDM for potential biomedical applications: A critical review. *Mater. Manuf. Process.* **2020**, 1–23. [CrossRef]

9. Al-Amin, M.; Abdul-Rani, A.M.; Abdu-Aliyu, A.A.; Bryant, M.G.; Danish, M.; Ahmad, A. Bio-ceramic coatings adhesion and roughness of biomaterials through PM-EDM: A comprehensive review. *Mater. Manuf. Process.* **2020**, *35*, 1157–1180. [CrossRef]

10. Devgan, S.; Sidhu, S.S. Surface modification of β-type titanium with multi-walled CNTs/μ-HAp powder mixed Electro Discharge Treatment process. *Mater. Chem. Phys.* **2020**, *239*, 122005. [CrossRef]

11. D'Urso, G.; Giardini, C.; Maccarini, G.; Quarto, M.; Ravasio, C. Analysis of the surface quality of steel and ceramic materials machined by micro-EDM. In Proceedings of the 18th International Conference of the European Society for Precision Engineering and Nanotechnology, EUSPEN, Venice, Italy, 4–8 June 2018; pp. 431–432.

12. Razavykia, A.; Yavari, M.R.; Iranmanesh, S.; Esmaeilzadeh, A. Effect of electrode material and electrical discharge machining parameters on machining of Co-Cr-Mo. *Int. J. Mech. Mechatron. Eng.* **2016**, *16*, 53–61.

13. Mahajan, A.; Sidhu, S.S. Potential of electrical discharge treatment to enhance the in vitro cytocompatibility and tribological performance of Co–Cr implant. *J. Mater. Res.* **2019**, *34*, 2837–2847. [CrossRef]

14. Philip, J.T.; Kumar, D.; Mathew, J.; Kuriachen, B. Experimental investigations on the tribological performance of electric discharge alloyed Ti–6Al–4V at 200–600 °C. *J. Tribol.* **2020**, *142*, 061702. [CrossRef]

15. Devgan, S.; Sidhu, S.S. Potential of electrical discharge treatment incorporating MWCNTs to enhance the corrosion performance of the β-titanium alloy. *Appl. Phys. A* **2020**, *126*, 211. [CrossRef]

16. Simao, J.; Lee, H.G.; Aspinwall, D.K.; Dewes, R.C.; Aspinwall, E.M. Workpiece surface modification using electrical discharge machining. *Int. J. Mach. Tools Manuf.* **2003**, *43*, 121–128. [CrossRef]

17. Singh, G.; Sidhu, S.S.; Bains, P.S.; Bhui, A.S. Improving microhardness and wear resistance of 316L by TiO_2 powder mixed electro-discharge treatment. *Mater. Res. Express.* **2019**, *6*, 086501. [CrossRef]

18. Pramanik, A.; Basak, A.K.; Dixit, A.R.; Chattopadhyaya, S. Processing of duplex stainless steel by WEDM. *Mater. Manuf. Process.* **2018**, *33*, 1559–1567. [CrossRef]

19. Alshemary, A.; Pramanik, A.; Basak, A.K.; Littlefair, G. Accuracy of duplex stainless steel feature generated by electrical discharge machining (EDM). *Measurement* **2018**, *130*, 137–144. [CrossRef]

20. Rajmohan, K.; Kumar, A.S. Experimental investigation and prediction of optimum process parameters of micro-wire-cut EDM of 2205 DSS. *Int. J. Adv. Manuf. Technol.* **2017**, *93*, 187–201. [CrossRef]

21. Rajaram, S.; Rajkumar, G.; Balasundaram, R.; Srinivasan, D. Experimental investigation of drilling small hole on duplex stainless steel (SS 2205) using EDM. *Mech. Mech. Eng.* **2019**, *23*, 98–102. [CrossRef]

22. Mahajan, A.; Sidhu, S.S.; Devgan, S. Examination of hemocompatibility and corrosion resistance of electrical discharge-treated duplex stainless steel (DSS-2205) for biomedical applications. *Appl. Phys. A* **2020**, *126*, 737. [CrossRef]

23. Khanra, A.K.; Sarkar, B.R.; Bhattacharya, B.; Pathak, L.C.; Godkhindi, M.M. Performance of ZrB_2-Cu composite as an EDM electrode. *J. Mater. Process. Technol.* **2007**, *183*, 122–126. [CrossRef]

24. Tsai, H.C.; Yan, B.H.; Huang, F.Y. EDM performance of Cr/Cu-based composite electrodes. *Int. J. Mach. Tools Manuf.* **2003**, *43*, 245–252. [CrossRef]

25. Grisharin, A.O.; Ogleznev, N.D.; Muratov, K.R.; Ablyaz, T.R.; Bains, P.S.; Sidhu, S.S. Investigation of the machinability of composite materials electrode-tools while EDM. *Conf. Ser. Mater. Sci. Eng.* **2019**, *510*, 012006. [CrossRef]

26. Teng, Y.L.; Li, L.; Zhang, W.; Wang, N.; Feng, C.C.; Ren, J.H. Machining characteristics of PCD by EDM with Cu-Ni composite electrode. *Mater. Manuf. Process.* **2020**, *35*, 442–448. [CrossRef]

27. Quarto, M.; Bissacco, G.; D'Urso, G. Machinability and energy efficiency in micro-EDM milling of zirconium boride reinforced with silicon carbide fibers. *Materials* **2019**, *12*, 3920. [CrossRef]

28. D'Urso, G.; Giardini, C.; Lorenzi, S.; Quarto, M.; Sciti, D.; Silvestroni, L. Micro-EDM milling of zirconium carbide ceramics. *Precis. Eng.* **2020**, *65*, 156–163. [CrossRef]

29. Wang, Z.L.; Fang, Y.; Wu, P.N.; Zhao, W.S.; Cheng, K. Surface modification process by electric discharge machining with a Ti powder green compact electrode. *J. Mater. Process. Technol.* **2002**, *129*, 139–142. [CrossRef]

30. Malayath, G.; Sidpara, A.M.; Deb, S. Fabrication of micro-end mill tool by EDM and its performance evaluation. *Mach. Sci. Technol.* **2019**, *24*, 169–194. [CrossRef]

31. Pratap, T.; Patra, K. Mechanical micro-texturing of Ti-6Al-4V surfaces for improved wettability and bio-tribological performances. *Surf. Coat. Technol.* **2018**, *349*, 71–81. [CrossRef]

32. Azevedo, A.F.; Corat, E.J.; Ferreira, N.G.; Trava-Airoldi, V.J. Wettability and corrosion tests of diamond films grown on Ti6Al4V alloy. *Surf. Coat. Technol.* **2005**, *194*, 271–275. [CrossRef]

33. Kliuev, M.; Florio, K.; Akbari, M.; Wegener, K. Influence of energy fraction in EDM drilling of inconel 718 by statistical analysis and finite element crater-modeling. *J. Manuf. Process.* **2019**, *40*, 84–93. [CrossRef]

34. Kumar, R.; Roy, S.; Gunjan, P.; Sahoo, A.; Sarkar, D.D.; Das, R.K. Analysis of MRR and surface roughness in machining Ti-6Al-4V ELI titanium alloy using EDM process. *Procedia Manuf.* **2018**, *20*, 358–364. [CrossRef]

35. Singh, G.; Sidhu, S.S.; Bains, P.S.; Bhui, A.S. Surface evaluation of ED machined 316L stainless steel in TiO_2 nano-powder mixed dielectric medium. *Mater. Today Proc.* **2019**, *18*, 1297–1303. [CrossRef]

36. Devgan, S.; Sidhu, S.S. Evolution of surface modification trends in bone related biomaterials: A review. *Mater. Chem. Phys.* **2019**, *223*, 68–78. [CrossRef]

37. Harcuba, P.; Bacakova, L.; Strasky, J.; Bacakova, M.; Novotna, K.; Janecek, M. Surface treatment by electric discharge machining of Ti–6Al–4V alloy for potential application in orthopaedics. *J. Mech. Behav. Biomed. Mater.* **2012**, *7*, 96–105. [CrossRef]

38. Mahajan, A.; Sidhu, S.S. Surface modification of metallic biomaterials for enhanced functionality: A review. *Mater. Technol.* **2018**, *33*, 93–105. [CrossRef]

39. Cogun, C.; Esen, Z.; Genc, A.; Cogun, F.; Akturk, N. Effect of powder metallurgy Cu-B_4C electrodes on workpiece surface characteristics and machining performance of electric discharge machining. *Proc. IMechE Part B J. Eng. Manuf.* **2016**, *230*, 2190–2203. [CrossRef]

40. Ji, R.; Liu, Y.; Diao, R.; Xu, C.; Li, X.; Cai, B.; Zhang, Y. Influence of electrical resistivity and machining parameters on electrical discharge machining performance of engineering ceramics. *PLoS ONE* **2014**, *9*, e110775. [CrossRef]

41. Devgan, S.; Sidhu, S.S. Enhancing tribological performance of β-titanium alloy using electrical discharge process. *Surf. Innov.* **2019**, *8*, 115–126. [CrossRef]

42. Hesketh, J.; Ward, M.; Dowson, D.; Neville, A. The composition of tribofilms produced on metal-on-metal hip bearings. *Biomaterials* **2014**, *35*, 2113–2119. [CrossRef] [PubMed]

43. Singh, G.; Sidhu, S.S.; Bains, P.S.; Singh, M.; Bhui, A.S. On surface modification of Ti alloy by electro discharge coating using hydroxyapatite powder mixed dielectric with graphite tool. *J. Bio Tribo Corros.* **2020**, *6*, 91. [CrossRef]

44. Buyukgoze-Dindar, M.; Tekbas-Atay, M. The effect of toothbrush abrasion on wear and surface roughness of direct and indirect composite laminate veneer restorations. *Surf. Topogr. Metrol. Prop.* **2020**, *8*, 035007. [CrossRef]

45. Chen, S.L.; Lin, M.H.; Huang, G.X.; Wang, C.C. Research of the recast layer on implant surface modified by micro-current electrical discharge machining using deionized water mixed with titanium powder as dielectric solvent. *Appl. Surf. Sci.* **2017**, *311*, 47–53. [CrossRef]

46. Gostimirovic, M.; Kovac, P.; Sekulic, M.; Skoric, B. Influence of discharge energy on machining characteristics in EDM. *J. Mech. Sci. Technol.* **2012**, *26*, 173–179. [CrossRef]

47. Sauli, Z.; Retnasamy, V.; Yeow, A.K.T.; Chui, G.S.; Anwar, K.; Abdullah, N. Surface roughness and wettability correlation on etched platinum using reactive ion ecthing. *Appl. Mech. Mater.* **2014**, *487*, 263–266. [CrossRef]

48. Bharathidasan, T.; Kumar, S.V.; Bobji, M.S.; Chakradhar, R.P.S.; Basu, B.J. Effect of wettability and surface roughness on ice-adhesion strength of hydrophilic, hydrophobic and superhydrophobic surfaces. *Appl. Surf. Sci.* **2014**, *314*, 241–250. [CrossRef]

49. Yang, S.; Xia, Q.; Zhu, L.; Xue, J.; Wang, Q.; Chen, Q.M. Research on the icephobic properties of fluoropolymer-based materials. *Appl. Surf. Sci.* **2011**, *257*, 4956–4962. [CrossRef]

50. Kumar, S.S.; Hiremath, S.S. Effect of surface roughness and surface topography on wettability of machined biomaterials using flexible viscoelastic polymer abrasive media. *Surf. Topogr. Metrol. Prop.* **2019**, *7*, 015004. [CrossRef]

51. Puckett, S.D.; Lee, P.P.; Ciombor, D.M.; Aaron, R.K.; Webster, T.J. Nanotextured titanium surfaces for enhancing skin growth on transcutaneous osseointegrated devices. *Acta Biomater.* **2010**, *6*, 2352–2362. [CrossRef]

 *micromachines*

Article

Surface Quality Improvement of 3D Microstructures Fabricated by Micro-EDM with a Composite 3D Microelectrode

Jianguo Lei, Kai Jiang, Xiaoyu Wu, Hang Zhao and Bin Xu *

Guangdong Provincial Key Laboratory of Micro/Nano Optomechatronics Engineering, Shenzhen University, Shenzhen 518060, China; ljg_sc111@163.com (J.L.); 13728644743@163.com (K.J.); wuxy@szu.edu.cn (X.W.); zh@szu.edu.cn (H.Z.)
* Correspondence: binxu@szu.edu.cn

Received: 21 August 2020; Accepted: 17 September 2020; Published: 19 September 2020

Abstract: Three-dimensional (3D) microelectrodes used for processing 3D microstructures in micro-electrical discharge machining (micro-EDM) can be readily prepared by laminated object manufacturing (LOM). However, the microelectrode surface always appears with steps due to the theoretical error of LOM, significantly reducing the surface quality of 3D microstructures machined by micro-EDM with the microelectrode. To address the problem above, this paper proposes a filling method to fabricate a composite 3D microelectrode and applies it in micro-EDM for processing 3D microstructures without steps. The effect of bonding temperature and Sn film thickness on the steps is investigated in detail. Meanwhile, the distribution of Cu and Sn elements in the matrix and the steps is analyzed by the energy dispersive X-ray spectrometer. Experimental results show that when the Sn layer thickness on the interface is 8 μm, 15 h after heat preservation under 950 °C, the composite 3D microelectrodes without the steps on the surface were successfully fabricated, while Sn and Cu elements were evenly distributed in the microelectrodes. Finally, the composite 3D microelectrodes were applied in micro-EDM. Furthermore, 3D microstructures without steps on the surface were obtained. This study verifies the feasibility of machining 3D microstructures without steps by micro-EDM with a composite 3D microelectrode fabricated via the proposed method.

Keywords: micro-EDM; composite 3D microelectrode; diffusion bonding; step; 3D microstructure

---

1. Introduction

Micro electrical discharge machining (micro-EDM), as a non-conventional machining technology, possesses unique advantages, such as non-contact machining nature and negligible cutting force, and is capable of machining any electrically conductive materials regardless of the hardness. Thus, micro-EDM was widely used to machine automotive, aerospace and surgical microstructures [1–4].

For the fabrication of three-dimensional (3D) microstructures, the current micro-EDM technology usually utilizes a layer-by-layer milling process with a cylindrical microelectrode. Yu et al. [5] developed a uniform wear method (UWM) for 3D micro-EDM. Based on the UWM, various complicated 3D microstructures were successfully fabricated via layer-by-layer scanning micro-EDM with a cylindrical electrode. Reynaerts et al. [6] fabricated a 3D microstructure consisting of two planes inclined under 45° in silicon by EDM milling, and found that the EDM process was independent of the silicon crystal orientation. Rajurkar et al. [7] integrated computer aided design and manufacturing (CAD/CAM) systems with micro-EDM, while accounting for tool wear utilizing UWM, and successfully generated various complex 3D micro cavities. Bleys et al. [8,9] introduced improved wear compensation and real-time wear sensing based on discharge pulse evaluation in EDM milling, achieving the accurate machining of square and hexagonal pockets. Zhao et al. [10] applied CAD/CAM system in micro-EDM to improve the machining quality, and successfully prepared a human face embossment with a dimension of 1 mm × 0.4 mm using a 30 μm diameter copper rod electrode.

To improve the machining accuracy and efficiency, based on the scanned area in each layer machining, Li et al. [11] proposed a new compensation method, which was integrated with a CAD/CAM system in 3D micro-EDM milling to generate 3D micro cavities. According to the proportional relationship between the removed workpiece volume and the number of discharge pulses, Jung et al. [12] developed a control method for a micro-EDM process using discharge pulse counting, improving the machining efficiency and accuracy without complex path planning to compensate electrode wear. Using a machine vision system, Yan et al. [13] proposed a multi-cut process planning method and an electrode wear compensation method for layer-by-layer 3D micro-EDM. Various 3D microstructures, such as pyramid cavity, hexagonal pyramid cavity, rectangular pyramid, and cone cavity, were prepared. This approach significantly improved machining accuracy and reduced machining time.

To improve the machining stability and the effective discharge ratio, Tong et al. [14] proposed workpiece vibration-assisted servo scanning 3D micro-EDM. With the optimized parameters, a series of 3D microstructures such as double-prism cavity with hemisphere, round cavity with hemisphere, camber and rectangular pyramid were prepared. To improve the machining surface quality and reduce electrode wear, Song et al. [15] developed a spray EDM milling method with a bipolar pulsed power source and deionized water. To improve the machining performance, such as material removal rate, electrode wear ratio and surface roughness, Yu et al. [16] combined the linear compensation method and the UWM to machine 3D microstructures in micro-EDM. Bissacco et al. [17,18] proposed a new method based on the discharge counting and discharge population characterization, effectively compensating tool electrode wear in micro-EDM milling.

Based on the established mathematical model of erosion depth, Li et al. [19] proposed a strategy of scanning speed adjusted with layer in 3D micro-EDM milling, effectively lowering the accumulative depth error to less than 1% under appropriate conditions. Tong et al. [20] proposed an on-machine process of rough-and-finishing servo scanning 3D micro-EDM to machine 3D micro cavities. In the rough machining process, high discharge energy and large-diameter tool electrodes were used to improve processing efficiency. In the finishing machining process, a small amount of material was removed by changing multi-factors of machining parameters. Based on the discharge pulse behavior, Wang et al. [21] combined off-line and in-line adaptive tool wear compensation to achieve effective and efficient tool wear compensation in micro-EDM milling, and prepared hemisphere, cone and pyramid cavities, with good form accuracy on the stainless steel. Tong et al. [22] developed a novel process of 3D servo scanning micro-EDM with the movement of two-axis linkage and one-axis servo, and efficiently and cheaply machined 3D complex micro driving structures pierced through thin-walled micro tube of NiTi SMA. D'Urso et al. [23] fabricated a micro-pocket on zirconium carbide ceramics by micro-EDM milling using a tungsten carbide cylindrical electrode of diameter 0.3 mm.

Micro-EDM milling simplifies the 3D machining into two-dimensional (2D) laminated scanning, however, the machining time is still long [5,10], which is mainly caused by the smaller cross section of the microelectrode and the milling strategy. To improve the anti-interference ability of the microelectrode, Xu et al. [24] introduced a foil queue microelectrode in micro-EDM to fabricate a 3D microstructure. The impacts of machining voltage and pulse width on rounded corner wear at the end of the foil microelectrode and the step effect on the 3D microstructure surface were investigated. The 3D microstructures with hemispherical shapes were successfully machined, and the dimensional errors of the 3D microstructures were less than 10 μm. Using the difference in the wear rates of different materials in EDM, Lei et al. [25] fabricated surface microstructures on Ti–6Al–4V alloy workpieces by EDM with laminated disc electrodes made of Cu and Sn foils, converting disadvantageous phenomenon of wear into a beneficial process. There was physical contact without metallurgical reaction between the foils. To improve the machining efficiency, based on the laminated object manufacturing (LOM), Xu et al. [26] fabricated a laminated 3D microelectrode with a complicated contour and shape via combining wire electrical discharge machining (WEDM) and thermal diffusion welding, and used the 3D microelectrode stacked by Cu foils to manufacture 3D microstructures by an up-and-down reciprocating machining method. To improve the machining accuracy, Xu et al. [27] prepared a 3D queue microelectrode by WEDM and vacuum pressure thermal diffusion welding, and applied it in micro-EDM to machine 3D microstructures. However, numerous and obvious ridges emerged on the machining surface of the 3D microstructure duo to imperfect thermal diffusion bonding between the Cu foils. To eliminate ridges on the machining surface, using Cu foils coated with Sn film as laminated materials, Lei et al. [28] manufactured laminated composite 3D microelectrodes by femtosecond laser cutting using a bending-and-avoidance mode and transient liquid phase bonding. Moreover, 3D microstructures without ridges on the surface were successfully machined by up-and-down reciprocating micro-EDM with the composite 3D microelectrodes.

LOM, as an additive manufacturing technology, is capable of fabricating various complex 3D functional micro parts. Without doubt, the LOM technology is suitable for preparing complex 3D microelectrodes, which are difficult to fabricate by traditional machining methods. However, when the 3D microelectrode possesses an inclined plane or curved surface, steps inevitably occur on the 3D microelectrode surfaces, due to the theoretical errors of LOM [29], and the steps are copied to the 3D microstructure surface on the workpiece.

To solve the problem above, the paper proposed a filling method to eliminate the steps on the 3D microelectrode surface. First, femtosecond laser was used for cutting Cu foils coated with Sn to obtain 2D layers. Then, 3D microelectrodes without steps were fabricated by applying the layers to thermal diffusion bonding, through which they were bonded and partially melt-softened (Cu) on the interface fills in the steps. Finally, 3D microstructures with no steps were successfully fabricated by micro-EDM using these electrodes.

2. Experimental Details

2.1. Experimental Setup

The main experimental setup for fabricating 3D microelectrodes and performing micro-EDM consists of a femtosecond laser, a PI motion platform, a high frequency pulse power supply, an oscilloscope, a control system and a vacuum furnace.

The titanium sapphire femtosecond laser, with a central wavelength of 800 ns, a pulse duration of 35 fs, a maximum repetition frequency of 1 kHz and pulse energy of 4 mJ, from Coherent Inc. was used to cut Cu foils coated with and without Sn to obtain 2D microstructures (Figure 1a) [30]. The high precision motion platform (model: M511.DD) made by Germany, PI was used to control the femtosecond laser cutting path and perform micro-EDM experiments. The maximum strokes of X/Y/Z-axis were 100 mm and the motion accuracy of each axis was of 0.2 μm. A wire EDM machine manufactured by HI-LINK Precision Machinery Co., Ltd., (Shenzhen, China) (model: H-CUT 32F), was used on fabricated samples for observation and characterization. The maximum stroke of X-axis, Y-axis and Z-axis were 400 mm, 320 mm and 400 mm, respectively. The location precision was 1 μm for all axes.

The schematic diagram of micro-EDM setup is shown in Figure 1b. A laminated 3D microelectrode was fixed on the PI platform to machine microstructures through a back and forth reciprocating machining strategy. The high frequency pulse power supply was used to generate ultra-short pulse signal and the machining process was real-time monitored using the oscilloscope. The pictures of experimental setup are displayed in Figure 2a–f.

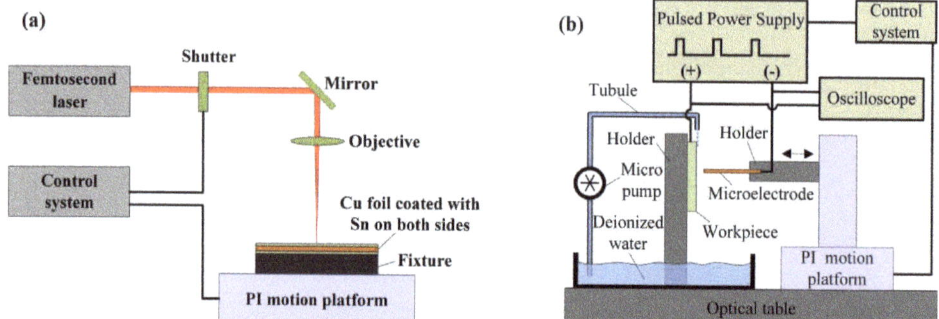

Figure 1. Schematic diagram of experimental setup: (**a**) Femtosecond laser cutting system; (**b**) Micro-electrical discharge machining (micro-EDM) system.

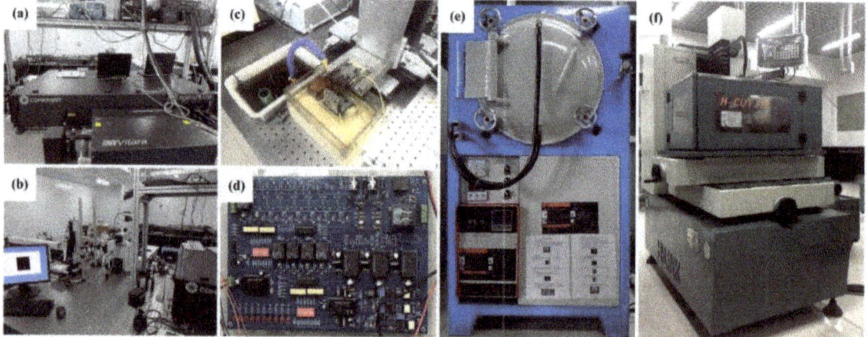

Figure 2. Pictures of experimental setup: (**a,b**) Femtosecond laser cutting system; (**c**) Micro-EDM platform; (**d**) Pulse power supply; (**e**) Vacuum furnace; (**f**) Wire EDM machine.

2.2. Experimental Materials and Conditions

Here, 50-μm-thick Cu foils coated with Sn on both sides were utilized to fabricate 3D microelectrodes. The thicknesses of Sn were 0 μm, 0.5 μm, 2.0 μm, and 4.0 μm, respectively. That is, the thickness of Sn layer in the sandwich structure was 0 μm, 1 μm, 4 μm, and 8 μm, respectively. Notably, 1 mm thick #304 stainless steel was used as a workpiece to machine microstructures. The working fluid used in the micro-EDM process was deionized water. According to the previous study [26–28], the machining conditions are selected to prepare the 3D microelectrodes and microstructures, as listed in Table 1.

Table 1. Machining conditions for femtosecond laser cutting, EDM and thermal diffusion bonding.

Parameters	Femtosecond Laser Cutting	EDM	Thermal Diffusion Bonding
Central wavelength	800 nm	-	-
Laser power	400 mW	-	-
Cutting speed	100 μm/s	-	-
Pulse duration	35 fs	800 ns	-
Pulse interval	-	4200 ns	-
Pulse frequency	1 kHz	0.2 MHz	-
Machining voltage	-	90 V	-
Peak current	-	0.1 A	-
Bonding temperature	-	-	900, 950, 1000 °C
Bonding time	-	-	15 h

2.3. Measurement and Evaluation

For observation and characterization, the fabricated 3D microelectrodes and the machined microstructures were cut by wire EDM and were polished afterwards. The morphology of the samples was observed by scanning electron microscopy (SEM) (S-3400N, Hitachi, Co., Ltd., Tokyo, Japan). To analyze the distribution of Cu and Sn in the steps and the matrix, energy dispersive X-ray spectrometer (EDS) point analysis was employed. The wear of the 3D microelectrodes and the surface roughness of the machined microstructures were measured by a laser scanning confocal microscope (VK-X250, KEYENCE, Osaka, Japan).

3. Results and Discussion

3.1. Fabrication of the Laminated 3D Microelectrode and the Microstructure

Based on the LOM technology, the process of fabricating 3D microelectrodes and microstructures is generally performed by the following steps:

(1)　CAD modeling and pre-process. Based on a 3D microstructure model, a 3D microelectrode model was prepared. Then, the 3D microelectrode model was sliced into several thin cross-sectional layers in a given direction by slicing software. After that, the total layer number and the profile data of each 2D cross-sectional layer were obtained, as shown in Figure 3a.

(2)　Femtosecond laser cutting. One end of the same multi-layer Cu foils coated with Sn on the both sides was clamped on a fixture, and the other end of the first layer was fixed on the fixture by a magnet. The other layers of the Cu foils coated with Sn were bent upward with a stopper to leave enough space for femtosecond laser cutting (Figure 3b). After completion of the first layer cutting (Figure 3c), it was bent downward with the stopper, and the former upward end of the second layer was fixed on the fixture (Figure 3d). With the above steps repeated (Figure 3d,e), multi-layer 2D microstructures were obtained (Figure 3f).

(3)　Diffusion bonding. The obtained multi-layer 2D microstructures were first cleaned using ethanol in an ultrasonic bath to remove contamination and dried with nitrogen at room temperature. After cleaning, they were sandwiched between two graphite blocks and placed in a vacuum furnace, where they were pressed by a constant pressure (36 KPa). Diffusion bonding was finally carried out at the required temperature at the rate of 20 °C/min to obtain 3D microelectrodes (Figure 3g,h).

(4)　Micro-EDM. The obtained 3D microelectrode was installed on the micro-EDM platform to machine 3D microstructures. The obtained 3D microstructure is shown in Figure 3i.

3.2. Bonding Temperature

Bonding temperature plays a great role on Cu-Sn interface reaction and the component of the compounds. To fully eliminate the steps on the surface of the 3D microelectrodes, Cu foils coated with 4-μm-thick Sn on both sides were chosen to fabricate V-shaped 3D microelectrodes. Figure 4 shows the cross-section profile of the V-shaped microelectrodes when bonding time is 15 h, bonding temperature is 900 °C, 950 °C and 1000 °C respectively. The bonding time of 15 h can guarantee that Sn and Cu atoms diffuse adequately [28]. It can be seen that, when bonding temperature was 900 °C, the steps were obvious, and no interface compounds was extruded at the steps (Figure 4a). EDS point analysis was carried out on the interface, which showed a content of nearly 5 wt.% Sn. Cu-Sn phase diagram (Figure 5) presented (Cu) was still in solid state under this temperature. Therefore, under constant pressure could not be squeezed into the steps. While temperature increased to 950 °C, the steps disappeared (Figure 4b). On the interface and the steps, the content of Sn was still around 5 wt.%, at which interface compound was on solid phase line (α + L), namely the critical melting point. Under the action of a constant pressure of 36 KPa, the interface compound (Cu) was gradually squeezed to the steps and after 15 h full diffusion reaction, thus V-shaped microelectrodes were with no steps.

As the temperature further went up to 1000 °C, interface compounds crossed the line and were totally in phase (α + L), featured by rheological behavior. With pressure working together, V-shaped microelectrodes were badly deformed (Figure 4c), which explained the reason that 950 °C was the optimized temperature.

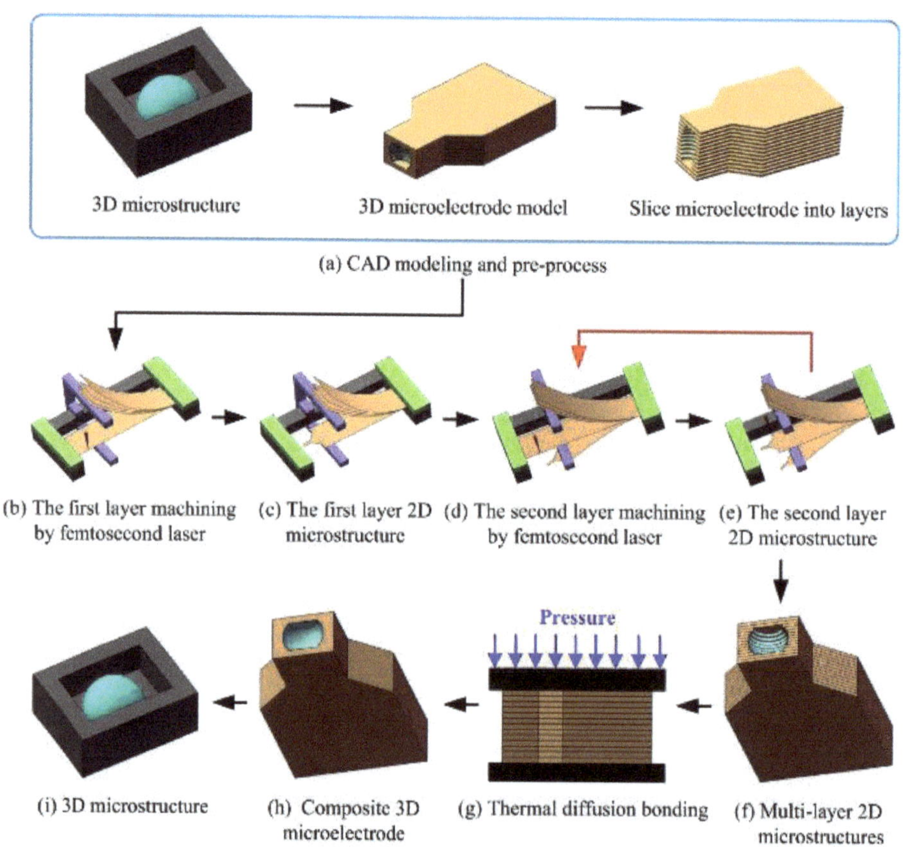

Figure 3. Schematic diagram for fabricating 3D microelectrode and microstructure.

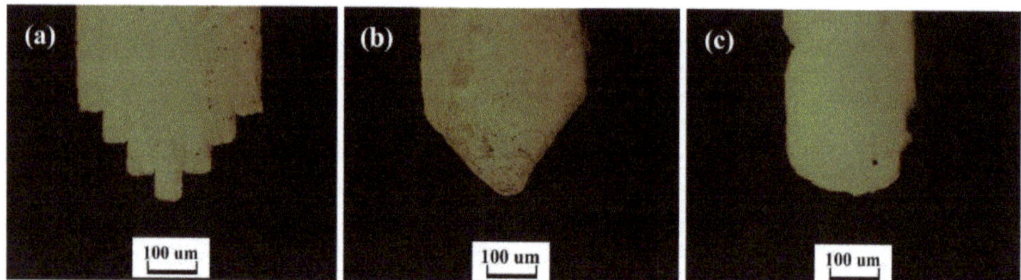

Figure 4. Sectional views of microelectrodes fabricated with different bonding temperatures: (**a**) 900 °C; (**b**) 950 °C; (**c**) 1000 °C.

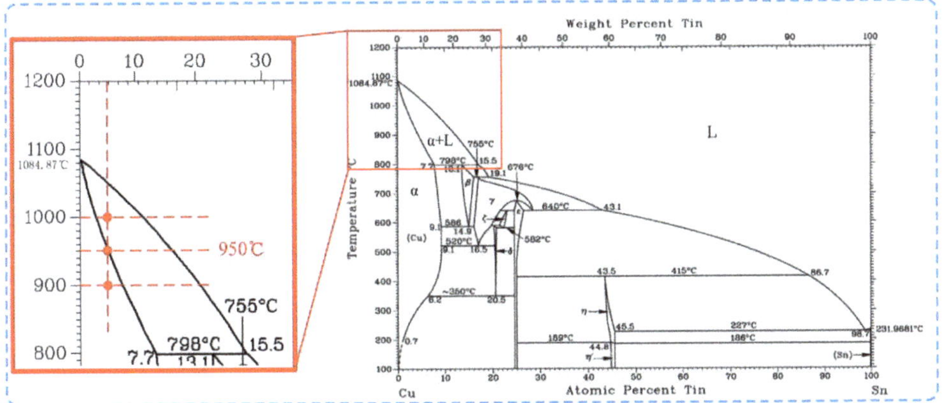

Figure 5. Cu-Sn phase diagram [30].

3.3. Thickness of the Sn Film

To study the effect of Sn layer thickness on the steps, Cu foils (50 µm in thickness) coated with Sn film in different thickness were chosen to machine V-shaped microelectrodes. Then, the prepared microelectrodes were carried out micro-EDM to fabricate V-shaped microstructures on a #304 stainless steel workpiece. The process parameters were: machining voltage of 90 V, pulse width of 800 ns, pulse interval 4200 ns.

When the thickness of Sn layer was 0 µm, the steps were obvious on the interface, bonding quality between Cu foils was poor (Figure 6a), and the microstructures machined using this microelectrode had clear steps on the surface (Figure 6e). As the thickness of Sn layer grew to 1 µm, though with a much better bonding of the Cu foils, the steps existed on the microelectrodes (Figure 6b) and were further copied to microstructures (Figure 6f). While the thickness of Sn continued to increase, partially melt-softened (Cu) was extruded out and flowed to the steps. After 15 h full diffusion reaction, the steps on the surface of the V-shaped microelectrodes decreased (Figure 6c,d) and so did those on the corresponding 3D microstructures (Figure 6g,h). However, on the other hand, if the Sn layer is too thick, the extruded (Cu) would be beyond the capacity of the steps, decreasing the quality of the microelectrodes. Therefore, as for Cu foils 50 µm thick, the optimized thickness of Sn layer on the interface would be 8 µm.

3.4. The Distribution of Elements

A uniform distribution of Cu and Sn elements on the 3D microelectrodes was beneficial for obtaining identical resistivity at each position of the microelectrode. With the bonding temperate of 950 °C and Sn layer thickness of 8 µm, a laminated composite 3D microelectrode was fabricated to study the distribution of Cu and Sn elements. EDS was chosen to analyze the steps and matrix of the microelectrodes.

Figure 7a illustrates section view of the V-shaped composite 3D microelectrode with no steps and EDS was carried out on point #1 to point #9. The experimental results are shown in Figure 7b. It could be seen from the diagram that all the points had uniform contributions of Cu and Sn, which indicated that after 15 h sufficient diffusion under 950 °C, Cu and Sn evenly distributed in the 3D microelectrodes. It effectively avoided the production of ridges due to the uneven distribution of Cu and Sn. Figure 7c,d respectively represent the energy spectra of point #1 and point #5 in Figure 7a. The composite 3D microelectrode mainly contained Cu and Sn. C and O may have resulted from the graphite blocks during the thermal diffusion bonding of multi-layer 2D microstructures. The content of C and O were both very low and did not affect the machining performance of the microelectrodes.

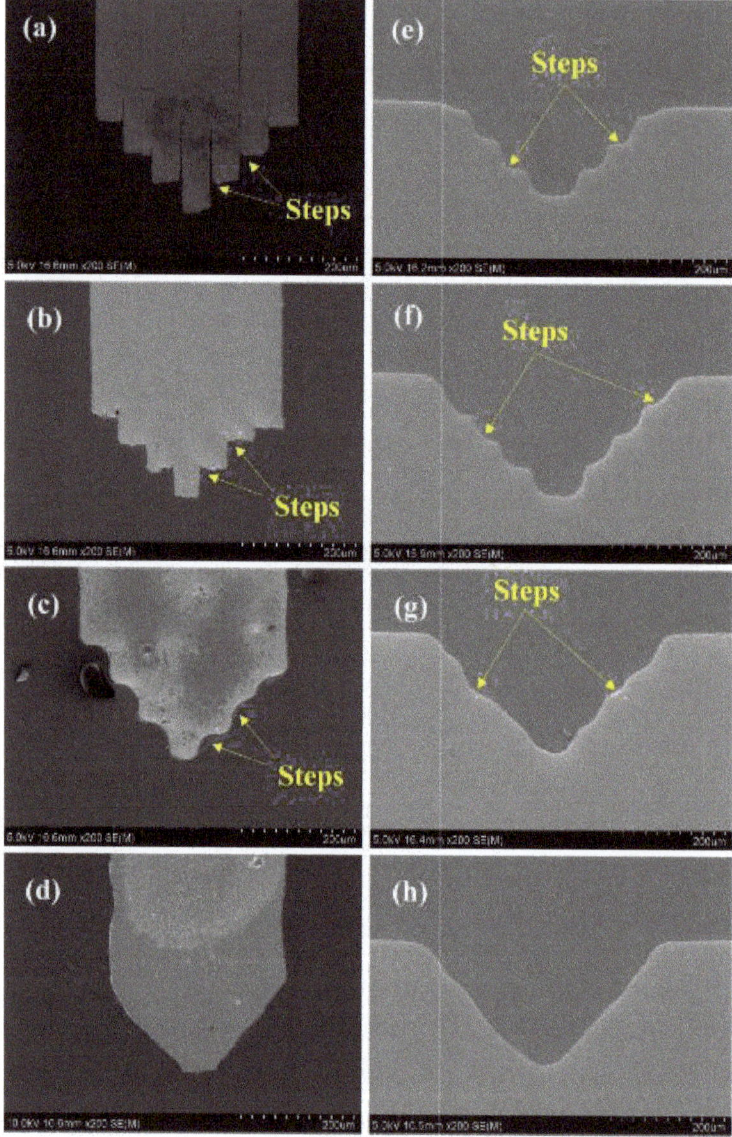

Figure 6. Sectional views of microelectrodes fabricated with different Sn layer thickness: (**a**) 0 μm; (**b**) 1 μm; (**c**) 4 μm; (**d**) 8 μm; (**e–h**) Corresponding sectional views of microstructures machined with these microelectrodes.

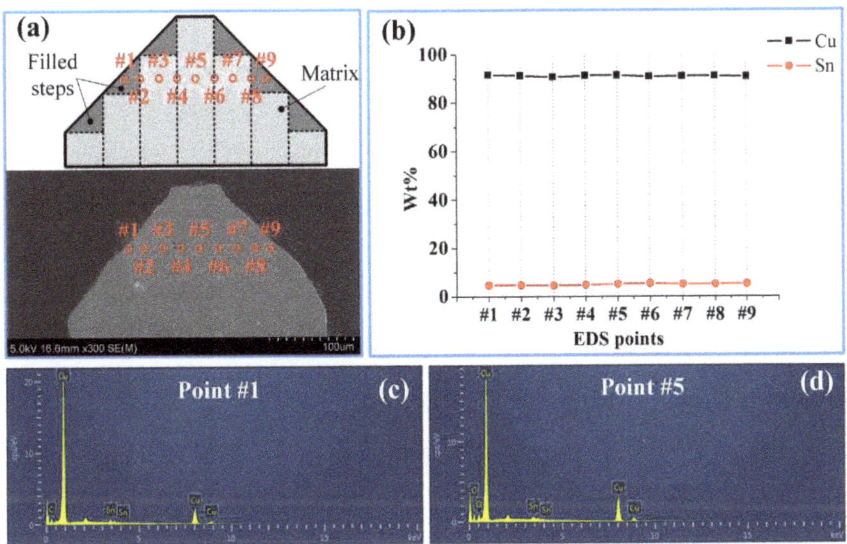

Figure 7. (**a**) Sectional views of microelectrodes; (**b**) The distribution of Cu and Sn along the laminated direction; (**c**) energy dispersive X-ray spectrometer (EDS) result of point #1; (**d**) EDS result of point #5.

3.5. Fabrication of the Microstructure with a Hemisphere

To verify the feasibility of the process, two kinds of 3D microstructures were designed, as shown in Figure 8a,c. It is well known that tool electrode wear inevitably significantly reduces the machining accuracy of microstructures. Therefore, the queued 3D microelectrode models were established according to the 3D microstructures (Figure 8b,d).

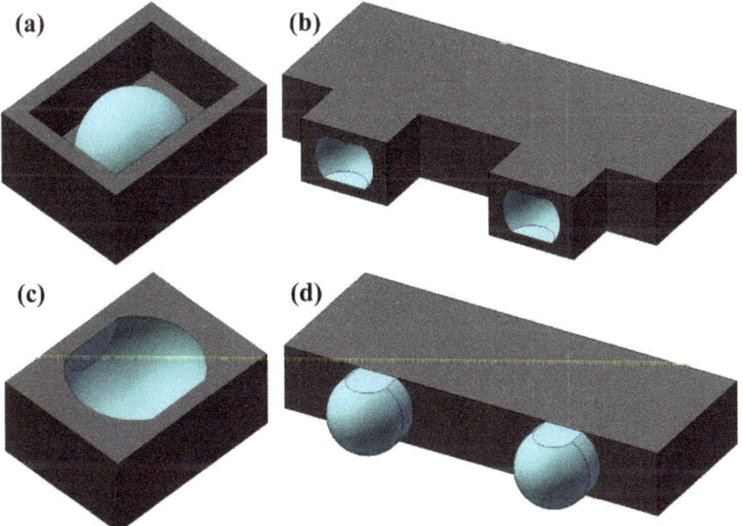

Figure 8. Models of (**a**,**c**) 3D microstructures and (**b**,**d**) queued 3D microelectrodes.

Notably, 50-μm-thick Cu foils coated with 4 μm thick Sn films on both sides were cut by a femtosecond laser beam to obtain multi-layer 2D microstructures. To obtain high quality 2D microstructures, the cutting parameters of femtosecond laser were a 400 mW laser power and a

100 µm/s cutting speed. The multi-layer 2D microstructures were heated up to 950 °C for 15 h with a pressure of 36 KPa in a vacuum furnace to obtain the queued 3D microelectrodes, as shown in Figure 9a,c. It can be observed that there were no steps on the surface of the 3D microelectrodes. This indicated that the steps had been filled effectively by the extruded melt-softened (Cu) during the diffusion bonding process.

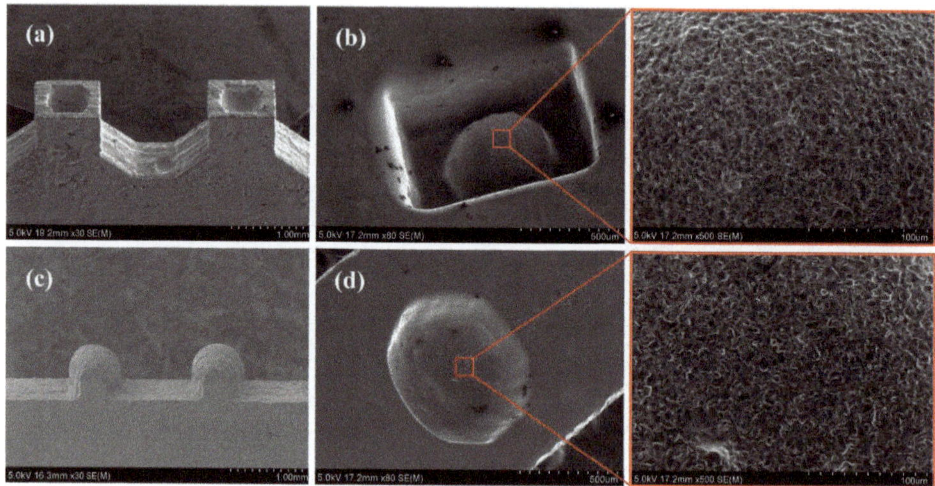

Figure 9. (**a,c**) Composite 3D microelectrodes; (**b,d**) 3D microstructures machined by EDM.

To further verify the elimination of the steps on the surface of the 3D microelectrode, the prepared queued 3D microelectrodes were applied to back-and-forth micro-EDM in sequence, after which 3D microstructures were gained, as shown in Figure 9b,d. The machining parameters were set as follows: voltage 90 V, pulse width 800 ns, pulse interval 4200 ns, pulse frequency 0.2 MHz. The working fluid was deionized water. It can be seen from Figure 9b,d that the arc surface of 3D microstructures had no steps, which indicated that the steps on the 3D microelectrode surface were eliminated. By laser scanning confocal microscopy, surface roughness analyses were carried out on the arc surface of Figure 9b,d. The results were $R_a = 1.365$ µm and $R_a = 1.51$ µm, respectively. The use of deionized water significantly reduced microelectrode wear. The wear of the first and the second microelectrode in the queued 3D microelectrodes were 10 µm and 5 µm, respectively.

For comparison, under the same process conditions, Cu foils coated with 0.5 µm thickness Sn film were used to fabricate queued 3D microelectrodes (Figure 10a,c), and the fabricated queued 3D microelectrodes were applied in micro-EDM to process 3D microstructures (Figure 10b,d). The experiment results show that the steps on the surfaces of 3D microelectrodes and microstructures were extremely obvious. It indicated that the amount of Sn on the interface was insufficient. Therefore, the steps on the surfaces of 3D microelectrodes still existed. As a result, the steps were copied to the machined surface of 3D microstructures from the 3D microelectrode surface.

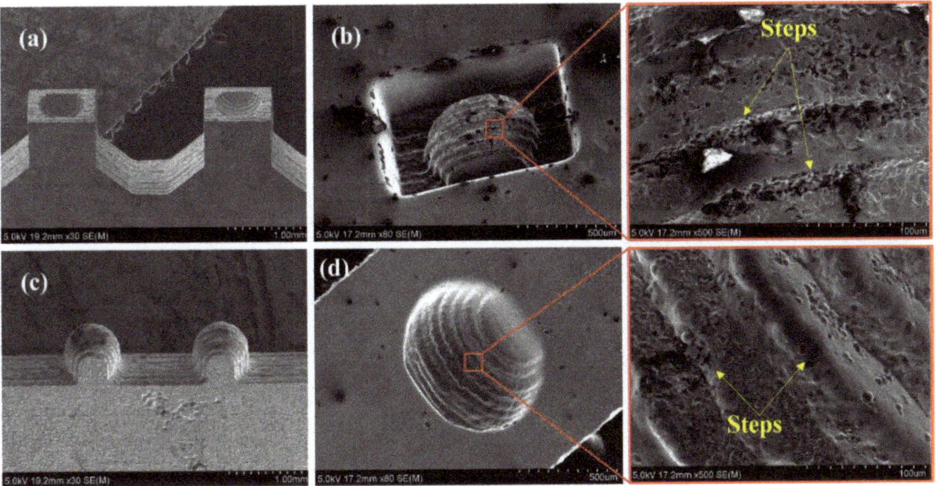

Figure 10. (**a,c**) Laminated 3D microelectrodes with steps; (**b,d**) Microstructures machined by micro-EDM.

4. Conclusions

In this paper, a filling method was proposed to eliminate the steps on the surface of laminated composite 3D microelectrodes using melt-softened (Cu). The microelectrodes were then used to machine 3D microstructures in stainless steel sheets by micro-EDM. The experimental results can be summarized as follows:

(1) The bonding temperature had a great influence on the state of Cu-Sn intermetallic compounds. The steps on the surface of laminated 3D microelectrodes were eliminated by controlling the bonding temperature to make the melt-softened (Cu) fill in the steps.

(2) When the thickness of Sn on Cu foil sides (50 μm thick) was 4 μm, bonding temperature was 950 °C and bonding time was 15 h, the composite 3D microelectrode without steps on the surface was successfully fabricated. Cu and Sn elements were uniformly distributed, both in the steps and matrix.

(3) The queued composite 3D microelectrodes fabricated with the optimized parameters were applied in micro-EDM, obtaining 3D microstructures without steps on the surfaces, significantly improving the machining surface quality. The surface roughnesses were $R_a = 1.365$ μm and $R_a = 1.51$ μm, respectively.

Author Contributions: Conceptualization, J.L. and X.W.; methodology, J.L.; validation, J.L., K.J. and B.X.; formal analysis, K.J.; investigation, J.L.; resources, H.Z.; data curation, K.J.; writing—original draft preparation, J.L.; writing—review and editing, J.L. and B.X.; visualization, K.J.; supervision, H.Z.; project administration, J.L.; funding acquisition, J.L. All authors have read and agreed to the published version of the manuscript.

Funding: This work is supported by the National Natural Science Foundation of China (Grant No. 51805333, 51975385), the Science and Technology Innovation Commission Shenzhen (Grant No. JCYJ20190808143017070, JCYJ20170817094310049, JCYJ20180305125118826 and JSGG20170824111725200), the Natural Science Foundation of Guangdong Province (Grant No. 2017A030313309 and 2018A030310512).

Acknowledgments: The authors are grateful to their colleagues for their essential contribution to the work.

Conflicts of Interest: The authors declare no conflict of interest.

References

1. Pei, J.; Zhou, Z.; Zhang, L.; Zhuang, X.; Wu, S.; Zhu, Y.; Qian, J. Research on the Equivalent Plane Machining with Fix-length Compensation Method in Micro-EDM. *Procedia CIRP* **2016**, *42*, 644–649. [CrossRef]
2. Xu, B.; Guo, K.; Zhu, L.; Wu, X.; Lei, J. Applying Foil Queue Microelectrode with Tapered Structure in Micro-EDM to Eliminate the Step Effect on the 3D Microstructure's Surface. *Micromachines* **2020**, *11*, 335. [CrossRef] [PubMed]
3. Liu, Y.; Chang, H.; Zhang, W.; Ma, F.; Sha, Z.; Zhang, S. A Simulation Study of Debris Removal Process in Ultrasonic Vibration Assisted Electrical Discharge Machining (EDM) of Deep Holes. *Micromachines* **2018**, *9*, 378. [CrossRef] [PubMed]
4. Sabotin, I.; Tristo, G.; Valentincic, J. Technical Model of Micro Electrical Discharge Machining (EDM) Milling Suitable for Bottom Grooved Micromixer Design Optimization. *Micromachines* **2020**, *11*, 594. [CrossRef] [PubMed]
5. Yu, Z.Y.; Masuzawa, T.; Fujino, M. Micro-EDM for Three-Dimensional Cavities—Development of Uniform Wear Method. *CIRP Ann. Manuf. Technol.* **1998**, *47*, 169–172. [CrossRef]
6. Reynaerts, D.; Meeusen, W.; Brussel, H.V. Machining of three-dimensional microstructures in silicon by electrodischarge machining. *Sens. Actuators A* **1998**, *67*, 159–165. [CrossRef]
7. Rajurkar, K.P.; Yu, Z.Y. 3D Micro-EDM Using CAD/CAM. *CIRP Ann. Manuf. Technol.* **2000**, *49*, 127–130. [CrossRef]
8. Bleys, P.; Kruth, J.P.; Lauwers, B. Sensing and compensation of tool wear in milling EDM. *J. Mater. Process. Technol.* **2004**, *149*, 139–146. [CrossRef]
9. Bleys, P.; Kruth, J.P.; Lauwers, B.; Zryd, A.; Delpretti, R.; Tricarico, C. Real-time Tool Wear Compensation in Milling EDM. *CIRP Ann. Manuf. Technol.* **2002**, *51*, 157–160. [CrossRef]
10. Zhao, W.; Yang, Y.; Wang, Z.; Zhang, Y. A CAD/CAM system for micro-ED-milling of small 3D freeform cavity. *J. Mater. Process. Technol.* **2004**, *149*, 573–578. [CrossRef]
11. Li, J.Z.; Xiao, L.; Wang, H.; Yu, H.L.; Yu, Z.Y. Tool wear compensation in 3D micro EDM based on the scanned area. *Precis. Eng.* **2013**, *37*, 753–757. [CrossRef]
12. Jung, J.W.; Jeong, Y.H.; Min, B.K.; Lee, S.J. Model-Based Pulse Frequency Control for Micro-EDM Milling Using Real-Time Discharge Pulse Monitoring. *J. Manuf. Sci. Eng.* **2008**, *130*, 031106. [CrossRef]
13. Yan, M.T.; Lin, S.S. Process planning and electrode wear compensation for 3D micro-EDM. *Int. J. Adv. Manuf. Technol.* **2011**, *53*, 209–219. [CrossRef]
14. Tong, H.; Wang, Y.; Li, Y. Vibration-assisted servo scanning 3D micro EDM. *J. Micromech. Microeng.* **2008**, *18*, 025011.
15. Song, K.Y.; Chung, D.K.; Park, M.S.; Chu, C.N. Micro electrical discharge milling of WC-Co using a deionized water spray and a bipolar pulse. *J. Micromech. Microeng.* **2010**, *20*, 045022. [CrossRef]
16. Yu, H.L.; Luan, J.J.; Li, J.Z.; Zhang, Y.S.; Yu, Z.Y.; Guo, D.M. A new electrode wear compensation method for improving performance in 3D micro EDM milling. *J. Micromech. Microeng.* **2010**, *20*, 055011. [CrossRef]
17. Bissacco, G.; Hansen, H.N.; Tristo, G.; Valentincic, J. Feasibility of wear compensation in micro EDM milling based on discharge counting and discharge population characterization. *CIRP Ann. Manuf. Technol.* **2011**, *60*, 231–234. [CrossRef]
18. Bissacco, G.; Tristo, G.; Hansen, H.N.; Valentincic, J. Reliability of electrode wear compensation based on material removal per discharge in micro EDM milling. *CIRP Ann. Manuf. Technol.* **2013**, *62*, 179–182. [CrossRef]
19. Li, Z.; Bai, J.; Zhu, X. Research on the Depth Error in Micro Electrical Discharge Milling. *Procedia. CIRP* **2016**, *42*, 638–643. [CrossRef]
20. Tong, H.; Li, Y.; Zhang, L. On-machine process of rough-and-finishing servo scanning EDM for 3D micro cavities. *Int. J. Adv. Manuf. Technol.* **2016**, *82*, 1007–1015. [CrossRef]
21. Wang, J.; Qian, J.; Ferraris, E.; Reynaerts, D. In-situ process monitoring and adaptive control for precision micro-EDM cavity milling. *Precis. Eng.* **2017**, *47*, 261–275. [CrossRef]
22. Tong, H.; Pu, Y.; Yang, J.; Li, Y.; Liu, X.; Liang, W.; Zhang, X. A special process of 3D servo scanning micro electro discharge machining for machining pierced micro structures of NiTi alloy tube. *J. Micromech. Microeng.* **2019**, *29*, 1–8. [CrossRef]

23. D'Urso, G.; Giardini, C.; Lorenzi, S.; Quarto, M.; Sciti, D.; Silvestroni, L. Micro-EDM milling of zirconium carbide ceramics. *Precis. Eng.* **2020**, *65*, 156–163. [CrossRef]

24. Xu, B.; Guo, K.; Wu, X.Y.; Lei, J.G.; Liang, X.; Guo, D.J.; Ma, J.; Cheng, R. Applying a foil queue micro-electrode in micro-EDM to fabricate a 3D micro-structure. *J. Micromech. Microeng.* **2018**, *28*, 055008. [CrossRef]

25. Lei, J.; Wu, X.; Wang, Z.; Xu, B.; Zhu, L.; Wu, W. Electrical discharge machining of micro grooves using laminated disc electrodes made of Cu and Sn foils. *J. Mater. Process. Technol.* **2019**, *271*, 455–462. [CrossRef]

26. Xu, B.; Wu, X.Y.; Lei, J.G.; Cheng, R.; Ruan, S.C.; Wang, Z.L. Laminated fabrication of 3D micro-electrode based on WEDM and thermal diffusion welding. *J. Mater. Process. Technol.* **2015**, *221*, 56–65. [CrossRef]

27. Xu, B.; Wu, X.Y.; Lei, J.G.; Cheng, R.; Ruan, S.C.; Wang, Z.L. Laminated fabrication of 3D queue micro-electrode and its application in micro-EDM. *Int. J. Adv. Manuf. Technol.* **2015**, *80*, 1701–1711. [CrossRef]

28. Lei, J.; Wu, X.; Xu, B.; Zhao, Z.; Ruan, S.; Cheng, R. Laminated fitting fabrication of Cu-Sn composite 3D microelectrodes and elimination of ridges on the machined surface of 3D micro-cavities. *J. Mater. Process. Technol.* **2015**, *225*, 24–31. [CrossRef]

29. Ahn, D.; Kweon, J.H.; Choi, J.; Lee, S. Quantification of surface roughness of parts processed by laminated object manufacturing. *J. Mater. Process. Technol.* **2012**, *212*, 339–346. [CrossRef]

30. Gagliano, R.A.; Ghosh, G.; Fine, M.E. Nucleation kinetics of Cu_6Sn_5 by reaction of molten tin with a copper substrate. *J. Electron. Mater.* **2002**, *31*, 1195–1202. [CrossRef]

Article

Enhancing Corrosion and Wear Resistance of Ti6Al4V Alloy Using CNTs Mixed Electro-Discharge Process

Gurpreet Singh [1], Timur Rizovich Ablyaz [2],*, Evgeny Sergeevich Shlykov [2], Karim Ravilevich Muratov [2], Amandeep Singh Bhui [1] and Sarabjeet Singh Sidhu [1]

[1]   Mechanical Engineering Department, Beant College of Engineering and Technology, Gurdaspur 143521, India; singh.gurpreet191@gmail.com (G.S.); meet_amandeep@yahoo.com (A.S.B.); sarabjeetsidhu@yahoo.com (S.S.S.)

[2]   Mechanical Engineering Faculty, Perm National Research Polytechnic University, 614000 Perm, Russia; kruspert@mail.ru (E.S.S.); karimur_80@mail.ru (K.R.M.)

*   Correspondence: lowrider11-13-11@mail.ru

Received: 26 August 2020; Accepted: 9 September 2020; Published: 12 September 2020

Abstract: This paper presents wear and corrosion resistance analysis of carbon nanotubes coated with Ti-6Al-4V alloy processed by electro-discharge treatment. The reported work is carried out using Taguchi's L18 orthogonal array to design the experimental matrix by varying five input process parameters i.e., dielectric medium (plain dielectric, multi-walled carbon nanotubes (MWCNTs) mixed dielectric), current (1–4 A), pulse-on-time (30–60 µs), pulse-off-time (60–120 µs), and voltage (30–50 V). The output responses are assessed in terms of microhardness and surface roughness of the treated specimen. X-ray diffraction (XRD) spectra of the coated sample reveal the formation of intermetallic compounds, oxides, and carbides, whereas surface morphology is observed using scanning electron microscopy (SEM) analysis. For the purpose of the in-vitro wear behavior of treated samples, the surface with superior microhardness values in plain dielectric and MWCNTs mixed dielectric is compared using a pin-on-disc type wear test. Furthermore, electrochemical corrosion test is also conducted to portray the dominance of treated substrate of Ti-6Al-4V alloy for biomedical applications. It is concluded that the wear-resistant and the corrosion protection efficiency of the MWCNTs treated substrate enhanced to 95%, and 96.63%, respectively.

Keywords: electro-discharge treatment; Ti-6Al-4V; MWCNTs; surface characterization; wear resistance; corrosion resistance

1. Introduction

Progress into medical diagnosis, therapy, and rehabilitation is not achievable without the persistent advances in the development of novel or refined materials. Evolutions in the field of metallic biomaterials have remarkably contributed to orthopedic, dental, cardiovascular, neural, and urological praxis [1,2]. Ti-6Al-4V alloy is very favorable in orthopedic applications owing to high weight-to-strength ratio, corrosion resistance, and biocompatibility. Yet, it possesses inferior wear and abrasion resilience, due to low hardness, and releases of toxic ions from the alloy surface within the human body environment [3]. The rather sub-standard tribo-characteristics and corrosion performance, possessed by Ti-6Al-4V alloy, has prompted the progress of various surface treatment approaches that include physical vapor deposition, chemical vapor deposition, dip coating, ion implantation, sol-gel, thermal treatments, laser alloying, etc. [4,5].

Among all the frequently-used techniques for surface-treatment, electro-discharge treatment (EDT) is a thermoelectric method that unfolded as a potential method for the surface modification process [6,7]. As a progressive technique in modifying substrates, EDT widely used to improve the surface characteristics, chemical, mechanical, corrosion, and tribological behavior of the base material [8]. In this process, the train of sparks within the electrodes resulted in the modification of substrate surface due to the material transfer mechanism [9–11]. During the EDT process, the particles of desired coating powder, added in the dielectric medium suspended around the discharge column, accelerated and gained sufficient velocity to penetrate to the molten pool before solidification by means of electrophoresis and negative pressure. This was induced after cessation of a discharge, which leads a surface embedded with added fine particles [12]. This technique significantly affects the surface characteristics of the modified surface, such as morphological structure and chemical compositional analysis [13]. Moreover, the surface produced by EDT exhibited superior corrosion-resistance, wear-resistance as compared to the untreated substrate material, and promoted bioactivity [14].

Various powders have been used by the researchers to investigate their machining performance and influence on the surface characteristics of the machined surface. The most commonly utilized powders are silicon, aluminum, chromium, silicon carbide, titanium, tantalum, boron carbide, and carbon nanotubes [15]. The selection of powder in the dielectric medium significantly depends on the final application of the product. For instance, carbon nanotubes exist in two variants, i.e., single-walled carbon nanotubes (SWCNTs) and multi-walled carbon nanotubes (MWCNTs). The MWCNTs are concentrically aligned sheets of graphene in the shape of cylinders having an outer diameter in the range of 20 to 50 nm [16]. These MWCNTs have been in the limelight in recent decades for exhibiting superior properties pertaining to chemical, electrical, mechanical durability, strength, and bioactivity [17,18]. As a result, carbon nanotubes prominently found applications in the area of medicine and biomedical industries [19,20].

Related Work

This section presents the overview of the research work associated with the surface modification of Ti-6Al-4V alloy for improved wear-resistance, microhardness, corrosion-resistance, and bioactivity. Sarraf et al. [21] reported the improved biological responses of Ti-6Al-4V alloy using the physical vapor deposition method. They had used the tantalum pentoxide nanotube for the surface modification of base metal and observed remarkable nanomechanical properties, wear-resistant, and corrosion-resistant surface with a protective efficiency of 95% in comparison to the base metal. In another study, plasma surface alloying was employed by Li et al. [22] to enhance the corrosion, and wear resistance of Ti-6Al-4V by forming a coated layer of ZrO_2/TiO_2 for biomedical applications. They validated the significance of the modified surface on their excellent performance in wear and corrosion testing. Pogrebnjak et al. [23] demonstrated that the implantation of W and Mo ions, followed by annealing of Ti-6Al-4V surface, resulted in improved nano-hardness and wear resilience, due to the formation of nitrides, carbo-nitrides, and inter-metalloids. Man et. al. [24] enhanced the tribological properties of Ti-6Al-4V alloy using laser diffusion nitriding. They found that the micro-hardness was increased by 2.3 times and wear resilience increased by 8 times compared to the substrate. Kgoete et al. [25] used the spark plasma sintering method to process the Ti-6Al-4V alloy reinforced with μ-sized Si_3N_4 powder to evaluate its corrosion behavior using electrochemical corrosion analysis. Numerous bioactive powders were used by the researchers to enhance the surface competency of Ti-6Al-4V alloy for biomedical applications [26]. However, in a review by Li et al. [27], it was concluded that the coating of CNTs on biomaterials works as a promising combination for achieving satisfactory tissue engineering, hence promoting cell attachment and proliferation. CNTs coating of Ti-6Al-4V alloy demonstrated enhanced fracture toughness, wear-resistance, and biocompatibility making this more favorable for load-bearing orthopedic implants. Deng et al. [28] concluded that the Ti-6Al-4V alloy surface treated with MWCNTs showed superior bio-tribological properties, which is the necessary condition for the bioimplant experiencing various loading conditions within the body. Similarly, Terada et al. [29] reported that the titanium alloy, coated with MWCNTs, improved surface roughness, which accelerated cell adhesion and proliferation.

The present article addresses the influence of electro-discharge process on the microhardness, surface roughness, surface morphology, phase transformation, wear, and corrosion behavior of Ti-6Al-4V surface. In the first step, the significant impact of the selected process parameters on the microhardness and surface roughness was analyzed. In the next step, the characterization of EDT modified surface in terms of morphological and compositional analysis was completed by using SEM, and XRD techniques. The Ti-6Al-4V alloy is well-established and a commonly used biomaterial for orthopedics and dental implants, where wear-resistance and corrosion-resistance play a vital role in long-term functioning of the implants. Therefore, pin-on-disc wear test and electrochemical corrosion behavior of treated samples were performed.

2. Material and Methods

2.1. Material

Commercial grade-5, Ti-6Al-4V ($\alpha + \beta$) titanium alloy was purchased in the form of a plate with dimensions $160 \times 80 \times 5$ (units: mm) from Baoji Fuyuantong Industry and Trade Co. Ltd., Baoji, China and used as workpiece material. Electrolytic pure graphite procured from Mersen Germany (Courbevoie, France), machined to 9.5 mm diameter was chosen as a tool electrode. Table 1 shows the physical properties of the workpiece and electrode used in the present experimentation. Table 2 listed the functionalized multi-walled carbon nanotubes (MWCNTs) properties, purchased from United Nanotech, Bangalore, India. These MWCNTs were mixed in commercial-grade hydrocarbon oil (EDM oil) to achieve the desired surface modification of Ti-6Al-4V alloy in EDT process.

Table 1. Physical properties of workpiece and electrode materials. Source: www.matweb.com.

Property	Ti-6Al-4V	Graphite
Chemical composition	Ti: 89.54%; Al: 6.1%; V: 4.2%; Fe: 0.09%; C: 0.03%; O: 0.03%; N: 0.003%; H: 0.001%	Pure carbon
Size (mm)	$70 \times 70 \times 5$	$\varnothing$ 9.5
Density (g/cm^3)	4.43	2.26
Melting temperature (°C)	1604–1660	3650
Thermal conductivity (W/m.K)	6.70	24.0
Specific heat (J/Kg °C)	526.3	0.7077
Electrical resistivity (Ω cm)	1.78×10^{-4}	6.0×10^{-3}

Table 2. Physical properties of MWCNTs. Source: Technical Data Sheet, provided with powder.

Property	Description
Production Method	Chemical Vapor Deposition
Available form	Black powder
Diameter	Outer Diameter: 10–30 nm
Length	10 microns
Nanotubes purity	>95%
Metal particles	<4%
Amorphous carbon	<1%
Specific surface area	330 m^2/g
Bulk density	0.04–0.06 g/cm^3

2.2. Experimental Work

In this experiment, Taguchi's L18 orthogonal array incorporating mixed-level design ($2^1 \times 3^4$) was selected. To draw valid conclusions, the machining parameters were current, pulse-on/off duration and voltage with three levels whereas two types of dielectric medium were selected. The different values for the levels were chosen from the pilot trials performed and reported in our earlier study [30]. Table 3 illustrates the machining parameters and the corresponding levels for the design of experiment.

Table 3. Experimental process parameters with their respective levels.

Parameter	Symbol	Units	Levels		
			Level 1	Level 2	Level 3
Dielectric type	A	–	Plain dielectric	MWCNTs mixed dielectric	–
Current	B	ampere	1	2	4
Pulse-on time	C	μ-seconds	30	45	60
Pulse-off time	D	μ-seconds	60	90	120
Voltage	E	volts	30	40	50

Table 4, illustrates the array of 18 experimental trials undertaken to investigate the potential of selected machining parameters on output responses of EDT process. For experimental trials, the numerical controlled electrical discharge machine with side flushing mechanism was used (make: OSCARMAX, Taichung City, Taiwan; model: S645 CMAX). The trials were conducted in doublet i.e., a total of 36 experiments (18 × 2 runs) were carried out for more precise output responses. The set of first nine experiments were conducted in plain dielectric type, i.e., EDM oil, followed by remaining experiments in MWCNTs mixed dielectric medium. The concentration of MWCNTs (7 g/L) in the dielectric medium was preferred on the basis of previously reported work by the authors, where higher concentration causes unstable machining [30]. Figure 1 illustrates the indigenously developed dielectric tank (12″ × 9″ × 9″) for the execution of MWCNTs mixed experiments. A circulation pump and stirrer were introduced for efficient flushing and suspension of nano-powder in the spark gap during machining. The polarity (i.e.,workpiece (–), tool (+)), and machining depth of 1 mm were kept constant for all the experimental runs.

Table 4. Taguchi's L18 experimental design matrix.

Exp. Trial	Levels of Process Parameters					Actual Values of Process Parameters				
	A	B	C	D	E	A	B	C	D	E
1.	1	1	1	1	1	Plain dielectric	1	30	60	30
2.	1	1	2	2	2	Plain dielectric	1	45	90	40
3.	1	1	3	3	3	Plain dielectric	1	60	120	50
4.	1	2	1	1	2	Plain dielectric	2	30	60	40
5.	1	2	2	2	3	Plain dielectric	2	45	90	50
6.	1	2	3	3	1	Plain dielectric	2	60	120	30
7.	1	3	1	2	1	Plain dielectric	4	30	90	30
8.	1	3	2	3	2	Plain dielectric	4	45	120	40
9.	1	3	3	1	3	Plain dielectric	4	60	60	50
10.	2	1	1	3	3	MWCNTs mixed dielectric	1	30	120	50
11.	2	1	2	1	1	MWCNTs mixed dielectric	1	45	60	30
12.	2	1	3	2	2	MWCNTs mixed dielectric	1	60	90	40
13.	2	2	1	2	3	MWCNTs mixed dielectric	2	30	90	50
14.	2	2	2	3	1	MWCNTs mixed dielectric	2	45	120	30
15.	2	2	3	1	2	MWCNTs mixed dielectric	2	60	60	40
16.	2	3	1	3	2	MWCNTs mixed dielectric	4	30	120	40
17.	2	3	2	1	3	MWCNTs mixed dielectric	4	45	60	50
18.	2	3	3	2	1	MWCNTs mixed dielectric	4	60	90	30

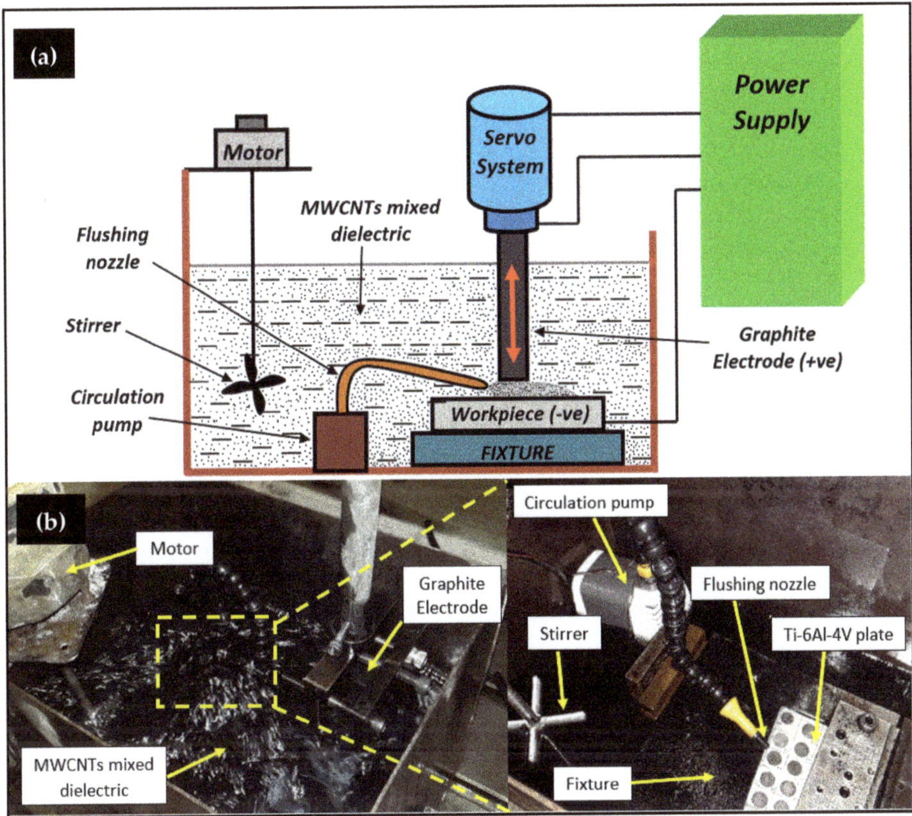

Figure 1. EDT line-diagram in; (**a**), and setup for MWCNTs-EDT in (**b**).

2.3. Measurement and Calculations of Output Responses

After the EDT, the output responses were assessed in the terms of surface roughness (SR) and microhardness (MH) of the machined substrate. Microhardness tester (model: HM-220, make: Mitutoyo, Neuss, Germany) and surface roughness tester (model: SJ-401, make: Mitutoyo, Germany) was used to measure the microhardness (units: HV) and surface roughness (units: µm) respectively, by taking three readings at distinct points. The microhardness was measured by profiling a pyramidal imprint on the sample surface using a diamond indenter under low-force hardness scale (HV 0.5) with a test force-load of 4.9 N and for a dwell time of 10 s.

For determining the roughness of EDT surface, stylus method was employed to find the roughness value (Ra) with a profile resolution of 12 nm, and range of 80 µm having resolution of 0.001 µm. It was reported in the literature that higher surface roughness participated in cell anchoring, facilities better bone-implant adhesion due to the presence of micro-crack and pores [31–33]. Therefore, for further testing, the samples with higher surface roughness were sectioned from the workpiece by using wire-EDM.

2.4. Surface Characterization of EDT specimen

The electro-discharge treated surface was thoroughly cleaned in acetone to remove any foreign particles or EDM oil entrapped in pores. The surface morphology of the sample depicting superior output responses was inspected using a scanning electron microscope (SEM; model: JSM-6610LV, make: Jeol Ltd., Tokyo, Japan) at an accelerating voltage of 20 kV. The phase composition of the EDT surface was examined by X-ray diffractometer (XRD; model: PANalytical X'Pert Pro MPD, make: Panalytical, Almelo, The Netherlands) with Cu-Kα X-ray radiations at λ = 1.5406 Å using 40 mA and 45 kV as generator settings.

2.5. Examination of Wear and Electrochemical Corrosion Behavior

Furthermore, in-vitro wear and corrosion tests were executed to scrutinize the wear resistance and corrosion resistance offered by the EDT samples. Two samples were selected in both cases, i.e., with maximum output values in MWCNTs mixed dielectric, in the plain dielectric, and it was further compared with the untreated substrate. The experimental trial depicting higher microhardness value in each dielectric medium (i.e., plain, MWCNTs mixed) was examined for the wear test, and similarly the two samples with higher surface roughness value in each dielectric medium was chosen for the electrochemical corrosion test. A pin-on-disk tribometer (Ducom instruments, Bangalore, India) was used to perform the wear test of specimens on EN31 stainless steel disk with diameter 120 mm, thickness 10 mm, and hardness of 63 HRC. The fixed parameters for the wear test were a steady load of 70 N, speed of rotating disc at 100 revolutions/min, and track diameter of 80 mm. The experimentation was conducted at room temperature using the ringer's solution (Nice Chemicals Pvt. Ltd., Cochin, India; pH value 7.2) as a lubricant to replicate the human body conditions during the investigation. The wear process of the pins was monitored by an incorporated computer system with TR-20LE software, and the combined plot of all samples was plotted using the reading values in OriginPro 8 software.

The potentiodynamic polarization technique was used to inspect the corrosion behavior of samples and the test was carried out on potentiostat/galvanostat electrochemical instrument (Metrohm Autolab, PGSTAT 302, Utrecht, The Netherlands) at ambient temperature. The equipment was an arrangement of three-electrodes, namely silver/silver chloride (Ag/AgCl) as a reference electrode, platinum rod (Pt) as a counter electrode, and workpiece sample (Ti-6Al-4V) as a working electrode. For better results, the surface of the specimen was covered with insulating tape to prevent the initiation of corrosion and exposed the fixed area of 0.32 cm^2 for testing. Prior to the testing, the samples were immersed in the solution for 24 h to stabilization. Likewise, for wear test analysis, ringer's solution is used as simulated body fluid (electrolyte in testing) to imitate the human body environment for the bio-implant. The results of the Tafel exploration plots were analyzed using the Nova software incorporated with the experimental arrangement, and the output values for anodic slope (βa), cathodic slope (βc), corrosion current density (i_{corr}), corrosion potential (E_{corr}) were recorded. Accordingly, the corrosion rate (CR) of the samples was then evaluated according to ASTM G102-89 [34], a standard for electrochemical measurements using Faraday's law (Equation (1)),

$$CR = K \frac{i_{corr}}{\rho} EW \tag{1}$$

where

CR is in mm/year,
i_{corr} is in μA/cm^2;
K = 3.27 × 10^{-3}, mm g/μA cm year;
ρ = density in g/cm^3;
EW = equivalent weight of the material.

3. Results and Discussion

3.1. Evaluation of the Output Responses

Table 5 demonstrates the measured values with three repetitions, and signal-to-noise (SN) ratios of microhardness and surface roughness associated with each experimental trial. The evaluation of microhardness and surface roughness values were as per the Taguchi's criterion "larger-is-better" (Equation (2)), where a higher value of SN ratio represents the favorable response value:

$$\text{Larger} - \text{is} - \text{better} : \eta = -10\log\left[\frac{1}{n}\sum_{i=1}^{n} yi^{-2}\right]. \tag{2}$$

where

'η' denotes the SN ratio (dB);
'yi' indicates the value of ith trial of experimental trial;
'n' is the repetition of the experiment.

Table 5. Response observations and SN ratios of microhardness (MH) and surface roughness (SR).

Exp. Trial	Output Responses						SN Ratio, dB	
	MH (HV)			SR (µm)			MH	SR
	Rep 1	Rep 2	Rep 3	Rep 1	Rep 2	Rep 3		
1.	897.8	937.0	925.7	0.053	0.061	0.087	59.2731	−24.0190
2.	1066.5	929.3	1025.7	0.138	0.094	0.105	60.0179	−19.3177
3.	1177.4	1098.1	1109.8	0.106	0.101	0.119	61.0373	−19.3390
4.	1091.9	1287.0	1259.3	0.802	0.657	0.883	61.6051	−2.3502
5.	1484.3	1561.2	1586.0	0.661	0.794	0.675	63.7615	−3.0605
6.	1497.0	1434.7	1205.9	0.698	1.021	0.976	62.6768	−1.3158
7.	1419.8	1459.2	1384.3	0.959	1.014	1.018	63.0465	−0.0358
8.	1736.8	1689.1	1656.5	0.855	0.954	0.969	64.5740	−0.7086
9.	1748.3	1761.2	1786.0	0.897	0.892	0.941	64.9347	-0.8266
10.	2140.5	2362.8	2306.1	0.523	0.591	0.615	67.0966	−4.8491
11.	2705.1	2653.1	3067.4	0.534	0.698	0.686	68.9165	−4.0849
12.	2901.8	3242.1	3263.2	0.412	0.469	0.501	69.8885	−6.8189
13.	3147.0	3242.9	3171.3	0.676	0.733	0.709	70.0657	−3.0383
14.	2890.0	2982.9	3329.9	0.885	0.974	0.874	69.6885	−0.8398
15.	3488.2	3792.7	3524.3	1.118	0.980	1.002	71.1123	0.2422
16.	4530.9	4346.8	4306.0	0.991	1.322	1.267	72.8520	1.3207
17.	4394.0	4494.0	4469.6	1.527	0.913	1.280	72.9709	1.2709
18.	3979.4	4283.0	4293.2	1.248	1.226	0.992	72.4180	1.1097

Rep: Repetitions

Additionally, the experimentally calculated values were subsequently analyzed via analysis of variance (ANOVA) to inspect the influential input parameters and their significance, in terms of percentage contribution for microhardness, and surface roughness, respectively.

3.1.1. Analysis of Microhardness

In electro-discharge treatment, the powder particles tend to enhance the microhardness by forming a coating layer of various intermetallic compounds, and carbides on the treated substrate [35,36]. Table 6 demonstrates the ANOVA results of microhardness of the machined surface. As shown in Table 6, dielectric type (MWCNTs mixed) noticed as the most eminent factor with the contribution of 83.019%, followed by current with the contribution of 13.698% affecting the microhardness of machined substrate.

Table 6. Analysis of variance for signal-to-noise ratios of Microhardness.

Source	DF	Seq SS	Adj MS	F-Value	*p*-Value	% Contribution
Dielectric type	1	304.897	304.897	571.58	0.000 *	83.019
Current	2	50.309	25.155	47.16	0.010 *	13.698
Pulse-on-time	2	5.919	2.959	5.55	0.031 *	1.612
Pulse-off-time	2	0.142	0.071	0.13	0.877	0.038
Voltage	2	1.727	0.863	1.62	0.257	0.472
Residual error	8	4.267	0.533			1.161
Total	17	367.261				100

DF: degrees of freedom; Seq SS: sequential sum of squares; Adj MS: adjusted mean sum of squares, * Significant at 95% confidence level, Rank 1: Dielectric type, Rank 2: Current, Rank 3: Pulse-on-time

Similar results were revealed by the signal-to-noise ratio plot (Figure 2), disclosing dielectric type, current, pulse-on-time as momentous parameters, and pulse-off-time, voltage as insignificant parameters for the microhardness of treated samples. The higher intensity of the current and pulse-on-time steeply promotes the deposition of particles, which improves the microhardness, and other changes in surface characteristics of the machining area during the EDT process [37–39]. Based on the results, higher values of current (4A) and MWCNTs, mixed dielectric medium, depicts utmost mean microhardness (4452.5 HV, trial 17) with an increase of 2.5 times comparative to the sample treated in plain dielectric (1765.2 HV, trial 9), and 10 times superior to the untreated material (435.4 HV), respectively.

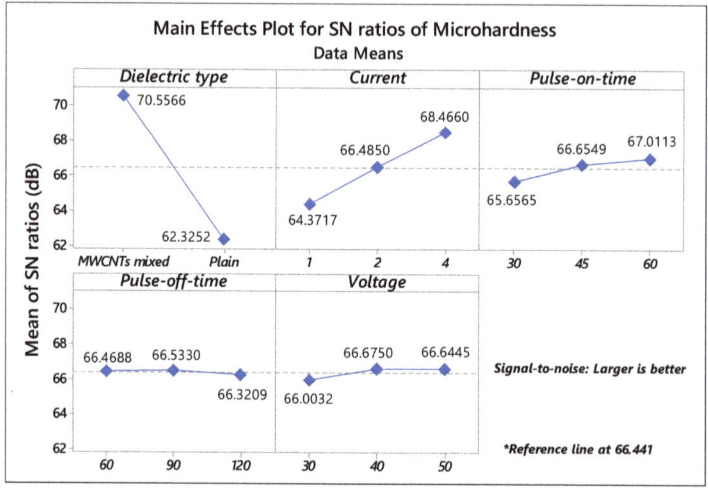

Figure 2. Main effects SN ratios plot of Microhardness.

3.1.2. Analysis of Surface Roughness

The ANOVA Table 7 manifested current (contribution: 61.076%) and dielectric type (contribution: 16.551%) as the prominent factors affecting the surface roughness of EDT processed Ti-6Al-4V substrate. As cleared from the SN ratios plot, shown in Figure 3, MWCNTs mixed dielectric and 4A of current intensity were the dominant parameters for the roughness of the EDT surface. In contrast, the variation of signal-to-noise ratios for pulse-on/off duration, voltage was not influential and termed as in-significant factors. Moreover, the interaction between dielectric type and current also noticed as significant, having *p*-value 0.001 and a contribution of 19.824%.

Table 7. Analysis of variance for signal-to-noise ratios of Surface Roughness.

Source	DF	Seq SS	Adj MS	F-Value	*p*-Value	% Contribution
Dielectric type	1	169.81	169.806	53.88	0.000 *	16.551
Current	2	626.60	313.299	99.41	0.000 *	61.076
Pulse-on-time	2	4.18	2.088	0.66	0.550	0.407
Pulse-off-time	2	2.65	1.325	0.42	0.675	0.258
Voltage	2	0.43	0.215	0.07	0.935	0.041
Dielectric type × current	2	203.38	101.689	32.27	0.001 *	19.824
Residual error	6	18.91	3.151			1.843
Total	17	1025.95				100

DF: degrees of freedom; Seq SS: sequential sum of squares; Adj MS: adjusted mean sum of squares, * Significant at 95% confidence level, Rank 1: Current, Rank 2: Dielectric type, Rank 3: Dielectric type × current

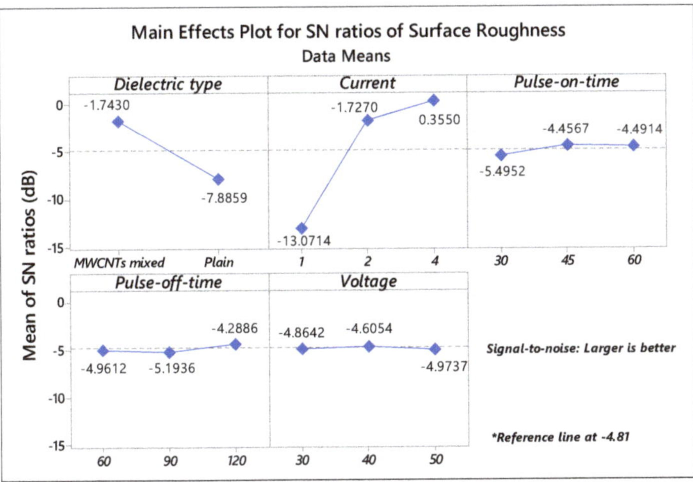

Figure 3. Main effects SN ratios plot of Surface Roughness.

Figure 4 portrays the variation of surface roughness in conjunction with the dielectric type and current. It was observed that current at level 3 (4 A) along with MWCNTs mixed dielectric increases the surface roughness of the treated samples. Whereas, at the lower level (1 A) of current intensity, changing the dielectric type from MWCNTs mixed to plain dielectric drastically decreases the surface roughness. Therefore, it can be revealed that the dielectric illustrates a substantial role in the surface roughness of the EDT surface. The maximum mean roughness of 1.240 μm (trial 17) was depicted at parametric settings of 4 A current, 45 μs pulse-on-time, 60 μs pulse-off-time, and 50V of voltage in MWCNTs mixed dielectric medium. Moreover, all the three trials (trial 16–18) in MWCNTs mixed dielectric at a discharge current of 4A exhibits good roughness values within the variation of 4% (approx.). The roughness value reported is acceptable and within the range, required for the bio-implants for proper cell proliferation, bone-implant fixation [40,41].

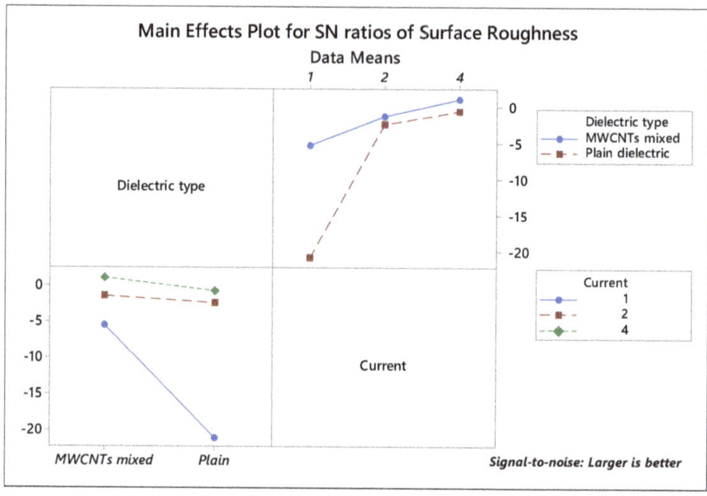

Figure 4. Interaction plot for Surface Roughness.

3.2. Morphology and Phase Composition Analysis of MWCNTs Treated Specimen

The supremacy of the carbon nanotubes mixed dielectric was investigated using scanning electron microscopy. Figure 5a,b manifests the morphological of untreated and treated substrate. For morphological examination, sample depicting superior microhardness and higher surface roughness (trial 17) was selected. It is evident from the micrograph (Figure 5b) that a higher spark energy, coupled with carbon particles, produce micro-macropores, small crests on the EDT surface. The existence of such pores, microcracks, and molten metal droplets on the surface facilitated good adhesion within the implant-bone interface, and offered cell growth for proper regeneration [42,43].

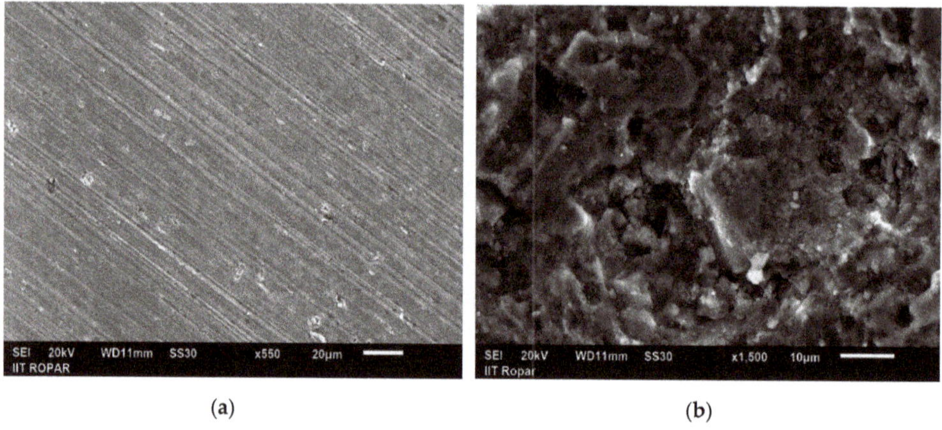

Figure 5. SEM showing; (**a**) substrate or untreated sample, and (**b**) surface treated in MWCNTs mixed dielectric (trial 17) representing micro cracks, porous surface and molten metal droplets.

Furthermore, the investigation of the electro-discharge-treated sample with maximum microhardness and surface roughness was carried using an X-ray diffractometer within the range of $20°$–$80°$. The XRD spectra in Figure 6 revealed the formation of various intermetallic compounds during the EDT process, which majorly contributes to improving the microhardness of the treated substrate. The newly formed compounds namely calcite ($C_1Ca_1O_3$), titanium nickel (Ni_1T_1) shows the hexagonal structure, molybdenum germanide (Ge$_2$Mo$_1$), molybdenum nickel (Mo$_1$Ni$_4$) represents a tetragonal structure, and aluminium nitride (Al$_1$N$_1$), chromium oxide (Cr$_3$O$_1$) disclosed as a cubic structure. Alongside, high peaks of titanium carbide (TiC), vanadium silicide (Si$_2$V$_1$) were witnessed on the surface, and the presence of such hard-bioactive layers exhibits better resistance to corrosion, wear by providing appropriate implant-bone bioactivity within the individual. Moreover, the MWCNTs-EDT surface was further examined for the valuation of in-vitro behavior to affirm the enhanced wear resistance and corrosion resistance by comparing the results with untreated surface and the sample treated in the plain dielectric.

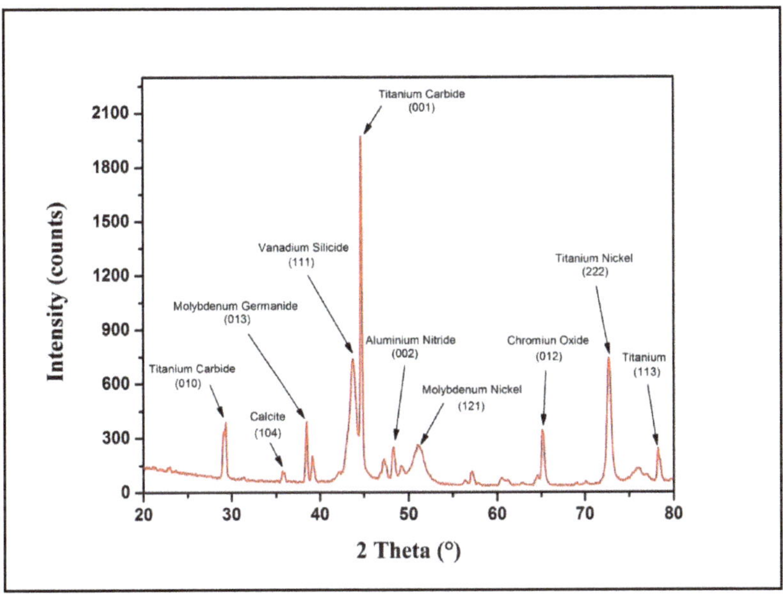

Figure 6. XRD spectra and crystallographic planes of MWCNTs treated sample (trial 17).

3.3. In-Vitro Wear and Corrosion Analysis of EDT Samples

The wear resistance proficiency of the treated Ti-6Al-4V surface was investigated using a pin-on-disc arrangement. Three samples were selected, i.e., untreated (as received), treated in the plain dielectric (trial 9) and in MWCNTs mixed dielectric (trial 17) for evaluation of wear performance. The samples were chosen on the basis of maximum mean microhardness in plain dielectric (1765.2 HV, trial 9) and MWCNTs mixed dielectric (4452.5 HV, trial 17).

Figure 7 unfolded the combined wear plot for three samples, and it was observed that the wear of substrate was nearly at an even rate of 450 µm. The wear of sample, treated in the plain dielectric, started after 150 µm in the beginning, due to the presence of ridges, then the wear rate became constant to 125 µm (approx.) up to 250 s until the machined or recast layer wore off. Afterwards, the wear rate was raised to 440 µm, which was close to the wear rate of the substrate material. The wear of MWCNTs-treated sample started initially at 50 µm, and once the surface became even from all edges, a constant wear rate of 22 µm was exhibited by the MWCNTs-EDT sample up to 900 s. This wear response was due to the modified surface having intermetallic compounds, carbides, and silicide, as discussed in the section of XRD analysis. Later, the wear rate was abruptly raised once the coated layer wore out, demonstrating an improved resistance to wear by 95% compared to substrate metal.

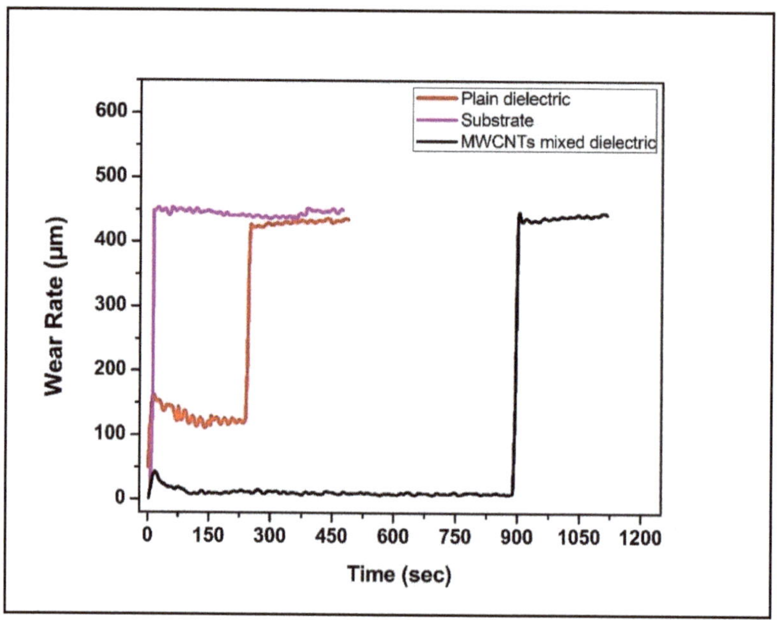

Figure 7. Comparison plot for the wear behavior of specimens.

In the electrochemical test analysis, the surface with maximum roughness value in the plain dielectric (trial 7), and MWCNTs mixed dielectric (trial 17), was compared with the untreated substrate to scrutinize the efficacy of the EDT method. The additional electrochemical characteristics attained via potentiodynamic polarization analysis were tabulated in Table 8.

Table 8. Polarization corrosion data of specimens in Ringer solution at 37 °C.

Sr. No.	Sample	Ecorr (mV)	i_{corr} (μA/cm^2)	βa (mV/dec)	βc (mV/dec)	Corrosion Rate (mm/y)	Protection Efficiency (Pe)
1.	MWCNTs mixed dielectric	3.50860	1.83	93.1230	187.420	0.03248	96.63%
2.	Plain dielectric	−82.4920	17.48	211.820	126.560	0.3091	67.83%
3.	Substrate	−160.640	54.35	234.610	292.470	0.9610	–

Figure 8 illustrates the combined polarization curves for the substrate and selected EDT samples. The plot showed that the MWCNTs-coated specimen exhibits a higher value of corrosion potential (E_corr), and thus, possesses a low corrosion rate compared to other samples. In contrast, the untreated substrate sample showed the lowest corrosion potential (E_corr), corrosion-resistance, and subsequently the higher corrosion current (i_{corr}), compared to all the EDT samples. The results of in-vitro corrosion behavior of the compared samples demonstrated that the EDT process uplifts the corrosion-resistant competences of the treated surface.

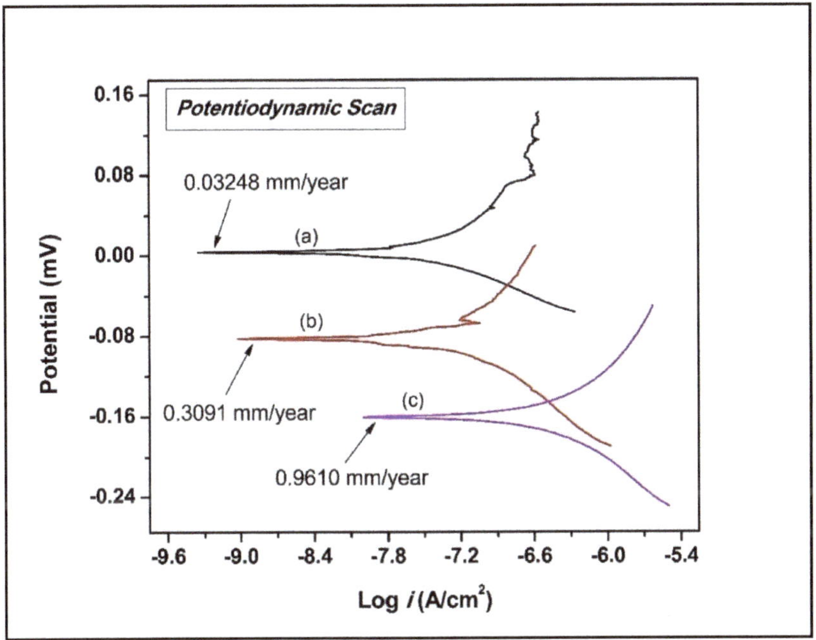

Figure 8. Combined polarization curves of samples (a) treated in MWCNTs, (b) treated in plain dielectric, (c) substrate.

In Table 8, MWCNTs-EDT specimen exhibit superior resistance to corrosion (CR) and a lesser value of corrosion current density (i_{corr}). These are the two main factors that directly influence the performance of a material under biological conditions. Also, carbon nanotubes-coated specimens depicted a higher value of corrosion potential (E_{corr} = 3.5086 mV) compared to another tested specimen, i.e., −82.492 mV is recorded for treated in plain dielectric and −160.640 mV for untreated substrate of Ti-6Al-4V alloy.

Additionally, the protection efficiency (P_e) of the EDT samples was calculated using the Equation (3),

$$P_e(\%) = \left[1 - \frac{i_{corr}}{i_{corr}{}^{substrate}}\right] \times 100 \tag{3}$$

where, i_{corr} denotes the corrosion current densities of EDT samples (in MWCNTs and plain dielectric) to the i_{corr} of the substrate sample, respectively.

The upmost protective efficiency of 96.63% was observed for MWCNTs-EDT on Ti-6Al-4V alloy. Finally, it was concluded that the MWCNT-modified Ti-6Al-4V alloy surface revealed enhanced bioactivity by minimizing the discharge of ions (Al and V) from the implant material causing toxicity. Therefore, the responses of in-vitro wear and corrosion analysis confirmed that the electro-discharge treated samples in MWCNT-mixed dielectric considerably discloses the surface for improved wear-resistance and corrosion-resistance.

4. Conclusions

The Ti-6Al-4V alloy surface was processed by electro-discharge treatment with the purpose of improving its surface hardness, surface characteristics, and to examine the in-vitro corrosion behavior and tribological performance of the modified surface. In the present work, the treated samples were compared with the substrate sample and the following conclusions can be drawn:

The surface characterization of the samples revealed that the surface-treated in MWCNTs medium demonstrated improved surface hardness to 10 times (4452.5 HV), compared with the untreated substrate sample (435.4 HV), thus, increasing wear-resistance to 95%.

The electrochemical corrosion analysis demonstrated that MWCNTs-EDT surface exhibited higher corrosion potential, and this non-reactive surface was attained because of the exceptional thermal properties and chemical stability of MWCNTs. The newly formed carbide- and oxide-rich layers stimulate the corrosion-resistance of the EDT surface.

The XRD pattern of the EDT sample substantiated the formation of intermetallic compounds (titanium nickel, molybdenum nickel, aluminium nitride, vanadium silicide), oxide (chromium oxide), and carbide (titanium carbide) on the treated surface that encourages the bioactivity, wear-resistance and corrosion-resistance of the EDT surface.

SEM analysis disclosed that MWCNTs-EDT surface with the presence of small craters, and evenly distributed micro-pores on the surface, which may facilitates the proper bone-implant adhesion and promotes cell proliferation.

The surface treatment of Ti alloy in MWCNTs mixed dielectric at 4 A current, 45 µs pulse-on-time, 60 µs pulse-off-time, and 50 V delivered the superlative results of tribological performance and corrosion analysis.

Therefore, the electro-discharge treated titanium-based implants can be further inspected for their practice in clinical applications. Moreover, the reported research can be further protracted to investigate the cytocompatibility, hemocompatibility, surface wettability, toxic ions degradation rate, etc., of the EDT surfaces.

Author Contributions: Conceptualization, S.S.S., T.R.A. and G.S.; methodology, S.S.S. and T.R.A.; software, G.S. and A.S.B.; validation, G.S., K.R.M. and E.S.S.; formal analysis, T.R.A., E.S.S. and S.S.S.; investigation, G.S., A.S.B.; resources, S.S.S.; data curation, E.S.S. and G.S.; writing—original draft preparation, G.S. and A.S.B.; writing—review and editing, G.S., S.S.S., T.R.A.; visualization, K.R.M. and E.S.S.; supervision, S.S.S. and T.R.A.; project administration, S.S.S. and G.S.; funding acquisition, T.R.A., E.S.S. and K.R.M. All authors have read and agreed to the published version of the manuscript.

Funding: This work was supported by Russian Science Foundation, grant number 20-79-00048.

References

1. Aleksandra, W.B.; Ewa, S.Z.; Wojciech, P.; Elzbieta, D.; Aleksandra, B.; Marta, B. A model of adsorption of albumin on the implant surface titanium and titanium modified carbon coatings (MWCNT-EPD) 2D correlation analysis. *J. Mol. Struct.* **2016**, *1124*, 61–70. [CrossRef]

2. Singh, G.; Sidhu, S.S.; Bains, P.S.; Singh, M.; Bhui, A.S. On surface modification of Ti alloy by electro discharge coating using hydroxyapatite powder mixed dielectric with graphite tool. *J. Bio- Tribo-Corros.* **2020**, *6*, 91. [CrossRef]

3. Aliyu, A.A.; Abdul-Rani, A.M.; Ginta, T.L.; Rao, T.V.V.L.N.; Axinte, E.; Ali, S.; Ramli, M. Hydroxyapatite electro discharge coating of Zr-based bulk metallic glass for potential orthopedic applications. *Key Eng. Mater.* **2019**, *796*, 123–128. [CrossRef]

4. Liu, X.; Chu, P.K.; Ding, C. Surface modification of titanium, titanium alloys, and related materials for biomedical applications. *J. Mater. Sci. Eng. R.* **2005**, *47*, 49–121. [CrossRef]

5. Wang, Z.L.; Fang, Y.; Wu, P.N.; Zhao, W.S.; Cheng, K. Surface modification process by electric discharge machining with a Ti powder green compact electrode. *J. Mater. Process. Technol.* **2002**, *129*, 139–142. [CrossRef]

6. Bui, V.D.; Mwangi, J.W.; Schubert, A. Powder mixed electrical discharge machining for antibacterial coating on titanium implant surfaces. *J. Manuf. Process.* **2019**, *44*, 261–270. [CrossRef]

7. Al-Amin, M.; Abdul-Rani, A.M.; Abdu-Aliyu, A.A.; Abdul-Razak, M.A.; Hastuty, S.; Bryant, M.G. Powder mixed-EDM for potential biomedical applications: A critical review. *Mater. Manuf. Process.* **2020**, 1–23. [CrossRef]

8. Umar-Farooq, M.; Pervez-Mughal, M.; Ahmed, N.; Ahmad-Mufti, N.; Al-Ahmari, A.M.; He, Y. On the investigation of surface integrity of Ti6Al4V ELI using Si-mixed electric discharge machining. *Materials* **2020**, *13*, 1549. [CrossRef]

9. Kumar, V.; Diyaley, S.; Chakraborty, S. Teaching-learning-based parametric optimization of an electrical discharge machining process. *FU Ser. Mech. Eng.* **2020**, *18*, 281–300. [CrossRef]

10. Tzeng, Y.F.; Lee, C.Y. Effects of powder characteristics on electro discharge machining efficiency. *Int. J. Adv. Manuf. Technol.* **2001**, *17*, 586–592. [CrossRef]

11. Furutania, K.; Saneto, A.; Takezawa, H.; Mohri, N.; Miyake, H. Accretion of titanium carbide by electrical discharge machining with powder suspended in working fluid. *Prec. Eng.* **2001**, *25*, 138–144. [CrossRef]

12. Singh, G.; Lamichhane, Y.; Bhui, A.S.; Sidhu, S.S.; Bains, P.S.; Mukhiya, P. Surface morphology and microhardness behavior of 316L in HAp-PMEDM. *FU Ser. Mech. Eng.* **2019**, *17*, 445–454. [CrossRef]

13. Al-Amin, M.; Abdul-Rani, A.M.; Abdu-Aliyu, A.A.; Bryant, M.G.; Danish, M.; Ahmad, A. Bio-ceramic coatings adhesion and roughness of biomaterials through PM-EDM: A comprehensive review. *Mater. Manuf. Process.* **2020**, *35*, 1157–1180. [CrossRef]

14. Moro, T.; Mohri, N.; Otsubo, H.; Goto, A.; Saito, N. Study on the surface modification system with electrical discharge machine in the practical usage. *J. Mater. Process. Technol.* **2004**, *149*, 65–70. [CrossRef]

15. Abdudeen, A.; Abu-Qudeiri, J.E.; Kareem, A.; Ahammed, T.; Ziout, A. Recent advances and perceptive insights into powder-mixed dielectric fluid of EDM. *Micromachines* **2020**, *11*, 754. [CrossRef]

16. Harris, P.J. Carbon nanotubes and related structures: New materials for the twenty-first century. *Am. J. Phys.* **2004**, *72*, 414–415. [CrossRef]

17. George, G.; Sisupal, S.B.; Tomy, T.; Kumaran, A.; Vadivelu, P.; Suvekbala, V.; Sivaram, S.; Ragupathy, L. Facile, environmentally benign and scalable approach to produce pristine few layers graphene suitable for preparing biocompatible polymer nanocomposites. *Sci. Rep.* **2018**, *8*, 11228. [CrossRef]

18. Kumar, S.; Parekh, S.H. Linking graphene-based material physicochemical properties with molecular adsorption, structure and cell fate. *Commun. Chem.* **2020**, *3*, 8. [CrossRef]

19. Han, Z.J.; Rider, A.E.; Fisher, C.; Van-der-Laan, T.; Kumar, S.; Levchenko, I.; Ostrikov, K. Biological Application of Carbon Nanotubes and Graphene. In *Carbon Nanotubes and Graphene*, 2nd ed.; Tanaka, K., Iijima, S., Eds.; Elsevier: Amsterdam, The Netherlands, 2014; pp. 279–312. [CrossRef]

20. Wang, W.; Yokoyama, A.; Liao, S.; Omori, M.; Zhu, Y.; Uo, M.; Akasaka, T.; Watari, F. Preparation and characteristics of a binderless carbon nanotube monolith and its biocompatibility. *Mater. Sci. Eng. C.* **2008**, *28*, 1082–1086. [CrossRef]

21. Sarraf, M.; Razak, B.A.; Nasiri-Tabrizi, B.; Dabbagh, A.; Kasim, N.H.A.; Basirun, W.J.; Bin-Sulaiman, E. Nanomechanical properties, wear resistance and in-vitro characterization of Ta_2O_5 nanotubes coating on biomedical grade Ti-6Al-4V. *J. Mech. Behav. Biomed. Mater.* **2017**, *66*, 159–171. [CrossRef]

22. Li, J.; He, X.; Zhang, G.; Hang, R.; Huang, X.; Tang, B.; Zhang, X. Electrochemical corrosion, wear and cell behavior of ZrO_2/TiO_2 alloyed layer on Ti-6Al-4V. *Bioelectrochemistry* **2018**, *121*, 105–114. [CrossRef] [PubMed]

23. Pogrebnjak, A.D.; Bratushka, S.N.; Uglov, V.V.; Rusakov, V.S.; Beresnev, V.M.; Anischik, V.M.; Malikov, L.V.; Levintant, N.; Zukovski, P. Structures and properties of Ti alloys after double implantation. *Vacuum* **2009**, *83*, S240–S244. [CrossRef]

24. Man, H.C.; Bai, N.; Cheng, F.T. Laser diffusion nitriding of Ti-6Al-4V for improving hardness and wear resistance. *Appl. Surf. Sci.* **2011**, *258*, 436–441. [CrossRef]

25. Kgoete, F.M.; Popoola, A.P.I.; Fayomi, O.S.I. Influence of spark plasma sintering on microstructure and corrosion behaviour of Ti-6Al-4V alloy reinforced with micron-sized Si_3N_4 powder. *Defence Technol.* **2018**, *14*, 403–407. [CrossRef]

26. Devgan, S.; Sidhu, S.S. Evolution of surface modification trends in bone related biomaterials: A review. *Mater. Chem. Phys.* **2019**, *223*, 68–78. [CrossRef]

27. Li, X.; Liu, X.; Huang, J.; Fan, Y.; Cui, F.-Z. Biomedical investigation of CNT based coatings. *Surf. Coat. Technol.* **2011**, *206*, 759–766. [CrossRef]

28. Deng, J.; Pang, S.; Wang, C.; Ren, T. Biotribological properties of Ti-6Al-4V alloy treated with self-assembly multi-walled carbon nanotube coating. *Surf. Coat. Technol.* **2020**, *382*, 125159. [CrossRef]

29. Terada, M.; Abe, S.; Akasaka, T.; Uo, M.; Kitagawa, Y.; Watari, F. Multiwalled carbon nanotube coating on titanium. *Biomed. Mater. Eng.* **2009**, *19*, 45–52. [CrossRef]

30. Bains, P.S.; Singh, G.; Bhui, A.S.; Sidhu, S.S. Parametric evaluation of medical grade titanium alloy in MWCNTs mixed dielectric using graphite electrode. In *Biomaterials in Orthopaedics and Bone Regeneration, Materials Horizons: From Nature to Nanomaterials*, 1st ed.; Bains, P., Ed.; Springer: Singapore, 2019; pp. 1–14. [CrossRef]

31. Buser, D.; Schenk, R.K.; Steinemann, S.; Fiorellini, J.P.; Fox, C.H.; Stich, H. Influence of surface characteristics on bone integration of titanium implants. A histomorphometric study in miniature pigs. *J. Biomed. Mater. Res.* **1991**, *25*, 889–902. [CrossRef]

32. Larsson, C.; Thomsen, P.; Aronsson, B.O.; Rodahl, M.; Lausmaa, J.; Kasemo, B.; Ericson, L.E. Bone response to surface-modified titanium implants: Studies on the early tissue response to machined and electropolished implants with different oxide thicknesses. *Biomaterials* **1996**, *17*, 605–616. [CrossRef]

33. Mour, M.; Das, D.; Winkler, T.; Hoenig, E.; Mielke, G.; Morlock, M.M.; Schilling, A.F. Advances in Porous Biomaterials for Dental and Orthopaedic Applications. *Materials* **2010**, *3*, 2947–2974. [CrossRef]

34. *ASTM G102-89, Standard Practice for Calculation of Corrosion Rates and Related Information from Electrochemical Measurements*; ASTM Int.: West Conshohocken, PA, USA, 2015. [CrossRef]

35. Gill, A.S.; Kumar, S. Surface Roughness and Microhardness Evaluation for EDM with Cu–Mn Powder Metallurgy Tool. *Mater. Manuf. Process.* **2015**, *31*, 514–521. [CrossRef]

36. Singh, G.; Sidhu, S.S.; Bains, P.S.; Bhui, A.S. Improving microhardness and wear resistance of 316L by TiO2 powder mixed electro-discharge treatment. *Mater. Res. Express.* **2019**, *6*, 086501. [CrossRef]

37. Zain, Z.M.; Ndaliman, M.B.; Khan, A.A.; Ali, M.Y. Improving micro-hardness of stainless steel through powder-mixed electrical discharge machining. *Proc. IMechE Part C J. Mech. Eng. Sci.* **2014**, *228*, 3374–3380. [CrossRef]

38. Batish, A.; Bhattacharya, A. Mechanism of material deposition from powder, electrode and dielectric for surface modification of H11 and H13 die steels in EDM process. *Mater. Sci. Forum.* **2012**, *701*, 61–75. [CrossRef]

39. Kumar, A.; Maheshwari, S.; Sharma, C.; Beri, N. Analysis of machining characteristics in additive mixed electric discharge machining of nickel-based super alloy Inconel 718. *Mater. Manuf. Process.* **2011**, *26*, 1011–1018. [CrossRef]

40. Melentiev, R.; Kang, C.; Shen, G.; Fang, F. Study on surface roughness generated by micro-blasting on Co-Cr-Mo bio-implant. *Wear* **2019**, *428–429*, 111–126. [CrossRef]

41. Rao, S.; Hashemiastaneh, S.; Villanueva, J.; Silva, F.; Takoudis, C.; Bijukumar, D.; Souza, J.C.M.; Mathew, M.T. In vitro osseointegration analysis of bio-functionalized titanium samples in a protein-rich medium. *J. Mech. Behav. Biomed. Mater.* **2019**, *96*, 152–164. [CrossRef]

42. Ryan, G.; Pandit, A.; Apatsidis, D. Fabrication methods of porous metals for use in orthopaedic applications. *Biomaterials* **2006**, *27*, 2651–2670. [CrossRef]

43. Zhu, X.; Chen, J.; Scheideler, L.; Altebaeumer, T.; Geis-Gerstorfer, J.; Kern, D. Cellular reactions of osteoblasts to micron- and submicron-scale porous structures of titanium surfaces. *Cells Tissues Organs* **2004**, *178*, 13–22. [CrossRef]

Article

Desktop Micro-EDM System for High-Aspect Ratio Micro-Hole Drilling in Tungsten Cemented Carbide by Cut-Side Micro-Tool

Yung-Yi Wu [1], Tzu-Wei Huang [2] and Dong-Yea Sheu [2,*]

[1] Graduate Institute of Mechanical & Electrical Engineering, CMEE, National Taipei University of Technology, Taipei 10608, Taiwan; wusiryy@gmail.com
[2] Graduate Institute of Manufacturing Technology, CMEE, National Taipei University of Technology, Taipei 10608, Taiwan; aspz519930@gmail.com
* Correspondence: dongyea@ntut.edu.tw; Tel.: +886-2-2771-2171 (ext. 2078)

Received: 15 June 2020; Accepted: 7 July 2020; Published: 11 July 2020

Abstract: Tungsten cemented carbide (WC-Co) is a widely applied material in micro-hole drilling, such as in suction nozzles, injection nozzles, and wire drawing dies, owing to its high wear resistance and hardness. Since the development of wire-electro-discharge grinding (WEDG) technology, the micro-electrical discharge machining (micro-EDM) has been excellent in the process of fabricating micro-holes in WC-Co material. Even though high-quality micro-holes can be drilled by micro-EDM, it is still limited in large-scale production, due to the electrode tool wear caused during the process. In addition, the high cost of precision micro-EDM is also a limitation for WC-Co micro-hole drilling. This study aimed to develop a low-cost desktop micro-EDM system for fabricating micro-holes in tungsten cemented carbide materials. Taking advantage of commercial micro tools in a desktop micro-EDM system, it is possible to reach half the amount of large-scale production of micro-holes. Meanwhile, it is difficult to drill the deep and high aspect ratio micro-holes using conventional micro-EDM, therefore, a cut-side micro-tool shaped for micro-EDM system drilling was exploited in this study. The results show that micro-holes with a diameter of 0.07 mm and thickness of 1.0 mm could be drilled completely by cut-side micro-tools. The roundness of the holes were approximately 0.001 mm and the aspect ratio was close to 15.

Keywords: Tungsten cemented carbide (WC-Co); desktop micro-electrical discharge machining (micro-EDM) system; cut-side micro-tool; micro-holes

---

1. Introduction

Tungsten carbide (WC) is a materiel which is widely applied in military industrial composite, metallurgy, aerospace, and other important fields because of its excellent physical and chemical properties [1]. Pure WC is very brittle. If doped with small amounts of titanium, cobalt, or other metals, the incorporated brittleness can be reduced. The interaction between Co-based binders and WC grains on early stages of liquid-phase sintering can be strongly affected by the carbon content in the binders [2]. Tungsten cemented carbide (WC-Co) has a series of excellent properties, such as hardness, strength, toughness, wear resistance, and corrosion resistance [3,4].

Micro-holes made from WC-Co are widely used as spraying nozzles, injection nozzles, and spinning nozzles, owing to their low wear and hardness [5]. Machining processes, such as micro-mechanical drilling, laser machining (LBM), and electron beam machining (EBM), are typically used for the mass- or semi-mass production of micro-holes in WC-Co materials. The micro-electrical discharge machining (micro-EDM) process is highly suitable for micro-hole fabrication because it is burr-free and efficient irrespective of the workpiece hardness, especially since the development of the WEDG technology [6–8]. However, micro-EDM still has some limitations for the mass production of micro-holes due to the low productivity of the micro-tools used for fabrication [9,10].

In order to achieve semi-mass production of micro-hole drilling using EDM in WC-Co material, a low-cost desktop micro-EDM system was developed in this study. Off-the-shelf spindle electrode tools with diameters of 0.15 mm were used directly as microelectrode tools. These tools have proven to be commercially successful in low-cost mechanical drilling. Using these commercially available micro-spindle tools, it was possible to achieve the semi-mass production of a WC-Co material with micro-holes drilled via a desktop micro-EDM system. However, small diameter and long electrode tools with diameters less than 0.1 mm are not commercially available. Therefore, in this study, a WEDG unit was attached to the desktop micro-EDM system to fabricate micro-spindle tools with diameters less than 0.01 mm by using commercially available 0.15 mm tools. In order to produce micro-holes with a high aspect ratio, a single-side notch electrode method was applied to flush debris. The machining parameters, such as machining time, aspect ratio, spindle tool wear, and micro-hole quality, were investigated in this paper. It is expected that the desktop micro-EDM system will be potentially useful for drilling micro-holes in tungsten carbide materials.

2. Structure of the Desktop Micro-EDM

2.1. Desktop Micro-EDM Structure

The micro-EDM with three axis computer numerical control (CNC) controllers has been commercialized in the industry market [11]. The high accuracy controlling system makes micro-EDM more expensive. However, most micro-EDM systems are still designed for micro-holes drilling in industry applications. Low productivity of micro-holes drilling is still the main challenge for micro-EDM due to the significant tool wear and micro-tools fabrication [12,13]. To achieve mass-production of micro-hole fabrication, these two factors need to be addressed in this paper.

The desktop micro-EDM system was developed and designed in this research. The block diagram of the operation relationship of each unit is shown in Figure 1. The system has only three axes, X, Y, Z for the micro-hole drilling, as shown in Figure 2. The X-Y stages were controlled manually with digital indicators, and the Z-axis was controlled by the microcontroller unit (MCU). The most important components of the desktop EDM system were the V-shaped block and the spindle electrode tools. The structure of the spindle tool is shown in Figure 3. The spindle tool was mounted on the V-shaped block and rotated by a direct current (DC) motor. The rotation speed was variable. The linear straightness and roundness of the spindles were important to ensure highly accurate micro-hole drilling. This machine was mainly suitable for micro-holes with diameters of 0.15 mm or less with micro resistor capacity electro discharge circuit (RC circuit). In this paper, the desktop EDM was designed for micro-holes drilling with diameters less than 0.15 mm. The desktop micro-EDM system was available for electro-conductive materials and used conductive materials such as tungsten (W), tungsten cemented carbide (WC), die steel (SKD), and stainless steel (SUS). The discharge energy of the desktop micro-EDM system simply adopts a RC discharge circuit with DC power supply of 80 to 100 V, and the workpiece cannot be touched during discharge machining. The discharge power must be turned off when replacing the microelectrode tool. The main specifications of the desktop micro-EDM system is shown in Table 1 [14].

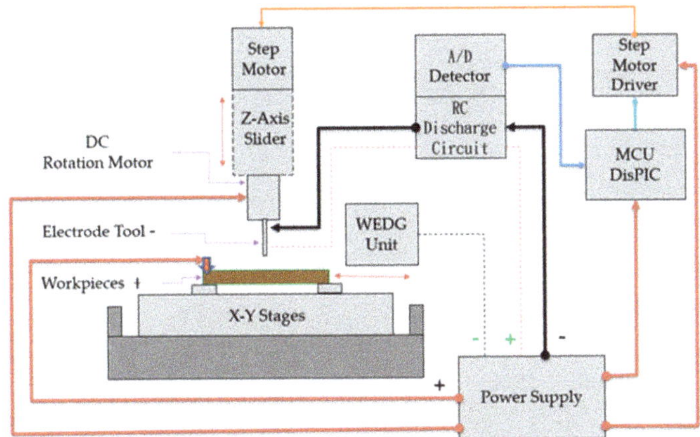

Figure 1. The block diagram sketch of the desktop micro-electrical discharge machining (micro-EDM) system.

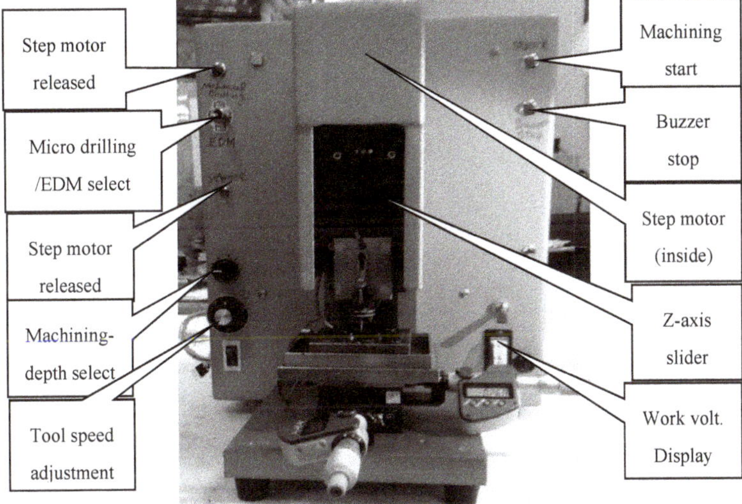

Figure 2. Complete structure of the desktop micro-EDM system.

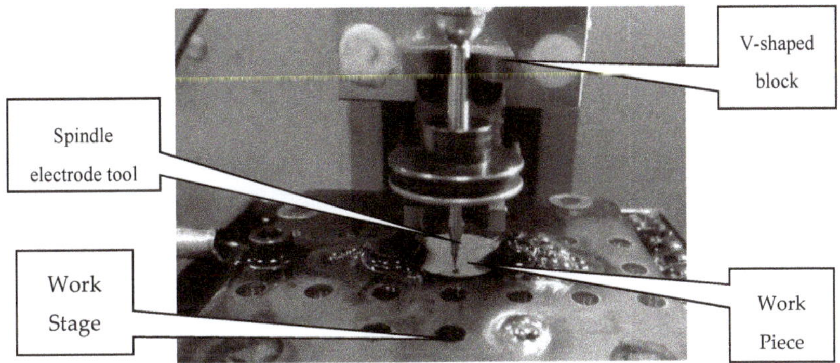

Figure 3. Structure of the spindle tools on the V-shaped block.

Table 1. Main specification of the desktop micro-EDM system.

Desktop Micro-EDM System Specification		
Machining dimensions (L × W × H)	450 mm × 320 mm × 360 mm	
Total weight	5 kg	
XY stage (L × W)	300 mm × 200 mm	
X, Y travel	25 mm × 25 mm	by manual process
Z-axis travel	150 mm	by microcontroller unit (MCU) control
Z-axis resolution	0.1 μm	By step motor
Spindle speed	6000 RPM	
Electrode diameter	35 up to 150 μm	

2.2. Commercially Available Spindle Tools

The desktop micro-EDM system uses off-the-shelf spindle electrode tools. The micro-tool length and diameter are 5 mm and 0.15 mm, respectively, as shown in Figure 4. It was successful and commercially available for the mass production of electrode tools and had a shank diameter of 3.0 mm and no screw slots due to the mechanical grinding process. The micro-tool accuracy of both the diameter and the length was approximately 0.003 mm for tungsten cemented carbide. Figure 5 shows the spindle micro-tool with pulley [15]. The pulley was fixed onto the spindle tool in the system, and it rotated directly on the V-shaped block. The rotation roundness of spindle tool was approximately 0.5 μm without any vibration. Due to a low tool wear ratio of the tungsten carbide by micro-EDM, the tool electrode is not available for commercial polishing process if it is made by tungsten. The WC-Co material demonstrates the possibility of mass production of the micro-tool by mechanical grinding process [16]. Even though it was possible to fabricate spindle micro-tools with a diameter of 0.05 mm by conventional grinding process, the aspect ratio was only 3 or 4 due to the mechanical grinding force. This low aspect ratio of spindle tools makes them unsuitable for mass-drilling micro-holes by micro-EDM. The other critical challenge involved in grinding is the fabrication of micro-tools with diameters less than 0.03 mm. In this paper, WEDG technology was used for micro-tools fabrication with diameters less than 0.03 mm by using commercial tools directly for high aspect ratio micro-hole drilling.

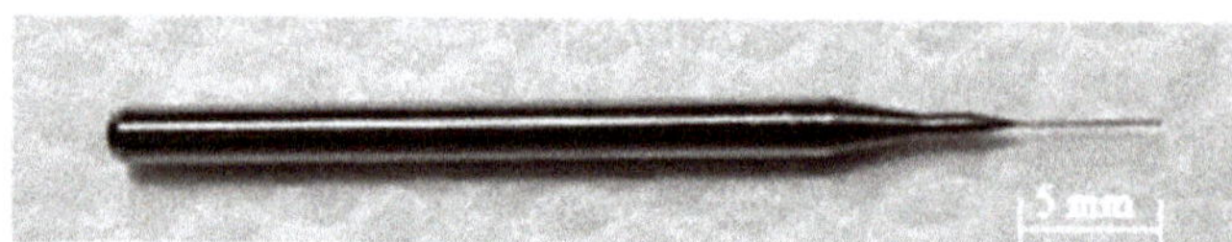

Figure 4. Commercially available spindle micro-tool without the screw slot. Diameter of shank: 3 mm. Diameter of tool: 0.15 mm.

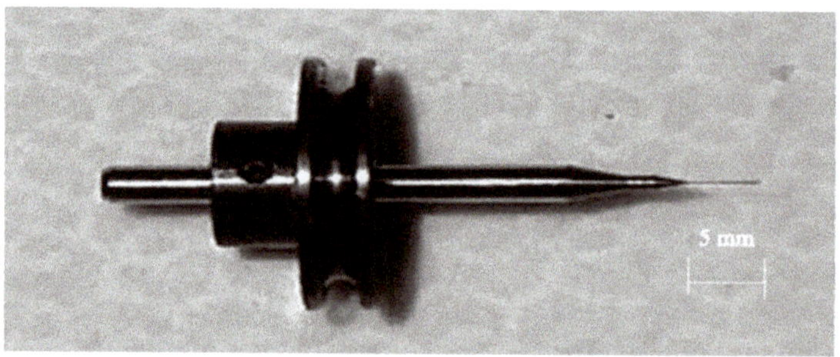

Figure 5. Spindle micro-tool with pulley.

2.3. WEDG Technology Unit

As described in the previous section, mechanical grinding process has been successful for the mass production of micro-spindle tools. The diameter was approximately 0.003 mm. However, it is still difficult to fabricate spindle tools with diameters less than 0.1 mm by using the grinding process, due to the mechanical grinding force [17]. It is possible to produce small spindle tools with a high precision grinding machine, however, the aspect ratio is only 3 or 4 with diameters less than 0.1 mm. Hence, in this study, in order to fabricate ultra-micro-holes with diameters of less than 0.05 mm, the wire-electro-discharge grinding (WEDG) technology unit was attached to the desktop micro-EDM system, as shown in Figure 6 [18,19]. The off-the-shelf spindle tools with diameters of 0.15 mm and lengths of 5 mm from the commercial market were used directly. The material of the spindle tool was tungsten carbide, which provides sufficient toughness and rigidness. Compared to the conventional WEDG process, the micro-electrode tool fabrication technique employed in this study will be more efficient, due to the finishing process post-machining. The x-axis was manually controlled by a digital micrometer head with the position control display resolutions as low as 1 μm, as shown in Figure 7. By aligning the position of the x-axis, the micro-electrode tools could easily be fabricated using WEDG technology. It is thus possible to shape the tools through one machining process only.

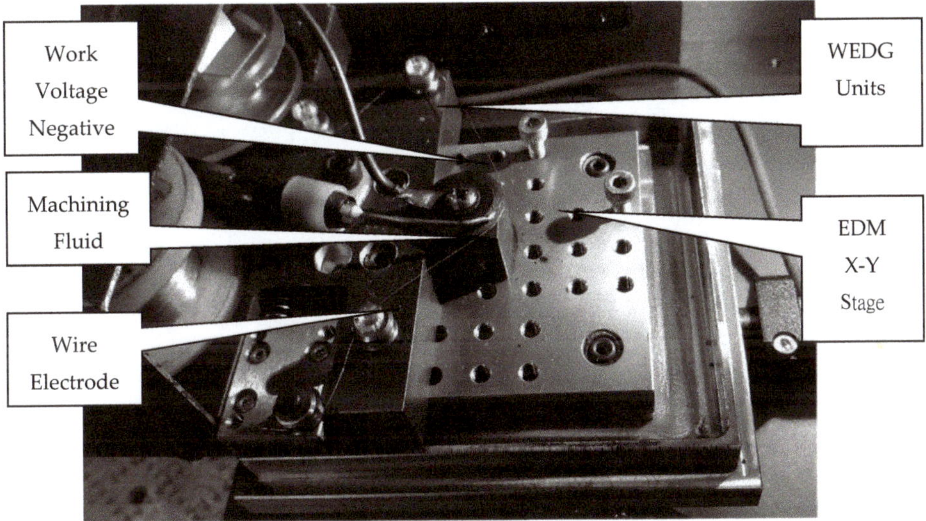

Figure 6. Wire-electro-discharge grinding (WEDG) technology unit attached to the desktop micro-EDM system.

Figure 7. X direction manual control by digital micrometer head.

2.4. Control Pad Micro-Controller Chips (dsPIC)

The conventional micro-EDM system is usually operated by numerical control (NC) or computer numerical control (CNC) controllers [20]. In addition, the position alignment and the scanning process with tool compensation are also possible. However, the large cost of conventional micro-EDM systems makes EDM unpopular for micro-hole drilling. To reduce the cost, this study used the MCU digital signal peripheral interface controller (dsPIC), to control the movement of the spindle tool and to detect the discharge gap [21]. The I/O connection provides drilling depth selection. The z-axis of the spindle micro-tool feeding rate is controlled by detecting voltage of the gap discharge. By only pushing the start button, the desktop micro-EDM system can automatically produce micro-holes. The diameter of the microelectrode tool could be manually adjusted using the x-axis position alignment with high resolution indicator. The micro-tools and micro-hole fabrication can be carried out on the same desktop micro-EDM system. Complete internal structure of the desktop micro-EDM system is as shown in Figure 8.

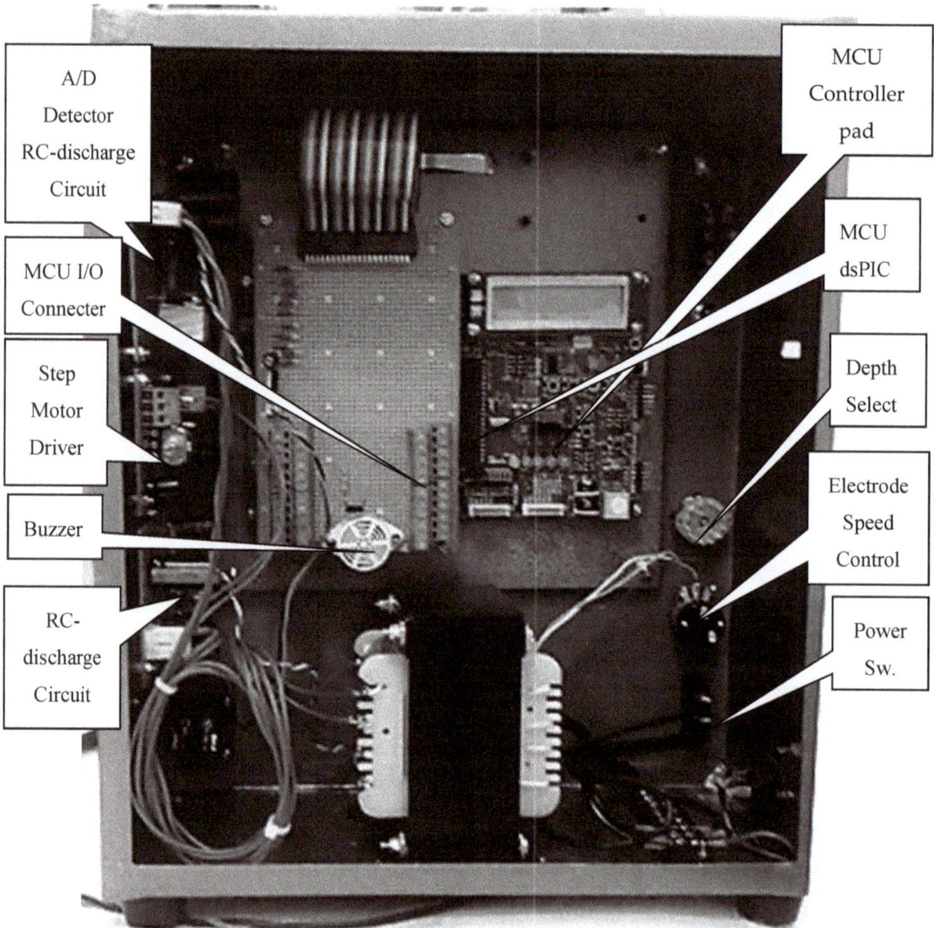

Figure 8. Internal structure of the desktop micro-EDM system.

3. Micro-Holes Drilling by Desktop Micro-EDM System

3.1. Micro-Electrode Tool Fabrication by WEDG

In this study, two types of micro-spindle tools with diameters 0.15 mm and 0.035 mm were used to fabricate the WC-Co micro-holes by micro-EDM drilling. To achieve the semi-mass production of micro-hole drilling using micro-EDM, commercial micro-spindle tools with a diameter of 0.15 mm can be used directly, but there is no supply for diameters less than 0.050 mm. However, fine micro-spindle tools may be produced by the WEDG grinding process, as shown in Figure 5. In order to fabricate micro-tools with diameters less than 0.035 mm, the WEDG unit mounted on the table was still used to reform the commercial spindle tools using only the finishing process. Figure 8 shows the potential of using the desktop EDM system to fabricate micro-tools with diameters of only 0.007 mm with tungsten carbide, as shown in Figure 9. It is thus possible to fabricate micro-holes with diameters less than 0.01 mm using this low-cost desktop micro-EDM system. By using commercial tools directly, the desktop EDM is able to produce micro-spindle tools more efficiently without rough machining.

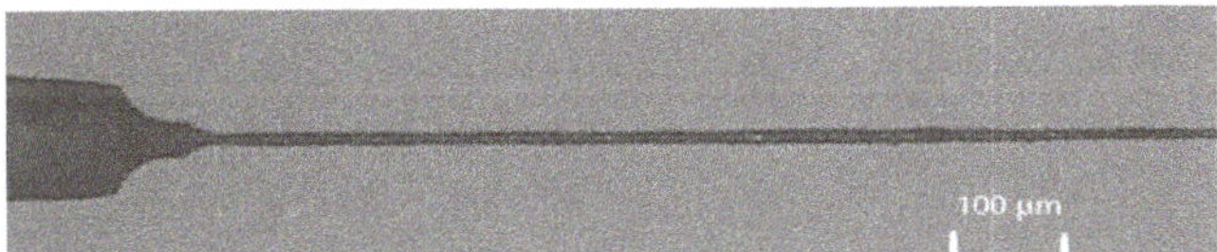

Figure 9. Microelectrode tool with diameter of 0.007 mm.

3.2. Micro-Hole Drilling by Micro-EDM System

The desktop micro-EDM system uses an RC discharge circuit and DC power with an open voltage of 80 V [22]. The machining conditions of the desktop micro-EDM system is as shown in Table 2. The V-shaped block was mounted onto the desktop micro-EDM system with high accuracy. Commercial micro-tools were used directly without the WEDG technology and the electrode tools are available from commercial market. However, for micro-tools with diameters less than 50 µm, the WEDG technology was still applied for reforming in this study. The main characteristic of the low-cost desktop micro-EDM system is that micro-tool and micro-hole convenience fabrication can be achieved conveniently on the same machine. After shaping the microelectrode tools, it is possible to drill micro-holes directly on the same micro-EDM. Experimental results of micro-hole drilling using tools with diameters of 150 µm and 35 µm are shown in Figure 10. It is facile to drill micro-holes on the WC-Co using this desktop micro-EDM system without any burr. It takes less than 2 min to drill a micro-hole of 150 µm diameter by the 1000 pf capacitor; and it takes approximately 5 min to drill a micro-hole of 40 µm when using discharge capacitor of 100 pf. Thus, when conducting less discharge capacitance, the machining time takes longer, and the electrode wear increases. Therefore, a small capacitance value will result in optimal micro-holes machining.

Table 2. Machining conditions of the desktop micro-EDM system.

Machining type	RC-EDM
Tool material	WC (150 µm)
Workpiece material	WC-Co
Machining open voltage	80 V
Discharge resistor	500 $\Omega \times 2$
Discharge capacitor	100 or 1000 pf
Workpiece thickness	0.5 or 0.3 mm
Machining depth	800 or 450 µm
Tool diameter by standard offer	150 µm
Tool diameter grind by WEDG	35 µm

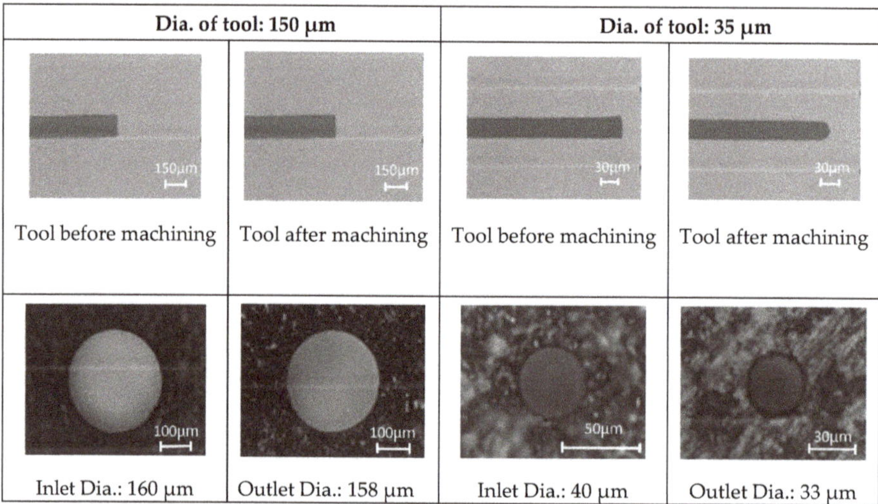

Dia. of tool: 150 μm		Dia. of tool: 35 μm	
Tool before machining	Tool after machining	Tool before machining	Tool after machining
Inlet Dia.: 160 μm	Outlet Dia.: 158 μm	Inlet Dia.: 40 μm	Outlet Dia.: 33 μm

Figure 10. Micro-hole drilling on tungsten cemented carbide (WC-Co) through a desktop micro-EDM system.

The surface observation of the micro-holes by Scanning Electron Microscope (SEM) is shown in Figure 11 [23]. Compared to another machining process, the desktop micro-EDM process can meet the requirement of micro-holes drilling on WC-Co material without any burrs. The experimental results show micro-tools with diameters less than 0.01 mm and micro-holes with diameters less than 0.05 mm. In micro-hole machining, the horizontal diameter (D_x) for a 0.3 mm thick workpiece is 39 μm, and the vertical diameter (D_y) is 40 μm; whereas the horizontal diameter (D_x) for 0.5 mm, the thick workpiece, is 66 μm, and the vertical diameter D_y is 66 μm. Therefore, the machining roundness is approximately 1 μm. For machining workpieces with apertures of 40 and 66 μm and thicknesses of 0.3 and 0.5 mm, the aspect ratio is about 8.

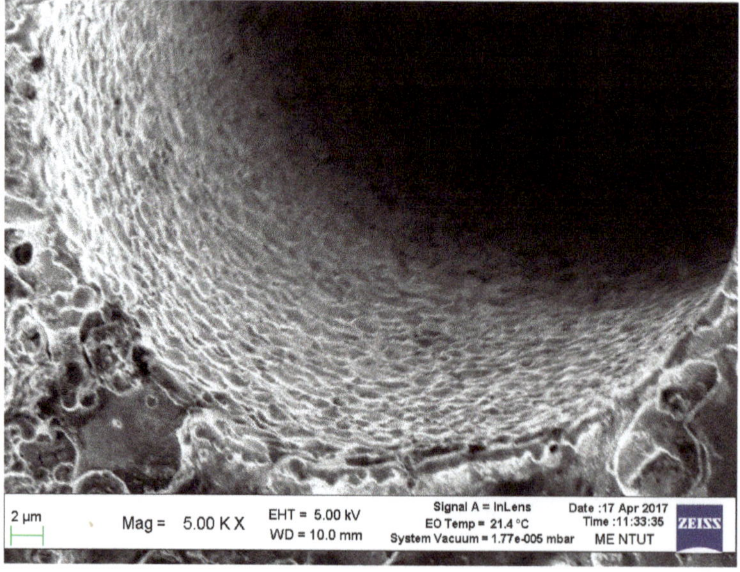

Figure 11. Surface of WC-Co micro-holes observed by Scanning Electron Microscope (SEM).

3.3. Electrode Tool Wear and Roundness

Figure 12 shows the length of tool wear of EDM micro-holes drilling with different electrical capacites. The micro-tools maintain their shape with only little tool wear by using small electric discharge capacities [24]. However, the tool wear increases significantly when the diameter is smaller than 0.05 mm. Even though the larger electric discharge capacities could increase the efficiency of micro-hole drilling, it leads to greater tool wear and deterioration of the micro-hole roundness. Therefore, it is necessary to consider all parameters to identify a suitable discharge capacity. In this study, for a work piece thickness of 0.3 mm and tool electrode diameter of 150 µm, approximately 1000 pF is the ideal drilling capacitance, whereas, for a tool electrode diameter of approximately 35 µm, a capacitance of 100 pF is the optimal value for micro-hole drilling.

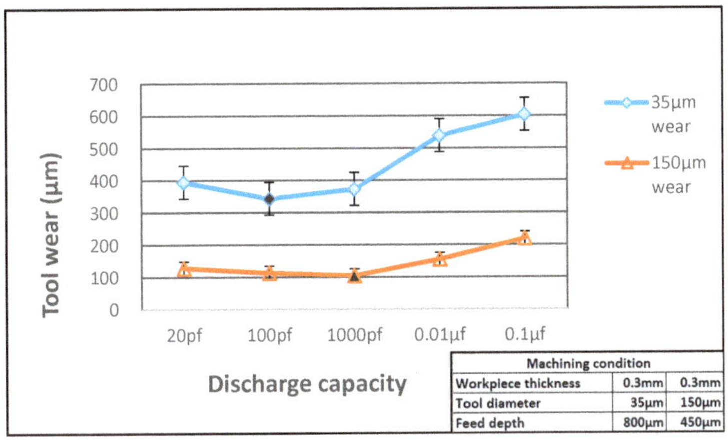

Figure 12. Tool wear in micro-hole drilling of WC-Co.

3.4. Limitations of the Machining Depth

Normally, transistor discharge circuits are used in commercial EDM because the discharge energy is controllable by adjusting the pulse generator and duty cycle [25]. However, micro-EDM requires extremely small electric discharge energy, especially for micro-tools with diameters less than 50 µm. The resistor capacity (RC) pulse generator is popular and widely used for the micro-EDM electric discharge energy [26]. The discharge energy depends on the capacity of the RC pulse generator. Therefore, the main parameter of the RC discharge circuit is the magnitude of capacity. Theoretically, it is possible to increase the machining speed with large electric capacity due to the large discharge energy. The large capacitors could cause bigger discharge sparks, however, the large discharge energy will lead to significant tool wear. The high aspect ratio of micro-holes will not be drilled to penetrate completely as the micro-tool wear will exceed the feed depth. As shown in Figure 13, the spindle feeding speed is efficient without any stagnation while the feeding depth is below 800 µm. However, the machining speed decreases significantly when the spindle feeding depth is larger than 800 µm due to insufficient debris flushing. Even the spindle feeding depth increases. However, this phenomenon means a large amount of micro-tool wear. It is clear that the micro-EDM drilling process is a method to fabricate micro-holes with aspect ratios less than 5 exclusively. The high aspect ratio micro-spindle tools are not possible for deep micro-hole drilling using micro-EDM due to the significant tool wear [27].

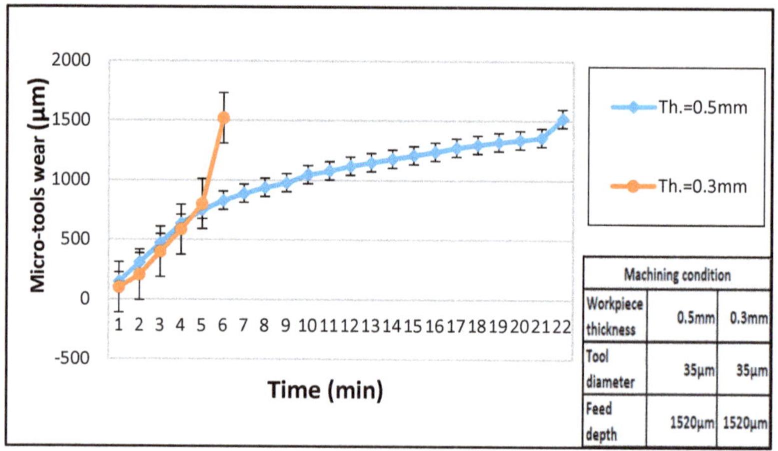

Figure 13. Micro-tools wear on WC-Co by the desktop micro-EDM.

3.5. High-Aspect Micro-Holes Fabrication

Micro-EDM encounters another critical problem for high-aspect ratio on deep micro-holes machining, especially when the diameter is less than 100 μm and the aspect ratio is larger than 10 [28]. The cut-side shaped micro-electrode tool was capable of fabrication using the desktop micro-EDM system. The purpose is to enhance the machining efficiency for micro-holes drilling with high aspect ratio. The 50 μm diameter tool-electrode can be ground out to 10 μm in depth by the WEDG technology machining unit without any tool spindle rotation, shown as in Figure 14. The side view and front view of cut-side shaped micro-tool after the WEDG process is shown as in Figures 15 and 16. The cut-side electrode tools of special shapes are able to improve the debris removal problem, but there is still no elevation for higher aspect ratio due to high tool wear [29,30]. Figure 17 shows the comparison between the feeding depth of the cylinder tool and cut-side tool. The experiment shows that initially the smaller cut-side electrode brought more electrode wear than the bigger cylindrical electrode. Later, the cut-side electrode contrarily brought less electrode wear than the bigger cylindrical electrode, for its larger space removed debris faster. Under such circumstances, to reduce high tools wear and increase machining efficiency of micro-holes EDM drilling, the cylinder and cut-side shaped micro-electrode tool should be shifted alternately for deep micro-holes drilling. In micro-holes machining with high-aspect ratio, the feeding depth was drilled alternately by cylindrical and cut-side micro-electrode tools. At first, the cylindrical micro-tool with less electrode wear started to drill micro-holes to reach the feeding depth limit area, and then shifted to the cut-side electrode with larger space to process it at the same speed without moving the workpiece. Continuing the process until the micro-holes completely drilled might improve the machining efficiency. The best machining method was achieved by using the cylinder- and cut-side shaped micro-tools alternately; a micro-hole with high aspect ratio will be drilled completely at 15 times larger in about 10 min of the machining time. The results show that it can be completely drilled by cut-side micro-tools, and high aspect ratio drilling can be performed on WC-Co material with a thickness of 1.0 mm, with an inlet diameter of 73 μm and an outlet diameter of 58 μm, as shown in Figure 18. The micro-holes surface with high aspect ratio on WC-Co was observed through SEM, as shown in Figure 19.

Figure 14. Cut-side shaped micro-tool process by WEDG.

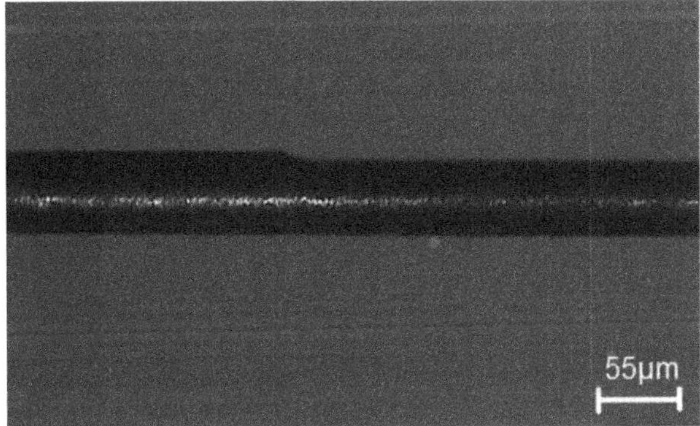

Figure 15. Side view of cut-side shaped micro-tool.

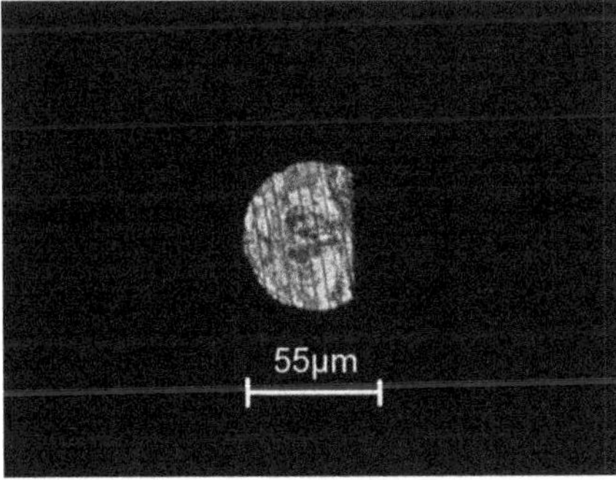

Figure 16. Front view of cut-side shaped micro-tool.

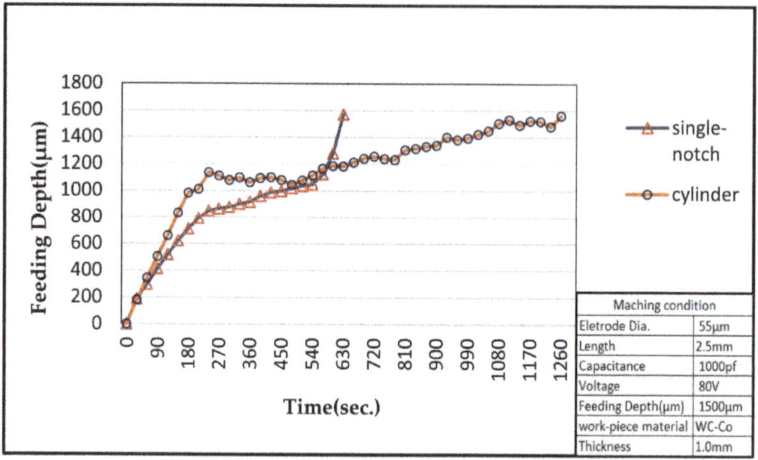

Figure 17. Comparison between the feeding depth of the cylinder tool and cut-side tool.

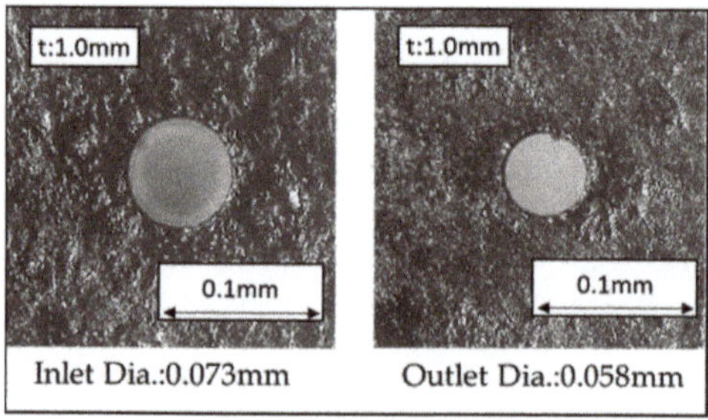

Figure 18. Micro-hole drilling of high aspect ratio on WC-Co through a desktop micro-EDM system.

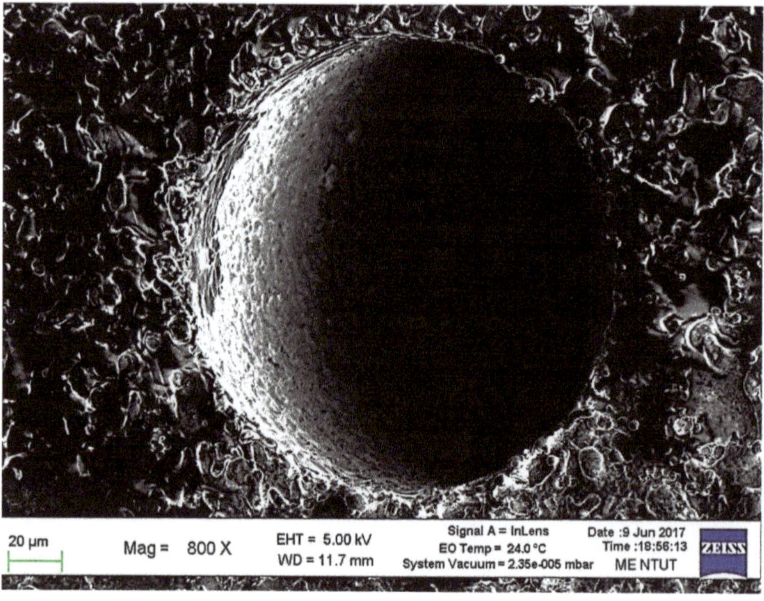

Figure 19. Surface of micro-hole drilling of high aspect ratio on WC-Co observed by SEM.

4. Conclusions

A review of the research trends in micro-EDM about various electrode tools and their effects in the characteristics of micro-EDM are presented. A low-cost desktop micro-EDM system was explored and developed for rapid drilling of micro-holes through tungsten cemented carbide in this study. Using commercial electrode tools of 50 up to 150 μm, for about 2 to 4 min, it is possible to achieve semi-large-scale production of micro-holes. The desktop micro-EDM system is also able to drill micro-holes by mechanical drilling using a micro-spindle tool with a screw slot. Besides, the WEDG technology can even be employed for more meticulous micro-electrode and shaping tool fabrication. In addition to superb microelectrode tool fabrication, micro-hole drilling could be also done by the same machine. Compared to the nearly one million dollars and high prices of commercial micro-EDM systems, there are more potential applications for the drilling of tungsten cemented carbide in the low-cost desktop micro-EDM system developed in this study, which is able to produce roundness of a micro-hole of approximately 1 μm and a 9 times standard aspect ratio, enhanced to 15 times in high-aspect ratio. We hope that research and analyses on machining characteristics of double cut-side tool electrodes will be continued in the future.

Author Contributions: Conceptualization, D.-Y.S. and Y.-Y.W.; methodology, Y.-Y.W., T.-W.H. and D.-Y.S.; software, Y.-Y.W. and T.-W.H.; validation, D.-Y.S.; software, Y.-Y.W. and T.-W.H.; formal analysis, D.-Y.S. and Y.-Y.W.; software, Y.-Y.W. and T.-W.H.; investigation, Y.-Y.W., T.-W.H. and D.-Y.S.; resources, D.-Y.S.; data curation, Y.-Y.W.; writing—original draft preparation, Y.-Y.W.; writing—review and editing, D.-Y.S.; visualization D.-Y.S. and Y.-Y.W.; supervision, D.-Y.S.; project administration, D.-Y.S. and Y.-Y.W.; funding acquisition, T.-W.H. and D.-Y.S. All authors have read and agreed to the published version of the manuscript.

Funding: This research was funded by the MOST of Taiwan grant number [MOST 107-2221-E-027-050-MY2].

Acknowledgments: The authors explain the gratefully acknowledge to the Ministry of Science and Technology (MOST) Taiwan.

Conflicts of Interest: The authors declare no conflict of interest

References

1. Malyshev, V.V.; Gab, A.I. Resource-saving methods for recycling waste tungsten carbide-cobalt cermets and extraction of tungsten from tungsten concentrates. *Theor. Found. Chem. Eng.* **2007**, *11*, 436 441. [CrossRef]
2. Konyashin, I.; Zaitsev, A.A.; Sidorenko, D.; Levashov, E.A.; Ries, B.; Konishev, S.N.; Sorokin, M.; Mazilkin, A.A.; Herrmann, M.; Kaiser, A. Wettability of tungsten carbide by liquid binders in WC-Co cemented carbides: Is it complete for all carbon contents. *Int. J. Refract. Met. Hard Mater.* **2017**, *62*, 134–148. [CrossRef]
3. Konyashin, I.; Riesa, B.; Hlawatscheka, D.; Zhukb, Y.; Mazilkincd, A.; Straumalcdef, B.; Dorng, F.; Park, D. Wear-resistance and hardness: Are they directly related for nanostructured hard materials. *Int. J. Refract. Met. Hard Mater.* 2015 49, 203–211. [CrossRef]
4. Singh, J.; Sharma, R.K. Assessing the effects of different dielectrics on environmentally conscious powder-mixed EDM of difficult-to-machine material (WC-Co). *Front. Mech. Eng.* **2016**, *11*, 374–387. [CrossRef]
5. Yan, B.H.; Huang, F.Y.; Chow, H.M.; Tsai, J.Y. Micro-hole machining of carbide by electric discharge machining. *J. Mater. Process. Technol.* **1999**, *87*, 139–145. [CrossRef]
6. Jahan, M.P.; Wong, Y.S.; Rahman, M. A study on the quality micro-hole machining of tungsten carbide by micro-EDM process using transistor and RC-type pulse generator. *J. Mater. Process. Technol.* **2009**, *209*, 1706–1716. [CrossRef]
7. Masuzawa, T.; Fujinoa, M.; Kobayashia, K.; Kinoshita, N. Wire electro-discharge grinding for micro-machining. *Cirp Ann. Manuf. Technol.* **1985**, *34*, 431–434. [CrossRef]
8. Zhao, W.; Wang, Z.; Di, S.; Chi, G.; Wei, H. Ultrasonic and electric discharge machining to deep and small hole on titanium alloy. *J. Mater. Process. Technol.* **2002**, *15*, 101–106.
9. Jahan, M.P.; Rahman, M.; Wong, Y.S. A review on the conventional and micro-electrodischarge machining of tungsten carbide. *Int. J. Mach. Tools Manuf.* **2011**, *51*, 837–858. [CrossRef]
10. Chavoshi, S.Z.; Goel, S.; Paul, M. Current trends and future of sequential micro-machining processes on a single machine tool. *Mater. Des.* **2017**, *127*, 37–53. [CrossRef]

11. Lee, L.W.; Yeh, S.S.; Lee, J.I. Application of Taguchi Method for Determining the Best-Fitted Control Parameters of CNC Machine Tools. In Proceedings of the International Conference on Mechatronics and Automation, Takamatsu, Japan, 6–9 August 2017.

12. Lee, P.A.; Kim, Y.; Kim, B.H. Effect of Low Frequency Vibration on Micro EDM Drilling. *Int. J. Precis. Eng. Manuf.* **2015**, *16*, 2617–2622. [CrossRef]

13. Lauwers, B.; Klocke, F.; Klink, A.; Tekkaya, A.E.; Neugebauer, R.; Mcintosh, D. Hybrid processes in manufacturing. *Cirp Ann. Manuf. Technol.* **2014**, *63*, 561–583. [CrossRef]

14. Jahan, M.P. Micro-Electrical Discharge Machining. In *Nontraditional Machining Processes*; Davim, J.P., Ed.; Springer Nature: Cham, Switzerland, 2013; Volume 4, pp. 115–151.

15. Muthuramalingama, T.; Mohanb, B. A review on influence of electrical process parameters in EDM process. *Arch. Civ. Mech. Eng.* **2015**, *15*, 87–94. [CrossRef]

16. Koyano, T.; Kunieda, M. Micro EDM by Stray-Capacitance-Coupled Electric Feeding. *Key Eng. Mater.* **2012**, *516*, 245–250. [CrossRef]

17. Cheng, J.; Yin, G.; Wen, Q.; Song, H.; Gong, Y. Study on grinding force modelling and ductile regime propelling technology in micro drill-grinding of hard-brittle materials. *J. Mater. Process. Technol.* **2015**, *223*, 150–163. [CrossRef]

18. Sheu, D.Y. Multi-spherical probe machining by EDM: Combining WEDG technology with one-pulse electro-discharge. *J. Mater. Process. Technol.* **2004**, *149*, 597–603.

19. Egashira, K.; Hosono, S.; Takemoto, S.; Masao, Y. Fabrication and cutting performance of cemented tungsten carbide micro-cutting tools. *Precis. Eng.* **2011**, *35*, 547–553. [CrossRef]

20. Natarajan, N.; Suresh, P. Experimental investigations on the microhole machining of 304 stainless steel by micro-EDM process using RC-type pulse generator. *Int. J. Adv. Manuf. Technol.* **2015**, *77*, 1741–1750. [CrossRef]

21. Raees, A.S.; Roger, I.G.; Paul, W.P. dsPIC-based advanced data acquisition system for Monitoring, Control and Security Applications. In Proceedings of the 12th International Bhurban Conference on Applied Sciences and Technology, Islamabad, Pakistan, 13–17 January 2015.

22. Singh, A.K.; Patowari, P.K.; Deshpande, N.V. Micro-Hole Drilling on Thin Sheet Metals by Micro-Electrical Discharge Machining. *J. Manuf. Technol. Res.* **2014**, *5*, 3–4.

23. Wu, Y.Y.; Sheu, D.Y. Investigating Tungsten Carbide Micro-Hole Drilling Characteristics by Desktop Micro-ECM with NaOH Solution. *Micromachines* **2018**, *9*, 512. [CrossRef]

24. Hourmand, M.; Sarhan, A.A.D.; Sayuti, M. Micro-electrode fabrication processes for micro-EDM drilling and milling: A state-of-the-art review. *Int. J. Adv. Manuf. Technol.* **2017**, *91*, 1023–1056. [CrossRef]

25. Koyano, T.; Kunieda, M. Achieving high accuracy and high removal rate in micro-EDM by electrostatic induction feeding method. *Cirp Ann. Manuf. Technol.* **2010**, *59*, 219–222. [CrossRef]

26. Liew, P.J.; Yan, J.; Kuriyagaw, T. Carbon nanofiber assisted micro electro discharge machining of reaction-bonded silicon carbide. *J. Mater. Process. Technol.* **2013**, *213*, 1076–1087. [CrossRef]

27. Urso, G.D.; Maccarini, G.; Ravasio, C. Process performance of micro-EDM drilling of stainless steel. *Int. J. Adv. Manuf. Technol.* **2014**, *72*, 1287–1298.

28. Urso, G.D.; Merla, C. Workpiece and electrode influence on micro-EDM drilling performance. *Precis. Eng.* **2014**, *38*, 903–914.

29. Barman, S.; Hanumaiah, N.; Puri, A.B. Investigation on shape, size, surface quality and elemental characterization of high-aspect-ratio blind micro holes in die sinking micro EDM. *Int. J. Adv. Manuf. Technol.* **2015**, *76*, 115–126. [CrossRef]

30. Chakraborty, S.; Dey, V.; Ghosh, S.K. A review on the use of dielectric fluids and their effects in electrical discharge machining characteristics. *Precis. Eng.* **2015**, *40*, 1–6. [CrossRef]

 *micromachines*

Article

Shaping Soft Robotic Microactuators by Wire Electrical Discharge Grinding

Edoardo Milana [1], Mattia Bellotti [1], Benjamin Gorissen [1,2], Jun Qian [1],
Michaël De Volder [1,3] and Dominiek Reynaerts [1,*]

[1] Department of Mechanical Engineering, KU Leuven and Flanders Make, Celestijnenlaan 300,
    3001 Leuven, Belgium; edoardo.milana@kuleuven.be (E.M.); mattia.bellotti@kuleuven.be (M.B.);
    benjamin.gorissen@kuleuven.be (B.G.); jun.qian@kuleuven.be (J.Q.); mfld2@cam.ac.uk (M.D.V.)
[2] Harvard John A. Paulson School of Engineering and Applied Sciences, Harvard University,
    Cambridge, MA 02138, USA
[3] Institute for Manufacturing, Department of Engineering, University of Cambridge,
    17 Charles Babbage Road, Cambridge CB3 0FS, UK
*   Correspondence: dominiek.reynaerts@kuleuven.be

Received: 13 June 2020; Accepted: 3 July 2020; Published: 4 July 2020

Abstract: Inflatable soft microactuators typically consist of an elastic material with an internal void that can be inflated to generate a deformation. A crucial feature of these actuators is the shape of ther inflatable void as it determines the bending motion. Due to fabrication limitations, low complex void geometries are the de facto standard, severely restricting attainable motions. This paper introduces wire electrical discharge grinding (WEDG) for shaping the inflatable void, increasing their complexity. This approach enables the creation of new deformation patterns and functionalities. The WEDG process is used to create various moulds to cast rubber microactuators. These microactuators are fabricated through a bonding-free micromoulding process, which is highly sensitive to the accuracy of the mould. The mould cavity (outside of the actuator) is defined by micromilling, whereas the mould insert (inner cavity of the actuator) is defined by WEDG. The deformation patterns are evaluated with a multi-segment linear bending model. The produced microactuators are also characterised and compared with respect to the morphology of the inner cavity. All microactuators have a cylindrical shape with a length of 8 mm and a diameter of 0.8 mm. Actuation tests at a maximum pressure of 50 kPa indicate that complex deformation patterns such as curling, differential bending or multi-points bending can be achieved.

Keywords: wire electrical discharge grinding (WEDG); micromoulding; soft microrobotics; electrical discharge machining (EDM)

---

1. Introduction

Soft robotic systems are capturing the interests of scientists and engineers with characteristics that are breaking with conventional robot traditions. Softness, compliancy and cost-effective manufacturing make soft robots preferable in applications where gentle manipulation and human interaction occur [1]. Thanks to their low mechanical stiffness, soft robots can safely operate in unstructured environments by adapting to unforeseen collisions and reduce the risk of harmful events [2]. For those reasons, soft robotic technology is extensively utilised to make universal grippers [3], some of which have been commercialised [4]. Other promising soft robotic applications include robots for search and rescue operations [5,6] as well as innovative techniques to make untethered and entirely soft machines [7]. There is an increasing interest in downscaling soft robotic technology to micrometre scales to drive advances in applications where the operational environment is unpredictable or extremely delicate, such as minimally invasive surgery and drug delivery [8–10]. Further, soft microrobotics has already been applied in microfluidics for making flexible active valves [11] and artificial cilia for biomimetic micromixing and micropumping [12].

Soft robots necessitate actuators that display large deformations as a response to a generalised force input. Typically, soft actuators are designed such that their deformation corresponds to a desired kinematic trajectory, behaving as a compliant mechanism with one degree of freedom. Many types of soft actuators can be distinguished according to the nature of the applied forces [8], which can vary from electrical (dielectric or ionic polymers) to magnetic fields (magnetic polymers), from solvent concentrations (hydrogels) to pressurised fluids (inflatable structures), etc.

Elastic inflatable actuators (EIAs) are one of the most widespread soft actuators, that rely for their motions on the morphology of the actuators [13]. Such actuators at small-scale were first introduced by Suzumori [14,15] and Konishi [9,16] for biomedical applications. For a comprehensive review of the different types of EIAs we referred to previous studies (see [13,17]). Current miniaturised EIAs are made with a single inflatable elastomeric cavity which leads to a simple motion that can be bending [18], twisting [19], contracting [20] or extending [21]. Large-scale actuators, on the other hand, have been presented with richer deformations, originating from a more complex design. However, these complex designs are challenging to copy at smaller scales due to manufacturing limitations. At a larger scale, inflatable actuators are typically made out of different elastomeric parts, which are subsequently bonded or glued together. This approach makes it possible to create intricate actuator geometries that drastically modify the actuator performance as shown by Mosadegh et al. [22] for pneumatic networks (PneuNets) bending actuators. However, at smaller scales, creating similar actuators by combining parts and bonding them is an uphill task, due to more stringent requirements on the tolerances of the parts to be assembled, possible misalignments and handling problems. A commonly used bonding process for sub-centimetre soft actuators in polydimethylsiloxane (PDMS) consists of oxygen plasma treatment to activate the PDMS surfaces before bonding [23]. However, these bonds are often the weakest part of the structure which eventually causes the actuator to rupture.

In order to circumvent these issues, we propose a bonding-free technique to fabricate millimetre-scaled soft bending microactuators, using out-of-plane moulding [18]. These microactuators consist of PDMS cylindrical structures with a simple cylindrical inflatable cavity, which is placed eccentric to the axis of the outer cylinder of the actuator. This eccentricity introduces an asymmetry in the cross-section of the actuator that causes the actuator to bend. The mould is composed of two micro-milled parts and a cylindrical microrod placed in between the two parts, where the shape of the microrod is replicated as the inflatable cavity. We used these devices for different applications such as artificial cilia [24] and flexible endoscopes [25]. However, given the simple morphology of the inflatable cavity, these microactuators have a limited operating range with a maximum bending angle of up to 45° [25].

In this paper we introduce an additional manufacturing step to this bonding-free technique in order to fabricate more complex microactuators. During this additional step, the cylindrical microrods are machined using a wire electrical discharge grinding (WEDG) process. WEDG is a manufacturing process in which material is removed from a rotating tool using a running wire through high-frequency sequences of electric discharges. WEDG, which was early developed by Masuzawa et al. [26], has been established over the past years as a proven technology for machining axisymmetric microrods down to less than 10 µm in diameter [27]. Nowadays, WEDG is used for machining not only cylindrical, but also tapered microrods [28] as well as microrods with a spherical tip [29]. WEDG'ed microrods find a wide variety of applications in industry such as touch probes for contact measurement systems [30] or microtools for drilling tapered microholes [31] and microhole arrays [32].

WEDG was already used for the fabrication of classic pneumatic microactuators [33], but not for soft actuators. Here, WEDG is used for machining axisymmetric microrods with more complex shapes to be placed in the internal cavity of inflatable soft robotic actuators in a micromoulding process. Dimensional and surface metrology are used to check that the accuracy and surface quality of the fabricated microrods are compatible with the moulding requirements. Further, an analytical model is used to predict the deformation of the moulded actuator. The model is validated by experimental tests on prototypes. These tests clearly show the benefits in terms of deformation patterns of using structured microrods over their unstructured counterparts. The paper is structured as follows: in the first section we report on three different morphologies of inflatable cavities and the details about the WEDG process to realise the respective internal shapes. Subsequently the manufacturing steps involved in the moulding process are described. Finally, the microactuators are characterised and compared to the state-of-the-art.

2. Materials and Methods

2.1. Inflatable Cavity Morphologies

All microactuators described in this work have the same general architecture and only differ in the shape of the inflatable cavities. Nevertheless, the deformation changes significantly as further reported. The microactuators are cylindrical pillars, 8 mm in length and 0.8 mm in outer diameter, with inflatable cavities that are placed at an eccentricity, e, of 110 µm. We designed the inflatable cavities to be compatible with the WEDG process as described in Section 2.2. Thus, all cavities are axisymmetric with local variations of the radius across the length. We identified three different actuators with distinct rod shapes that lead to different functional deformations. In the paper we refer to them as: (i) *Saw* (Figure 1a); (ii) *Totem* (Figure 1b); (iii) *Halter* (Figure 1c).

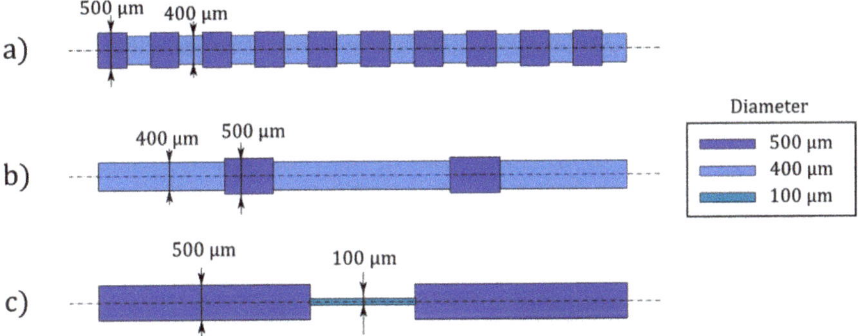

Figure 1. Inflatable cavity morphologies: (**a**) *Saw*, (**b**) *Totem*, (**c**) *Halter*. The cavities are axisymmetric, and the colour code corresponds to the local diameter of each segment. The segment lengths are reported in Table 1.

Table 1. Inflatable cavities segments lengths.

Morphology	Segments	Lengths (mm)
Saw	20	0.4–0.35 (×10)
Totem	5	1.8–0.7–2.5–0.7–1.8
Halter	3	3–1.5–3

The inflatable cavities of the three actuators can be divided into segments of different length and diameter. The segment lengths are reported in Table 1, while the diameters can be found in Figure 1, where the colour codes distinguish the different segments. In total, the combined length of the inflatable cavities equals 7.5 mm.

Given the eccentricity of the cavity with respect to the axis of symmetry of the structure, all actuators undergo a bending motion upon pressurisation, as explained in our previous paper [18]. Due to the asymmetric placement of the central void (Figure 2a), the centroid of pressure is shifted with respect to the neutral (bending) axis of the structure, resulting in a large bending deformation. In Figure 2b,c,

intricate shape, the microrods are stuck in the microactuators after demoulding (Figure 4e). Removing the microrods is the last and most delicate step, which requires the use of ethanol to slightly swell the silicone and allow a safe removal of the rod (Figure 4f). Since the rubber cures around the microrod, the absence of air generates a negative relative pressure that locks the microrod inside the microactuator. The function of ethanol is to both lubricate and swell the rubber so that air can penetrate the cavity and eliminate the negative relative pressure. This effect combined with the compliancy of the soft material enables the removal of the microrod. After removing the microrod, the microactuators are dried at room temperature to fully evaporate the ethanol. Then, they can be connected to pressure supply tubing.

Microactuators of the *Halter* type require a different process to remove the microrod. Indeed, due to the large diameter variation (from 500 μm to 100 μm), the swelling-induced removal of the microrod is not effective. Alternatively, the microrod is broken at the thin section and the two parts are extracted from the base and the tip of the microactuator. The tip is subsequently sealed with a drop of uncured rubber.

2.4. Analytical Model

In order to evaluate the deformation patterns associated with the different inflatable cavity morphologies of the actuators, we apply a multi-segment linear analytical model consisting of Euler–Bernoulli beam segments that are each loaded with a constant bending moment. For one segment, this model has already been applied to capture the overall bending deformation of a soft bender [18,37,38], while here we introduce a segmented approach. The different actuator types that we analyse in this work can be divided into n segments (Table 1), where each radial variation r_i is considered as the i-th segment ($i = 1, \dots, n$) of length l_i. In our model, each segment is subjected to a constant moment M_i

$$M_i = p_E \pi r_i^2 d_i, \tag{1}$$

where d_i is the distance between the centre of the cavity section and the neutral bending axis, passing through the centre of mass of the section (Figure 2 and Equation (2)), while p_E is a non-dimensional parameter corresponding to the normalised pressure with respect to the Young's modulus of the material ($p_E = p/E$).

$$d_i = e + \frac{e r_i^2}{R^2 - r_i^2} \tag{2}$$

Therefore, the curvature of the i-th segment is equal to

$$k_i = \frac{M_i}{I_i}, \tag{3}$$

where I_i is the second moment of area of the section:

$$I_i = \frac{\pi\left(R^4 - r_i^4\right)}{4} + \pi R^2 (d_i - e)^2 - \pi r_i^2 d_i^2 \tag{4}$$

The curved profiles of the n segments are numerically computed and assembled in MATLAB to show the overall deformed configuration of the actuator.

2.5. Microactuators Experimental Setup

The experimental setup to characterise the microactuator response consists of a pressure regulator valve (Festo LR-D-7-I-Mini) fed with compressed air coupled to a manometer (Festo FMAP-63-1-1/4-EN) and connected to the tested microactuator. A 500 μm outer diameter (OD) tube is inserted in the inflatable cavity of the microactuator and fixed with uncured silicone rubber. In the experiment, the pressure input is manually increased by 10 kPa increments, while a camera (Nikon 1 V3) captures the actuator deformation in the bending plane once a static equilibrium is reached at each pressure increment. As the actuators have different curvatures in accordance to their segmentation, the deformed configurations are characterised using the tip trajectory as parameter.

3. Results and Discussions

3.1. WEDG Accuracy and Quality

In order to evaluate the accuracy of the WEDG process for the fabrication of soft robotic microactuators, the WEDG'ed sections of the microrods are measured by means of a ZEISS® SteREO Discovery V20 microscope. The measurement results for ten different sections, which were machined on microrods with a *Saw* morphology, are shown in Figure 5. The target diameter for the measured sections is 400 µm.

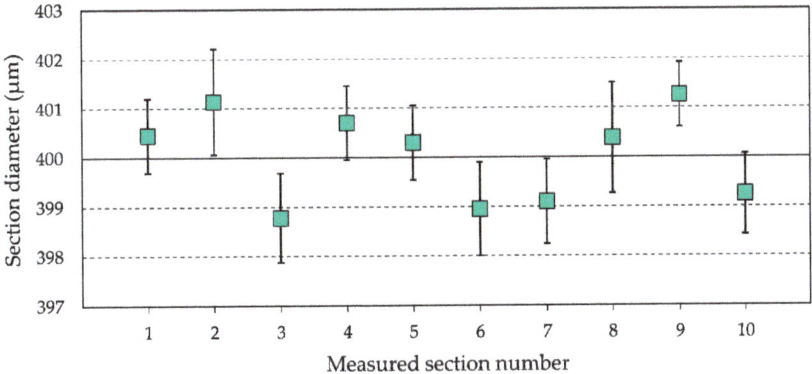

Figure 5. Measurement of the diameter of the WEDG'ed sections of the microrods. Ten different sections are measured, and eight measurements are carried out per section. The data points refer to the mean diameter of each measured section, while the error bars indicate the standard deviation of the eight measurements.

From Figure 5, it can be observed that the WEDG'ed sections deviate less than ±3 µm from the target diameter. In particular, the average diameter of the measured sections is 400.4 µm, while the standard deviation is 2.59 µm. These results highlight the high precision and accuracy of the proposed WEDG processing method. In order to study the effect of the processing accuracy on the performance of soft robotic microactuators, the analytical model presented in Section 2.4 is used to analyse the influence of variations to the diameter of the inflatable cavity of a segment on the curvature relative error (CRE). Equation (5) is used to compute the CRE from the curvature k_0 of a segment of nominal diameter and the curvature k_e of a segment of diameter affected by a machining error.

$$\text{CRE} = \frac{|k_e - k_0|}{k_0} \tag{5}$$

In Figure 6, the effect of the machining error on the CRE is shown. Segments of diameter equal to 100 µm and 400 µm are considered, which are the diameters of the segments of the three types of microactuators presented in this work. It can be seen that the CRE linearly increases with the machining error. WEDG processing, which allows the machining of segments having a deviation of less than ±3 µm from the nominal diameter, results in maximum CREs of about 2.2% and 5.3% for segments of diameter equal to respectively 400 µm and 100 µm. These maximum errors are relatively small and confirm that WEDG processing can be considered as a viable technique for machining axisymmetric microrods to be used in the bonding-free fabrication process of soft robotic microactuators. The trends shown in Figure 6 also suggest that the finishing step is crucial and unavoidable for ensuring high performance of the soft robotic microactuators since deviations from the nominal diameter in the order of 4–10 µm are observed after roughing. These deviations would result in an increase of the CRE up to 20% for a section diameter of 100 µm.

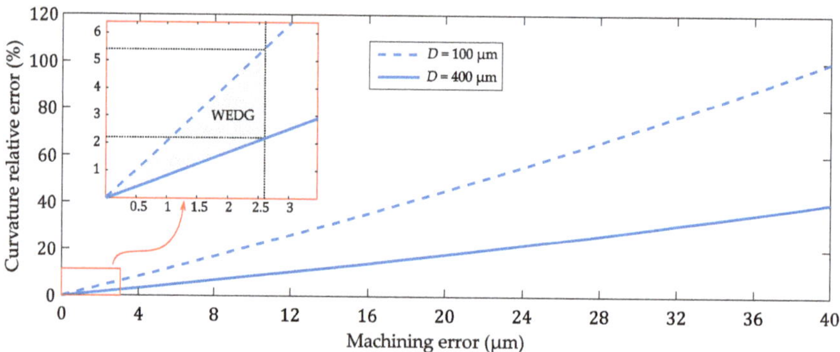

Figure 6. Effect of the machining error on the curvature relative error (CRE). In the graph, the calculated CRE of segments with inner cavity diameter D equal to 100 µm and 400 µm is shown.

The main drawback of the proposed WEDG method is the relatively long processing time. For instance, it takes approximately 5 min for machining a section of 400 µm diameter and 0.4 mm length on a cylindrical microrod of 500 µm diameter. In this case, the two roughing steps take approximately 3 min in total, while roughly 40% of the total machining time is spent in finishing. However, long processing time are acceptable, taking into consideration that a microrod can be used for moulding multiple soft robotic microactuators. A possible solution for reducing the WEDG processing time could be to increase the discharge energy during the roughing steps. This can be accomplished, for example, by increasing the capacitance or open voltage parameters [39]. Nevertheless, an increase of the discharge energy is likely to result in a more aggressive and less repeatable removal of material by electric discharges, thus decreasing the overall accuracy and precision of the WEDG process.

The surface roughness of the WEDG'ed sections of the microrods is analysed qualitatively by means of scanning electron microscopy (SEM) and confocal microscopy. In order to study the effects of finishing on the surface quality, the surface analysis is carried out on the same sections considered in Figure 5 and on other sections, which were machined by interrupting the WEDG process before performing the final finishing step. The benefits of the finishing step are clear when observing the SEM micrographs of the microrods before and after the finishing step (Figure 7). In particular, it can be seen that a less uneven surface morphology can be achieved once the single-step finishing is performed after roughing. The observed difference corresponds to a decrease of the surface roughness from $S_a = 0.84$ µm to $S_a = 0.37$ µm. These values refer to the average values of the S_a surface parameter, which are computed from 20 samples measured on 10 different grooves by a Sensofar® S lynx microscope in confocal mode (magnification: ×50, field of view: 350 × 260 µm). The S_a parameter represents the arithmetical mean height of a surface. It is the extension of the R_a parameter (arithmetical mean height of a line) to the surface. In light of the bonding-free fabrication process of the microactuators, a reduction in surface roughness is advantageous, since the demoulding forces in microreplication processes depend on friction [40]. Despite the relatively long machining time, it can be concluded that the finishing step is crucial not only to achieve the required processing accuracy, but also to facilitate the removal of the microrod after moulding.

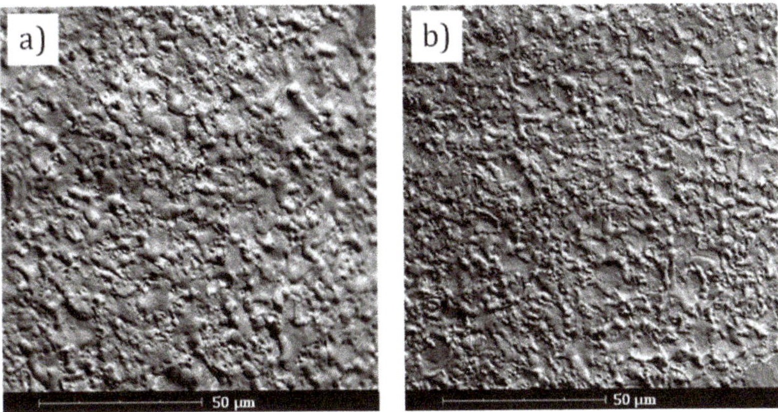

Figure 7. Surface morphology of a microrod (**a**) before and (**b**) after the finishing step. The images are taken by means of Phenom® Pro scanning electron microscope.

Figure 8 shows a microrod after WEDG processing. The enlarged views taken by SEM reveal that a flat tip and straight edges can be achieved by WEDG. It can also be seen that the WEDG'ed sections have round chamfers, of which the radius depends on the radius of the wire which is used in the WEDG process. Round chamfers are crucial for demoulding the microactuators. Very sharp edges should indeed be avoided as they can damage the microactuators.

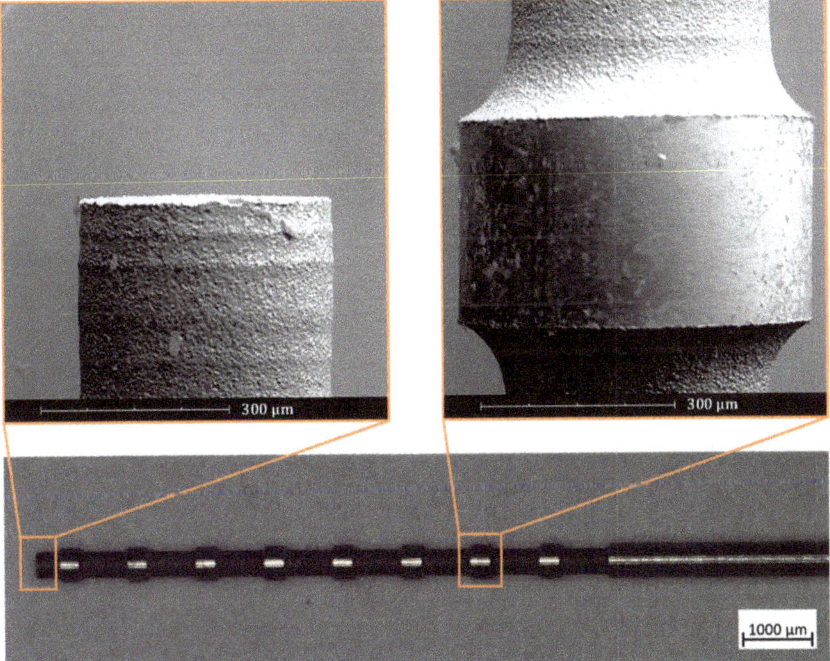

Figure 8. WEDG'ed microrod for moulding of soft pneumatic microactuators. The image of the microrod is taken using a ZEISS® SteREO Discovery V20 optical microscope, while the enlarged views are taken by means of Phenom® Pro scanning electron microscope.

3.2. Microactuators Analytical Model

The analytical model described in Section 2.4 is solved for the three different inflatable cavities. The normalised input pressure (p_E) is equal for the three actuators, varying linearly from 0 to a maximum of 0.2. Figure 9 displays the deformation of the three actuators for six equidistant pressure values along the input ramp.

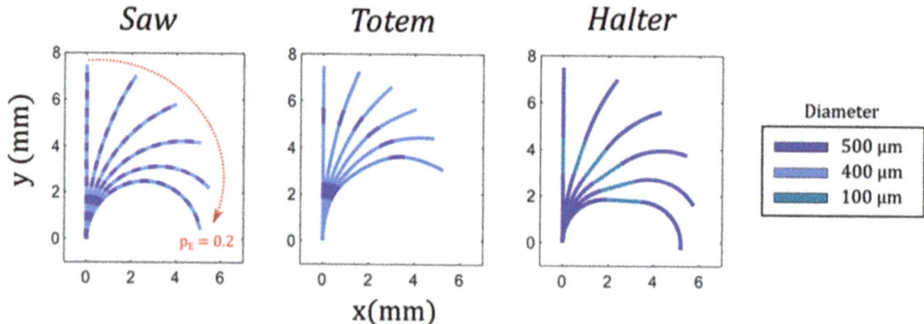

Figure 9. Analytical model results. Six configurations are depicted for each actuator at the same normalised pressure inputs (p_E) varying from 0 (undeformed) to 0.2 (maximum deformation). Segments are distinguished with the same colour code as used in Figure 1.

The different segments of each inflatable cavity are distinguished using the same colour code as for Figure 1. The sections of the inflatable cavity with a reduced diameter (light blue) undergo a lower curvature for two reasons. First of all, the bending stiffness is higher due to the increase of the second moment of area as the cross-sectional void (Figure 2) is smaller. Secondly, the normal force driving the bending moment scales linear with this cross-sectional area (Equation (1)).

The diverse responses of the segments along the microactuators determine the complex deformation patterns that we aim to achieve. As such, each low-stiffness segment acts as a compliant joint. For example, we expect *Saw* to achieve a full-curled configuration due to the higher distribution of joints, whereas *Totem* has a discrete deformation, with only two low-stiffness segments working as the main bending points. On the other hand, *Halter* bends only at the extremities of the microactuators while the central part stays undeformed.

3.3. Microactuators' Characterisation

The three microactuators are experimentally characterised using the setup described in Section 2.5. Figure 10 shows the deformation of the three microactuators at a pressure of 20, 40 and 50 kPa. The deformed shapes are in agreement with the results of the analytical model, showing the predicted segmented curvatures according to the shape of the inflatable cavity, as discussed in the previous paragraph.

However, the experimental displacement starts to deviate from the model at large displacements. This is more significant for actuators *Saw* and *Halter* as they have a higher distribution of thinner membranes that undergo large strains. Indeed, the assumptions of the linear model (such as linear elasticity and undeformed cross-sections) hold at small deformations, but for large displacements silicone rubbers follow hyperelastic models, where stress and strain are nonlinearly related, and cross sections deform. Moreover, circumferential strains become important at large deformations, leading to nonlinear phenomena that occur in rubber structures, such as ballooning and elastic instabilities [41]. This is one of the main reasons why soft bending actuators are manufactured with fibre-reinforcements or bellows shapes to limit circumferential strains [22,38]. Given the asymmetric geometry of the cross-section of our microactuators, the analytical formulation of the nonlinear problem is not trivial, and finite element method (FEM) is commonly used to deal with these nonlinearities [42].

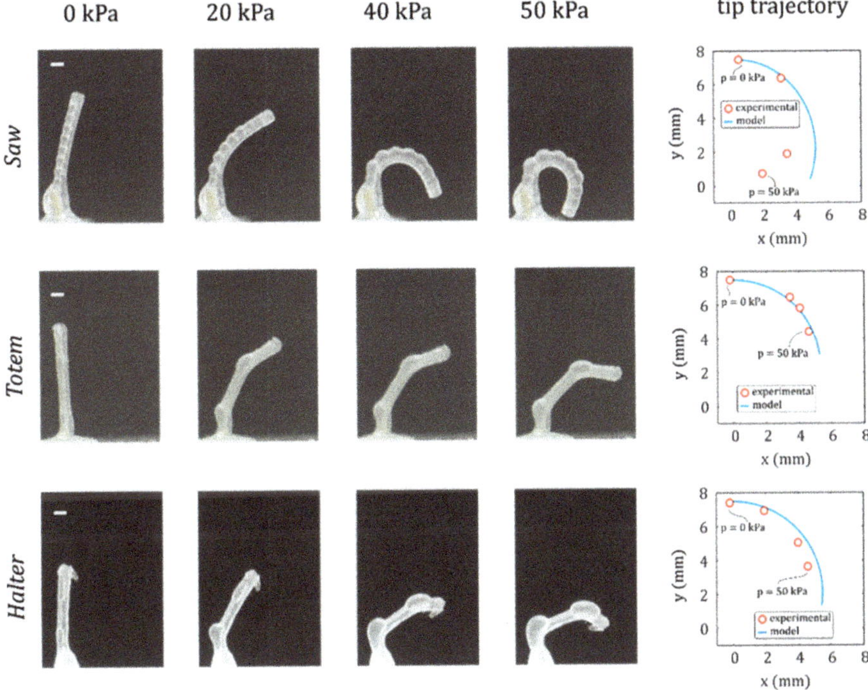

Figure 10. Microactuator inflation tests. Deformed configuration at four different pressures is reported for each microactuator. The white dash in the subfigures of the first column corresponds to a length of 0.8 mm. The experimental and modelled tip trajectory with respect to the initial position at 0 pressure is reported in the graphs.

Thanks to the segmented shape of the inner chamber introduced with the WEDG process, the ballooning effect is limited to the segments with lower stiffness and does not affect the whole actuator. For example, the 400 µm segments in *Saw* work as circumferential strain limiters and prevent the propagation of the ballooning from the 500 µm segments. This decreases the risk of bursting and allows the microactuator to safely achieve a fully curled configuration.

Another interesting feature that we obtained with WEDG can be observed in the *Totem* deformation (Figure 10, second row), where the low-stiffness segments balloon and bend while the rest of the cavity is less deformed, resulting in a finger-like motion. Therefore, larger bending deformations can be locally concentrated in the actuator.

4. Conclusions

In this paper we investigated a new production technology to improve the design of soft inflatable bending microactuators and achieving more complex deformations. WEDG processing was shown to accurately machine micromoulds whose shapes are replicated in the soft actuators through a bonding-free micromoulding process. As demonstrators, we proposed three different actuators that share the same global geometry and material except for the shape of the inflatable cavity. The actuators showed very different kinematics. To predict the response, we applied a simple analytical model based on a multi-segment approximation and linear beam theory. Experimental results agreed with the model-based predictions, within the limits posed by the linear approximation. We manufactured microactuators that exhibit application-relevant behaviours such as full curl, flexible joint-like fingers and undeformed segments. Indeed, this type of actuator has been used to develop flexible microgrippers [36], as well as biomedical devices [25]. In the future we envision a reverse kinematics approach in the form of an optimisation algorithm that, starting from a given trajectory of the end effector, is able to deduce the right morphology of the inflatable cavity which is compatible with WEDG manufacturing.

Author Contributions: E.M. and M.B. conceived the research and wrote the manuscript; M.B. characterised the WEDG process parameters and manufactured the microrods; E.M. created the analytical model, manufactured and characterised the microactuators; B.G. and J.Q. critically analysed the experimental results; M.D.V. and D.R. supervised and supported the research. All authors revised the manuscript. All authors have read and agreed to the published version of the manuscript.

Funding: This research was supported by the Fund for Scientific Research-Flanders (FWO). The authors also gratefully acknowledge the financial support from the European Union's Horizon 2020 projects ProSurf (grant agreement number 767589) and Microman (Project ID: 674801) and from Flanders Make vzw.

Conflicts of Interest: The authors declare no conflicts of interest.

References

1. Rus, D.; Tolley, M.T. Design, fabrication and control of soft robots. *Nature* **2015**, *521*, 467–475. [CrossRef]
2. Kim, S.; Laschi, C.; Trimmer, B. Soft robotics: A bioinspired evolution in robotics. *Trends Biotechnol.* **2013**, *31*, 287–294. [CrossRef]
3. Shintake, J.; Cacucciolo, V.; Floreano, D.; Shea, H. Soft Robotic Grippers. *Adv. Mater.* **2018**, *30*, 1707035. [CrossRef]
4. Soft Robotics Inc. Available online: https://www.softroboticsinc.com (accessed on 5 March 2020).
5. Hawkes, E.W.; Blumenschein, L.H.; Greer, J.D.; Okamura, A.M. A soft robot that navigates its environment through growth. *Sci. Robot.* **2017**, *2*, eaan3028. [CrossRef]
6. Shepherd, R.F.; Ilievski, F.; Choi, W.; Morin, S.A.; Stokes, A.A.; Mazzeo, A.D.; Chen, X.; Wang, M.; Whitesides, G.M. Multigait soft robot. *Proc. Natl. Acad. Sci. USA* **2011**, *108*, 20400–20403. [CrossRef]
7. Wehner, M.; Truby, R.L.; Fitzgerald, D.J.; Mosadegh, B.; Whitesides, G.M.; Lewis, J.A.; Wood, R.J. An integrated design and fabrication strategy for entirely soft, autonomous robots. *Nature* **2016**, *536*, 451–455. [CrossRef]
8. Hines, L.; Petersen, K.; Lum, G.Z.; Sitti, M. Soft Actuators for Small-Scale Robotics. *Adv. Mater.* **2017**, *29*, 1603483. [CrossRef] [PubMed]
9. Konishi, S. Small, soft, safe micromachine for minimally invasive surgery. In Proceedings of the IMFEDK International Meeting for Future of Electron Devices, Osaka, Japan, 9–11 May 2012.
10. De Greef, A.; Lambert, P.; Delchambre, A. Towards flexible medical instruments: Review of flexible fluidic actuators. *Precis. Eng.* **2009**, *33*, 311–321. [CrossRef]
11. Mosadegh, B.; Tavana, H.; Lesher-Perez, S.C.; Takayama, S. High-density fabrication of normally closed microfluidic valves by patterned deactivation of oxidized polydimethylsiloxane. *Lab Chip* **2011**, *11*, 738–742. [CrossRef] [PubMed]
12. Den Toonder, J.M.J.; Onck, P.R. Microfluidic manipulation with artificial/bioinspired cilia. *Trends Biotechnol.* **2013**, *31*, 85–91. [CrossRef] [PubMed]
13. Gorissen, B.; Reynaerts, D.; Konishi, S.; Yoshida, K.; Kim, J.-W.; De Volder, M. Elastic Inflatable Actuators for Soft Robotic Applications. *Adv. Mater.* **2017**, 1604977. [CrossRef] [PubMed]
14. Suzumori, K.; Iikura, S.; Tanaka, H. Development of flexible microactuator and its applications to robotic mechanisms. In Proceedings of the IEEE International Conference on Robotics and Automation, Sacramento, CA, USA, 9–11 April 1991; pp. 1622–1627.
15. Wakimoto, S.; Ogura, K.; Suzumori, K.; Nishioka, Y. Miniature soft hand with curling rubber pneumatic actuators. In Proceedings of the IEEE International Conference on Robotics and Automation, Kobe, Japan, 12–17 May 2009; pp. 556–561.
16. Konishi, S.; Kawai, F.; Cusin, P. Thin flexible end-effector using pneumatic balloon actuator. *Sens. Actuators A Phys.* **2001**, *89*, 28–35. [CrossRef]
17. Polygerinos, P.; Correll, N.; Morin, S.A.; Mosadegh, B.; Onal, C.D.; Petersen, K.; Cianchetti, M.; Tolley, M.T.; Shepherd, R.F. Soft Robotics: Review of Fluid-Driven Intrinsically Soft Devices; Manufacturing, Sensing, Control, and Applications in Human-Robot Interaction. *Adv. Eng. Mater.* **2017**, *19*, 1–22. [CrossRef]
18. Gorissen, B.; Vincentie, W.; Al-Bender, F.; Reynaerts, D.; De Volder, M. Modeling and bonding-free fabrication of flexible fluidic microactuators with a bending motion. *J. Micromech. Microeng.* **2013**, *23*, 45012. [CrossRef]

19. Connolly, F.; Walsh, C.J.; Bertoldi, K. Automatic design of fiber-reinforced soft actuators for trajectory matching. *Proc. Natl. Acad. Sci. USA* **2016**, *114*, 51–56. [CrossRef]
20. Chou, C.P.; Hannaford, B. Measurement and modeling of McKibben pneumatic artificial muscles. *IEEE Trans. Robot. Autom.* **1996**, *12*, 90–102. [CrossRef]
21. Overvelde, J.T.B.; Kloek, T.; D'haen, J.J.; Bertoldi, K. Amplifying the response of soft actuators by harnessing snap-through instabilities. *Proc. Natl. Acad. Sci. USA* **2015**, *112*, 10863–10868. [CrossRef]
22. Mosadegh, B.; Polygerinos, P.; Keplinger, C.; Wennstedt, S.; Shepherd, R.F.; Gupta, U.; Shim, J.; Bertoldi, K.; Walsh, C.J.; Whitesides, G.M. Pneumatic networks for soft robotics that actuate rapidly. *Adv. Funct. Mater.* **2014**, *24*, 2163–2170. [CrossRef]
23. Gorissen, B.; De Volder, M.; De Greef, A.; Reynaerts, D. Theoretical and experimental analysis of pneumatic balloon microactuators. *Sens. Actuators A Phys.* **2011**, *168*, 58–65. [CrossRef]
24. Milana, E.; Gorissen, B.; Peerlinck, S.; De Volder, M.; Reynaerts, D. Artificial Soft Cilia with Asymmetric Beating Patterns for Biomimetic Low-Reynolds-Number Fluid Propulsion. *Adv. Funct. Mater.* **2019**, *29*, 1900462. [CrossRef]
25. Gorissen, B.; De Volder, M.; Reynaerts, D. Chip-on-tip endoscope incorporating a soft robotic pneumatic bending microactuator. *Biomed. Microdevices* **2018**, *20*, 73. [CrossRef] [PubMed]
26. Masuzawa, T.; Fujino, M.; Kobayashi, K.; Suzuki, T.; Kinoshita, N. Wire Electro-Discharge Grinding for Micro-Machining. *CIRP Ann. Manuf. Technol.* **1985**, *34*, 431–434. [CrossRef]
27. Fleischer, J.; Masuzawa, T.; Schmidt, J.; Knoll, M. New applications for micro-EDM. *J. Mater. Process. Technol.* **2004**, *149*, 246–249. [CrossRef]
28. Wang, Y.-Q.; Bellotti, M.; Li, Z.; Qian, J.; Reynaerts, D. Twin-wire electrical discharge grinding for shaping tapered micro rods. In Proceedings of the 19th International Conference and Exhibition of the European Society for Precision Engineering and Nanotechnology, EUSPEN 2019, Bilbao, Spain, 3–7 June 2019.
29. Wang, Y.; Chen, X.; Gan, W.; Wang, Z.; Guo, C. Complex Rotary Structures Machined by Micro-WEDM. *Procedia CIRP* **2016**, *42*, 743–747.
30. Sheu, D.Y. Study on an evaluation method of micro CMM spherical stylus tips by μ-EDM on-machine measurement. *J. Micromech. Microeng.* **2010**, *20*, 075003. [CrossRef]
31. Diver, C.; Atkinson, J.; Helml, H.J.; Li, L. Micro-EDM drilling of tapered holes for industrial applications. *J. Mater. Process. Technol.* **2004**, *149*, 296–303. [CrossRef]
32. Zhang, L.; Tong, H.; Li, Y. Precision machining of micro tool electrodes in micro EDM for drilling array micro holes. *Precis. Eng.* **2015**, *39*, 100–106. [CrossRef]
33. De Volder, M.; Peirs, J.; Reynaerts, D.; Coosemans, J.; Puers, R.; Smal, O.; Raucent, B. Production and characterization of a hydraulic microactuator. *J. Micromech. Microeng.* **2005**, *15*, S15. [CrossRef]
34. Bellotti, M.; Milana, E.; Gorissen, B.; Qian, J.; Reynaerts, D. Wire electrical discharge grinding of micro rods for bonding-free fabrication of soft pneumatic microactuators. In Proceedings of the 18th International Conference and Exhibition of the European Society for Precision Engineering and Nanotechnology, EUSPEN 2018, Venice, Italy, 4–8 June 2018.
35. Bellotti, M.; Qian, J.; Reynaerts, D. Enhancement of the micro-EDM process for drilling through-holes. *Procedia CIRP* **2018**, *68*, 610–615. [CrossRef]
36. Milana, E.; Bellotti, M.; Gorissen, B.; De Volder, M.; Reynaerts, D. Precise bonding-free micromoulding of miniaturized elastic inflatable actuators. In Proceedings of the RoboSoft 2019–2019 IEEE International Conference on Soft Robotics, Seoul, Korea, 14–18 April 2019; pp. 768–773.
37. Paek, J.; Cho, I.; Kim, J. Microrobotic tentacles with spiral bending capability based on shape-engineered elastomeric microtubes. *Sci. Rep.* **2015**, *5*, 10768. [CrossRef]
38. Polygerinos, P.; Wang, Z.; Overvelde, J.T.B.; Galloway, K.C.; Wood, R.J.; Bertoldi, K.; Walsh, C.J. Modeling of Soft Fiber-Reinforced Bending Actuators. *IEEE Trans. Robot.* **2015**, *31*, 778–789. [CrossRef]
39. Jahan, M.P.; Rahman, M.; Wong, Y.S. A review on the conventional and micro-electrodischarge machining of tungsten carbide. *Int. J. Mach. Tools Manuf.* **2011**, *51*, 837–858. [CrossRef]
40. Delaney, K.D.; Bissacco, G.; Kennedy, D. Demoulding force prediction for micro polymer replication: A review of relevant literature. In Proceedings of the 6th International Conference on Multi Material Micro Manufacture (4M), Bourg en Bresse and Oyonnax, France, 17–19 November 2010.

41. Gent, A.N. Elastic instabilities in rubber. *Int. J. Non-Linear Mech.* **2005**, *40*, 165–175. [CrossRef]
42. Moseley, P.; Florez, J.M.; Sonar, H.A.; Agarwal, G.; Curtin, W.; Paik, J. Modeling, Design, and Development of Soft Pneumatic Actuators with Finite Element Method. *Adv. Eng. Mater.* **2016**, *18*, 978–988. [CrossRef]

Article

Applying Foil Queue Microelectrode with Tapered Structure in Micro-EDM to Eliminate the Step Effect on the 3D Microstructure's Surface

Bin Xu, Kang Guo, Likuan Zhu, Xiaoyu Wu and Jianguo Lei *

Guangdong Provincial Key Laboratory of Micro/Nano Optomechatronics Engineering, Shenzhen University, Shenzhen 518060, China; binxu@szu.edu.cn (B.X.); gk_szu@163.com (K.G.); zhulikuan@szu.edu.cn (L.Z.); wuxy@szu.edu.cn (X.W.)
* Correspondence: leijg@szu.edu.cn

Received: 29 February 2020; Accepted: 20 March 2020; Published: 24 March 2020

Abstract: When using foil queue microelectrodes (FQ-microelectrodes) for micro electrical discharge machining (micro-EDM), the processed results of each foil microelectrode (F-microelectrode) can be stacked to construct three-dimensional (3D) microstructures. However, the surface of the 3D microstructure obtained from this process will have a step effect, which has an adverse effect on the surface quality and shape accuracy of the 3D microstructures. To focus on this problem, this paper proposes to use FQ-microelectrodes with tapered structures for micro-EDM, thereby eliminating the step effect on the 3D microstructure's surface. By using a low-speed wire EDM machine, a copper foil with thickness of 300 µm was processed to obtain a FQ-microelectrode in which each of the F-microelectrodes has a tapered structure along its thickness direction. These tapered structures could effectively improve the construction precision of the 3D microstructure and effectively eliminate the step effect. In this paper, the effects of the taper angle and the number of microelectrodes on the step effect were investigated. The experimental results show that the step effect on the 3D microstructure's surface became less evident with the taper angle and the number of F-microelectrodes increased. Finally, under the processing voltage of 120 V, pulse width of 1 µs and pulse interval of 10 µs, a FQ-microelectrode (including 40 F-microelectrodes) with 10° taper angle was used for micro-EDM. The obtained 3D microstructure has good surface quality and the step effect was essentially eliminated.

Keywords: foil queue microelectrode; micro-EDM; step effect; tapered structure

1. Introduction

Micro electrical discharge machining (micro-EDM) is a non-contact machining technology, which has the advantage of a small cutting force. In view of this advantage, micro-EDM has been widely used in the processing of micro structures [1,2].

For achieving three-dimensional (3D) micro-EDM, Yu et al. [3] proposed the uniform wear method (UWM) and applied it to process 3D micro-cavities through layer-by-layer micro-EDM of a micro-electrode with a simple cross-section. In order to process complex 3D structures with arbitrary components, Rajurkar et al. [4] combined UWM and the computer-aided design (CAD)/computer-aided manufacturing (CAM) system in micro-EDM. With the purpose of investigating the effects of alternating-current on the energy usage and the erosion efficiency in the micro-EDM process, Yang et al. [5] built an electrical model and provided a theoretical analysis.

To improve the efficiency of 3D micro-EDM, Tong et al. [6] proposed the servo scanning 3D micro-EDM (3D SSMEDM) method based on the macro/micro-dual-feed spindle. In order to improve the machining efficiency and reduce the electrode wear, Fu et al. [7] proposed piezoelectric self-adaptive micro-EDM based on the inverse piezoelectric effect. To improve machining quality and machining efficiency of 3D micro-EDM, Yu et al. [8] proposed a new electrode wear compensation method, which combined the linear compensation method (LCM) with the uniform wear method (UWM). In order to fabricate micro groove arrays and columnar microstructures, Wu et al. [9] applied rotary dentate disc foil electrodes in micro-EDM in #304 stainless steel workpieces.

To improve machining accuracy of 3D micro-EDM, Nguyen et al. [10] identified and analyzed the error components of 3D micro-EDM milling process, which found that the corner radius of virtual electrode is also important to determine the machining accuracy. In order to achieve the high-precise machining of the micro rotating structure, Wang et al. [11] proposed micro reciprocated wire electrical discharge machining (wire-EDM) to fabricate micro-rotating structures. Using the low speed wire electrical discharge turning (LS-WEDT) method combined with the numerical control technology, Sun et al. [12] manufactured the microelectrodes and micro-cutting tools with good surface quality and high machining accuracy. To further study the discharge characteristics and machining mechanism of micro-EDM, Liu et al. [13] studied the variation trends of the discharge energy and discharge crater size in single-pulse experiments. Focus on the optimization of the processing parameters and quality control in micro-EDM, Bellotti et al. [14] applied a process fingerprint approach in micro-EDM drilling. In order to process micro-holes on tungsten carbide plates, D'Urso et al. [15] applied tubular electrodes in micro-EDM and evaluated the influence of variable process parameters on process performance. For achieving high precision machining of cubic boron nitride, Wyszynski et al. [16] described an application of EDM for drilling micro holes in cubic boron nitride and determined a set of parameters and technical specifications. Focus on the fabrication of deep micro-channels, Ahmed et al. [17] used wire-cut electrical discharge machining (EDM) to fabricate deep micro-channels with thin inter-channels fins.

Focus on the optimal selection of machining parameters, Świercz et al. [18] performed an analytical and experimental investigation of the influence of the EDM parameters. In order to understand the debris movement in high aspect ratio hole EDM machining, Liu et al. [19] developed a model to simulate the distribution and removal of debris in different machining conditions in ultrasonic assisted EDM with side flushing. To study the hydrogen dielectric strength forces in the EDM, António Almacinha et al. [20] applied electro-thermal model to simulate a single discharge in an electric discharge machining process. For obtaining an array micro-grooves, Wang et al. [21] developed a manufacturing method by applying disk electrode in micro electrochemical machining. For getting the high-efficiency removal, Zhang et al. [22] adopted a tool electrode with an optimized helical structure in tube electrode high-speed electrochemical discharge machining (TSECDM). Focus on the problem of the current micro-EDM pulse generator, Wang et al. [23] designed a micro-energy pulse source with narrow pulse width and high-voltage amplitude for getting more fine-etching ability. For fabricating micro punching mold with complex cross-sectional shape, Yu et al. [24] developed a micro punching system with a micro electrical discharge machining (EDM) module. To optimize the process parameters for micro EDM of Ti-6Al-4V alloy, Huang et al. [25] used the Taguchi method to determine the performance characteristics in micro EDM milling operations. In order to flush the debris generated in micro-EDM, Beigmoradi et al. [26] proposed a new numerical approach for enhancing flushing.

The above studies did good work on 3D micro-EDM and promoted the development of micro-EDM. For improving machining efficiency of 3D micro-EDM, Xu et al. [27,28] proposed a novel process to fabricate 3D micro-electrodes by superimposing multilayer 2D micro-structures and applied it in micro-EDM. However, the fabrication process of 3D micro-electrode is complicated and has a low success rate (30%).

Focusing on the complexity and low fabrication success rate of 3D microelectrodes, Xu et al. [29,30] discretized 3D micro-electrodes into several foil micro-electrodes and these foil micro-electrodes composed foil queue micro-electrode (FQ-microelectrode). According to the planned process path, each foil micro-electrode in FQ-microelectrode (Figure 1a) was sequentially applied in micro-EDM and processed results of each foil microelectrode (F-microelectrode) can be stacked to construct 3D microstructures. However, the surface of the 3D microstructure obtained from this process will have a step effect (Figure 1b), which will affect the surface quality and shape accuracy of the 3D microstructure. Focus on this problem, this paper used FQ-microelectrode with tapered structure for micro-EDM processing. These taper structures can effectively improve the construction accuracy of the 3D microstructure and effectively eliminate the step effect.

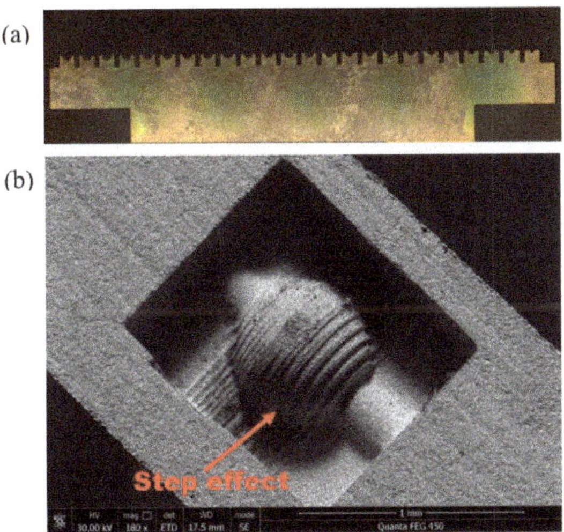

Figure 1. (**a**) Foil queue (FQ)-microelectrode; (**b**) 3D microstructure fabricated by the micro-electrical discharge machining (EDM) of FQ-microelectrode.

2. Method

Firstly, the 3D microstructure model was established using 3D modelling software. According to the 3D microstructure model, the corresponding 3D microelectrode model was obtained. Then, the 3D microelectrode model was sliced along its thickness direction to obtain a number of F-microelectrode models and thus machining data of each F-microelectrode can be obtained. Based on the machining data, the copper foil was processed by low-speed wire EDM machine to obtain each F-microelectrode and these F-microelectrodes composed the FQ-microelectrode.

FQ-microelectrode was applied for micro-EDM in sequence and processed results of each F-microelectrode can be stacked to construct 3D microstructures. Similar to the 3D printing process, the surface of 3D microstructure has a step effect (Figure 1), which seriously affects the surface quality and shape accuracy of the 3D microstructure.

When the electrode wear factor was not considered, the 3D microstructure obtained from micro-EDM of FQ-microelectrode was formed by many step superpositions. In this case, the step effect on the surface of the 3D microstructure was evident (Figure 2a). When the F-microelectrode had a tapered structure, the processing contour of the 3D microstructure was composed of oblique lines, which could better fit the design contour (Figure 2b), thereby reducing the step effect and improving the shape accuracy of the 3D microstructure.

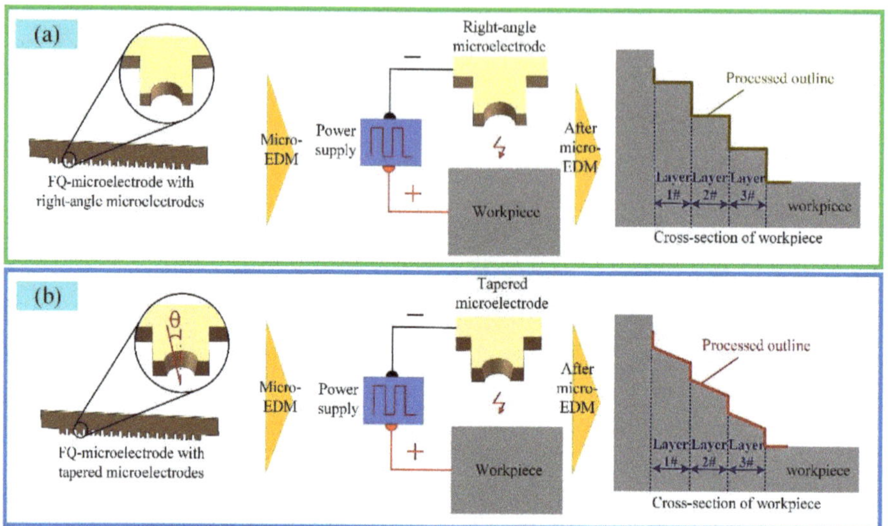

Figure 2. (**a**) Fabricating 3D microstructure based on the FQ-microelectrode without tapered structure; (**b**) Fabricating 3D microstructure based on the FQ-microelectrode with tapered structure.

3. Experimental Materials and Equipment

The FQ-microelectrode with taper structures was machined from a 300 µm thick copper foil using a LS-WEDM machine (Sodick company, Model: AP250LS, Suzhou, China), and then 3D micro-EDM was performed in cemented carbide. The FQ-microelectrode was observed by laser scanning confocal microscopy (Keyence company, model: VK-X250, Osaka, Japan). The surface topography of the 3D microstructure was observed by scanning electron microscopy (FEI company, Model: Quanta FEG 450, Hillsboro, OR, USA).

4. Experimental Results and Discussion

To eliminate the step effect on the 3D microstructure surface, this paper used FQ-microelectrode with tapered structures for micro-EDM to process 3D microstructure. This paper studied in detail the influence of different taper angles and numbers of F-microelectrodes on the step effect. Under the voltage of 72 V, the pulse width of 0.5 µs and the pulse interval of 5 µs, the FQ-microelectrodes were machined from copper foil with thickness of 300 µm by using the LS-WEDM machine.

4.1. Influence of Taper Angle on the Elimination of Step Effect

To study the effect of the taper angle on the step effect, FQ-microelectrodes with different taper angles were used for micro-EDM. Due to the limitations of the processing equipment, FQ-microelectrodes with a taper angle of more than 10° cannot be machined. Therefore, the taper angle of the FQ-microelectrodes was set to 0°, 2°, 4°, 6°, 8° and 10°. The workpiece material was cemented carbide, and the FQ-microelectrode containing 16 F-microelectrodes was fabricated in 300 µm thick copper foil. The processing object was 1/4 sphere with diameter of 600 µm. Based on the previous studies [17,18], under the processing voltage of 120 V, the pulse width of 1 µs and the pulse interval of 10 µs, the processing object fabricated by micro-EDM had well surface morphology.

The experimental results are shown in Figure 3. When the taper angle of the F-microelectrode is 0°, the number of steps on the 3D microstructure surface is highest and the step effect is evident (Figure 3a). When the taper angle of the F-microelectrode is 10°, the number of steps on the 3D microstructure surface is 2 and the step effect is not evident. When the F-microelectrode has a taper structure, the processing contour of the 3D microstructure is composed of oblique lines, which effectively improves the shape precision of the 3D microstructure and reduces the step effect. Therefore, as the taper angle of the F-microelectrode increases, the number of steps on the 3D microstructure surface gradually decreases and the step effect becomes increasingly less evident.

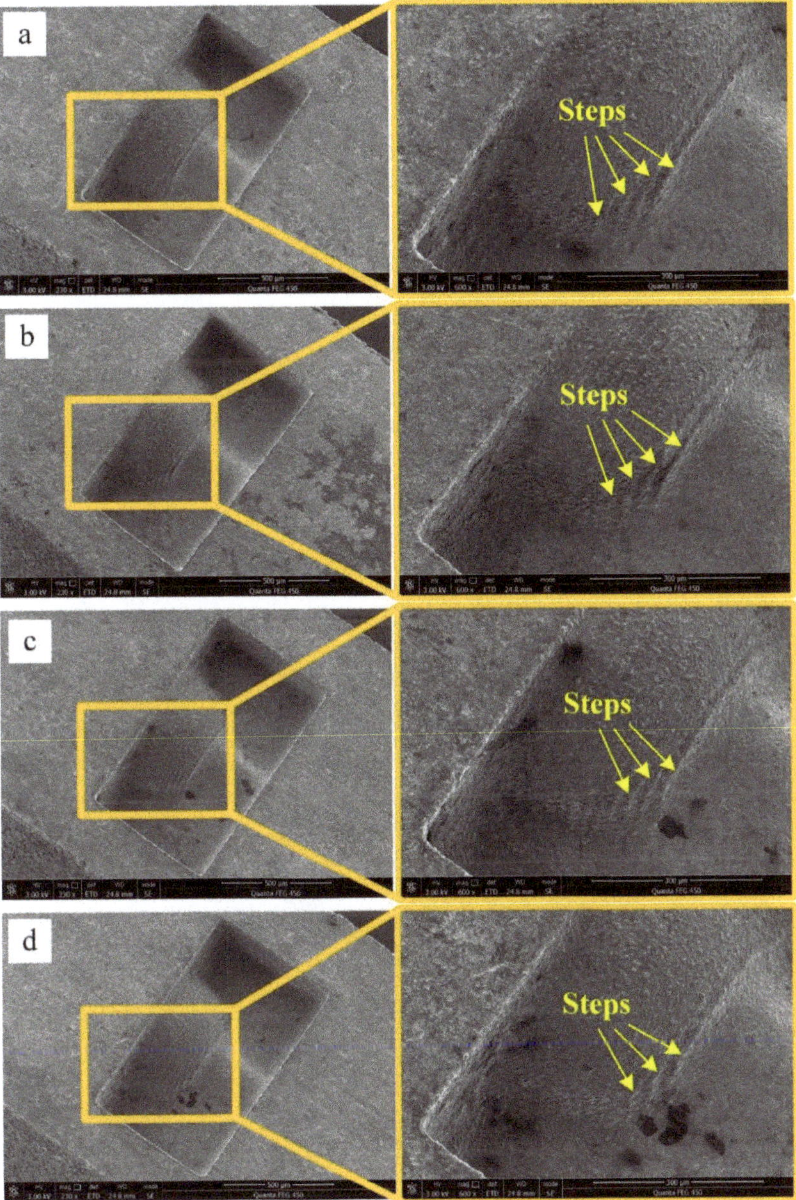

Figure 3. *Cont.*

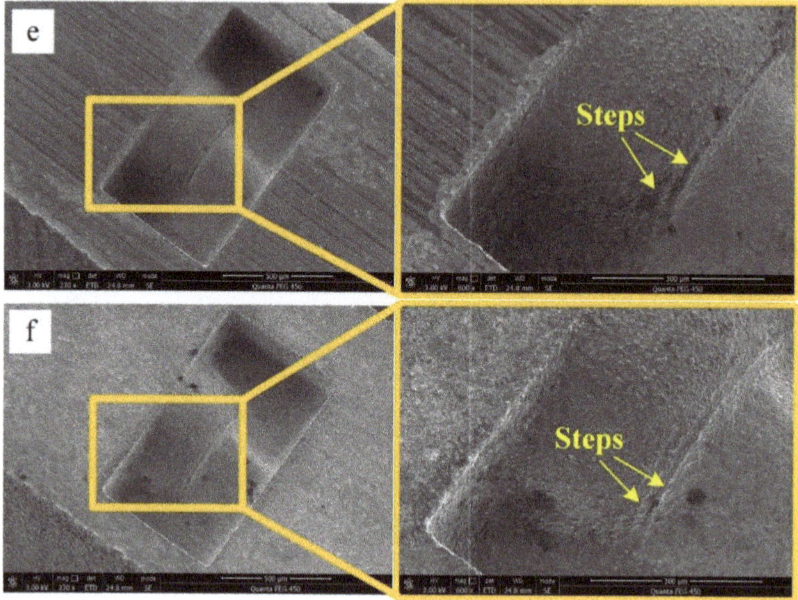

Figure 3. Micro-EDM results of FQ-microelectrodes with different taper angles: (**a**) 0°; (**b**) 2°; (**c**) 4°; (**d**) 6°; (**e**) 8°; (**f**) 10°.

To further clarify the position of the steps, the cross-sectional profile of the 3D microstructure was measured by laser confocal microscopy and the experimental results are shown in Figure 4.

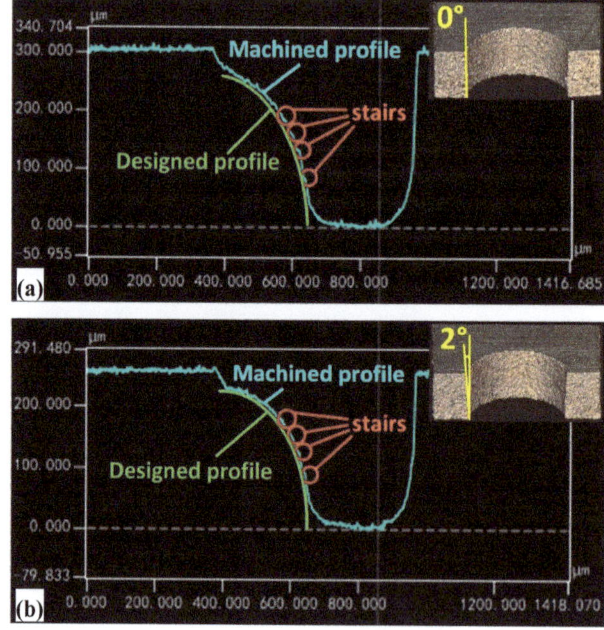

Figure 4. *Cont.*

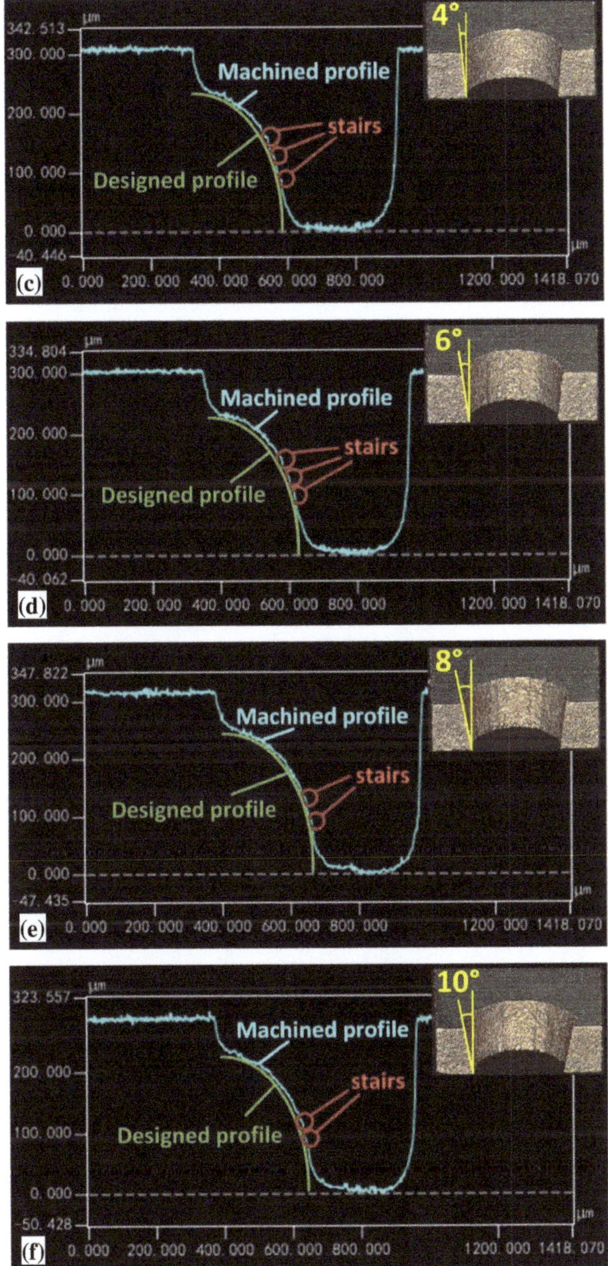

Figure 4. Micro-EDM results of FQ-microelectrode with different tapers observed by laser scanning confocal microscopy: (**a**) 0°; (**b**) 2°; (**c**) 4°; (**d**) 6°; (**e**) 8°; (**f**) 10°.

When the taper angle of the F-microelectrode is 0°, the steps are mainly distributed in the middle and tail of the spherical surface. When the taper angle of the F-microelectrode gradually increases from 0° to 10°, the steps in the middle of the spherical surface are gradually eliminated, and the steps in the tail of the spherical surface are somewhat attenuated. When the F-microelectrode had a tapered structure, the processing contour of the 3D microstructure was composed of oblique lines. If the slope of the oblique line was close to the slope of the processing results of the adjacent F-microelectrodes, the steps on this position can be substantially eliminated. From the processing position of the first F-microelectrode to the last F-microelectrode, the height difference of the processing results of the adjacent F-microelectrodes is continuously increased (Figure 5), so the slope of the processing results of the adjacent F-microelectrodes is continuously increased. Therefore, when the taper angle of the F-microelectrode is close to the slope of the processing results of the adjacent F-microelectrodes, the step effect can be effectively eliminated (Figure 5). Thus, in the middle of the spherical surface, when the taper angle of the F-microelectrode is 10°, the taper angle is relatively close to the slope of the processing results of the adjacent F-microelectrodes. Therefore, the steps on this position is substantially eliminated. In the tail of the sphere surface, the taper of the F-microelectrode differs greatly from the slope of the processing results of the adjacent F-microelectrodes, which results in a more pronounced step at that location.

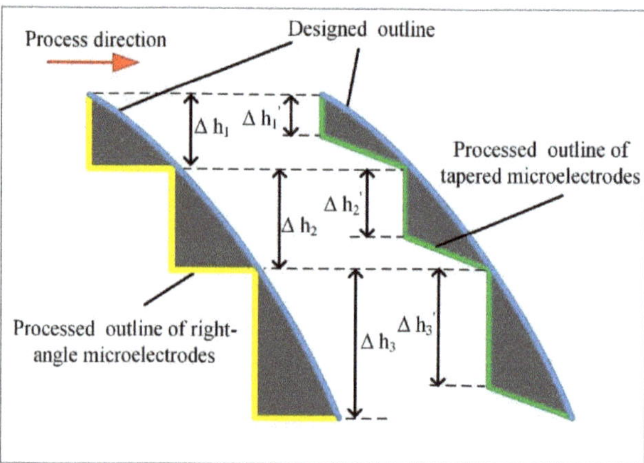

Figure 5. The height difference of the processing results of the adjacent F-microelectrodes.

4.2. Influence of Numbers of F-Microelectrode on the Elimination of Step Effect

Due to the limitation of processing equipment, FQ-microelectrodes with a taper angle of more than 10° cannot be machined. Therefore, it is difficult to eliminate the steps in the tail of the sphere surface. To focus on this problem, this paper proposes to eliminate the steps in the tail of the sphere surface by reducing slice thickness of the 3D microelectrode model and thereby increasing the number of the F-microelectrodes.

To study the effect of the number of F-microelectrodes on the step effect, FQ-microelectrodes with different numbers of F-microelectrodes were applied for micro-EDM. The FQ-microelectrodes had 16, 25, 33 and 40 F-microelectrodes. The workpiece material was cemented carbide. The FQ-microelectrode was fabricated in 300 µm thick copper foil and every F-microelectrode had a taper angle of 10°. The processing object was 1/4 sphere with diameter of 600 µm, the processing voltage was 120 V, the pulse width was 1 µs and the pulse interval was 10 µs.

As shown in Figure 6, when the number of F-microelectrodes is 16, the 3D microstructure surface has a small number of steps and these steps are located in the tail of the sphere surface. As the number of F-microelectrodes increasing, the steps on the 3D microstructure surface become fewer and fewer. When the number of F-microelectrodes increases to 40, the steps on the 3D microstructure surface are essentially eliminated (Figure 6d). These experimental results prove that the step effect of the 3D microstructure surface can be effectively eliminated by increasing the number of the F-microelectrodes. To further clarify the position of the steps, the cross-sectional profile of the 3D microstructure was measured by laser confocal microscopy and the experimental results are shown in Figure 7.

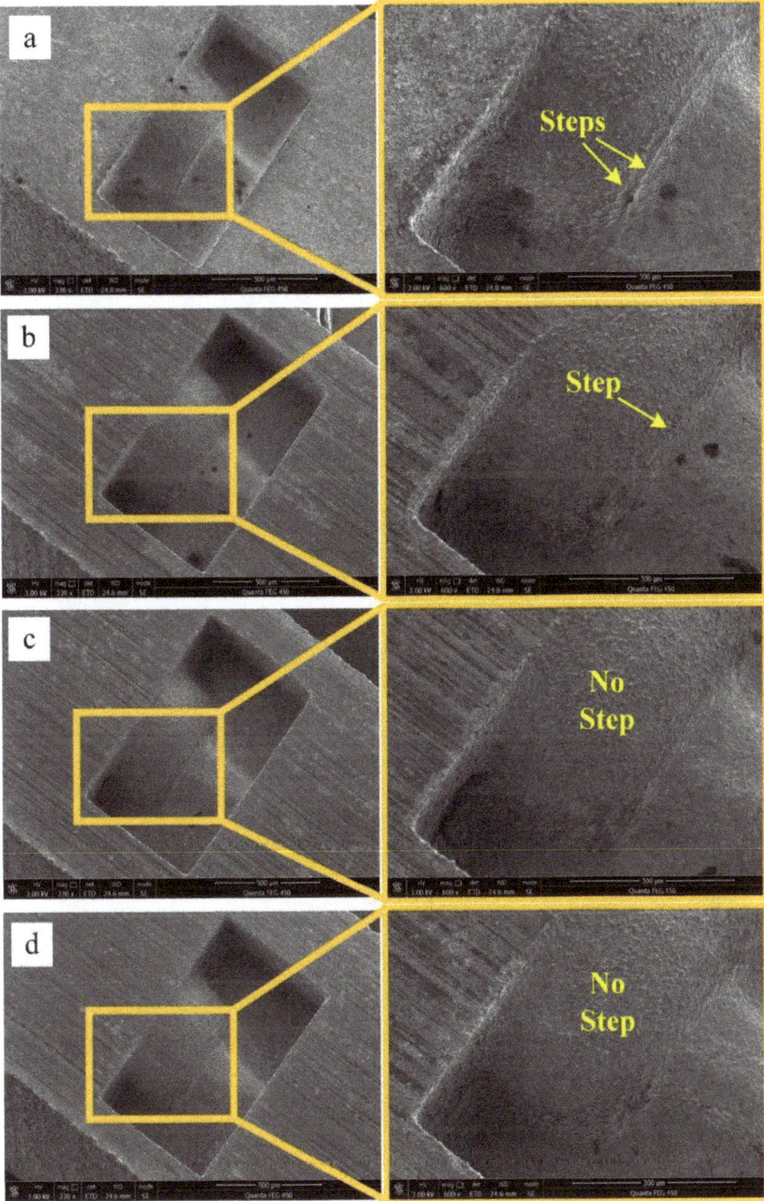

Figure 6. Micro-EDM results of FQ-microelectrodes with different number of F-microelectrodes: (**a**) 16; (**b**) 25; (**c**) 33; (**d**) 40.

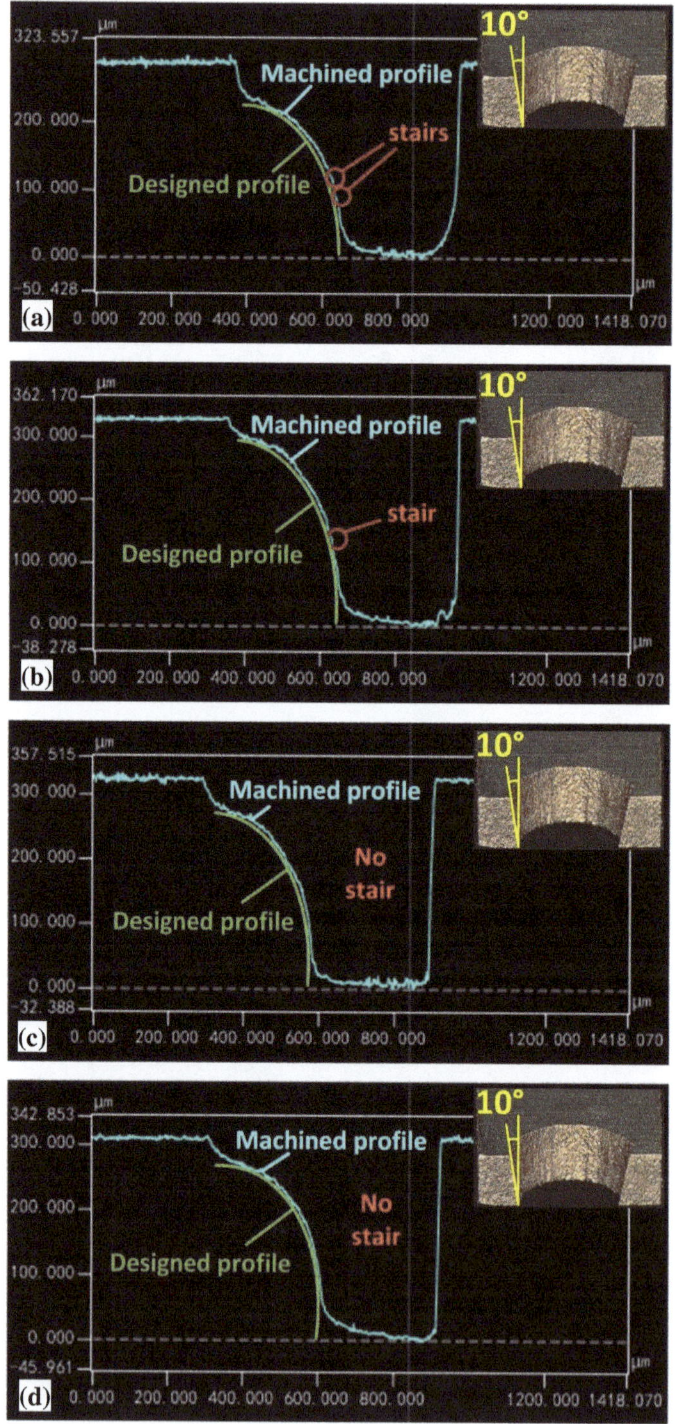

Figure 7. Micro-EDM results of FQ-microelectrodes with different number of F-microelectrodes observed by laser scanning confocal microscopy: (**a**) 16; (**b**) 25; (**c**) 33; (**d**) 40.

As shown in Figure 7, when the number of F-microelectrodes is 16, the step is mainly distributed at the tail of the spherical surface. With the number of F-microelectrodes gradually increasing from 16 to 40, the position at which the steps occur gradually moves to the tail of the sphere, until it is eliminated. As slice thickness of the 3D microelectrode model decreasing, the number of F-microelectrodes will increase, which could improve the fitting precision of 3D microstructure. Therefore, in the tail of the sphere, the step can be eliminated through increasing the number of F-microelectrodes. In addition, during the micro-EDM, the wear of the F-microelectrode is unavoidable. Under the effect of micro-EDM, the vertical angle at the end face of the F-microelectrode is worn and becomes rounded corner. Under the influence of these factors, the surface of the processing results is processed into a corresponding curved surface and thus the steps on the 3D microstructure surface are gradually eliminated.

5. Application of FQ-Microelectrode with Tapered Structures in Micro-EDM

To verify the feasibility of the proposed process, an FQ-microelectrode was prepared using a 300 μm thick copper foil. The FQ-microelectrode contained 40 F-microelectrodes, each of which had taper structure with taper angle of 10° (Figure 8a). The process parameters of the FQ-microelectrode preparation were described as follows: wire cutting voltage of 72 V, the pulse width of 0.5 μs and the pulse interval of 5 μs. The FQ-microelectrode was applied in micro-EDM and its process parameters were described as follows: pulse width of 1 μs, pulse interval of 10 μs and voltage of 120 V. The processing object was a hemisphere with diameter of 600 μm (Figure 9) and the workpiece material was cemented carbide.

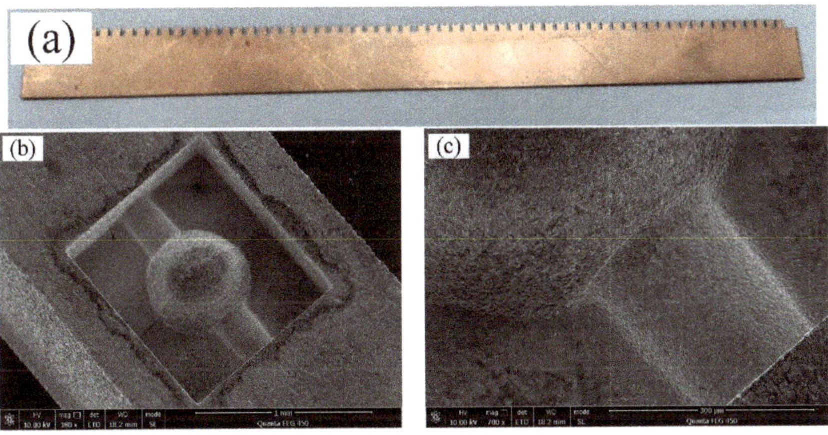

Figure 8. (**a**) FQ-microelectrode with tapered structures; (**b**,**c**) Micro-EDM results of FQ-microelectrodes observed by scanning electron microscopy.

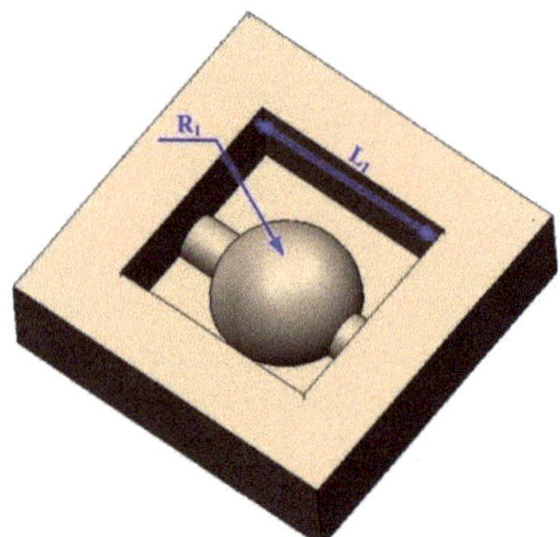

Figure 9. The computer-aided design (CAD) model of 3D microstructure.

The 3D microstructure was observed and measured by using scanning electron microscopy and laser scanning confocal microscopy. The experimental results are shown in Figure 8 and Table 1. From the experimental results, it can be seen that the surface quality of the 3D microstructure is good (Figure 8b) and the surface step effect is eliminated (Figure 8c). The dimensional accuracy of 3D microstructure is well and the maximum dimensional error is within 10 μm.

Table 1. Dimensional comparison between computer-aided design (CAD) model shown in Figure 9 and micro electrical discharge machining (micro-EDM) result shown in Figure 8.

Dimensional Symbols	Dimensional Values (μm)			Processing Time
	CAD Model Shown in Figure 9	**Micro-EDM Result Shown in Figure 8**	**Errors**	
L1	1156	1150	6	
R1	300	295	5	120 min
Depth	330	322	8	

6. Conclusions

Using FQ-microelectrodes for micro-EDM, processed results of each F-microelectrode can be stacked to construct 3D microstructures. However, the surface of the 3D microstructure obtained from this process will have step effect, which will affect the surface quality and shape accuracy of the 3D microstructure. To focus on this problem, this paper proposed to eliminate the step effect by using FQ-microelectrodes with tapered structures for micro-EDM and increasing the number of F-microelectrodes. Through the detailed study, the following conclusions can be drawn:

(1) When the FQ-microelectrodes with tapered structures were used for micro-EDM, the step effect on the surface of the 3D microstructure can be significantly weakened and the shape accuracy can be improved.

(2) By reducing the slice thickness of the 3D microelectrode model and thereby increasing the number of F-microelectrodes, the step effect on the 3D microstructure surface can be further eliminated. When the taper angle of the F-microelectrode is 10° and the number of F-microelectrodes is 40, the step effect on the surface of the 3D microstructure can be essentially eliminated.

(3) Under the processing voltage of 120 V, pulse width of 1 μs and pulse interval of 10 μs, the FQ-microelectrode (including 40 F-microelectrodes) with 10° taper angle was used for micro-EDM. The obtained 3D microstructure has good surface quality and the step effect was essentially eliminated. The dimensional accuracy of 3D microstructure is well and the maximum dimensional error is within 10 μm.

Author Contributions: Conceptualization, J.L. and B.X.; methodology, B.X.; validation, J.L., X.W. and B.X.; formal analysis, K.G. and L.Z.; investigation, K.G.; resources, X.W.; data curation, L.Z.; writing—original draft preparation, B.X. and K.G.; writing—review and editing, J.L.; visualization, K.G.; supervision, B.X.; project administration, B.X. and K.G.; funding acquisition, J.L. All authors have read and agree to the published version of the manuscript.

Funding: This work is supported by the National Natural Science Foundation of China (Grant Nos. 51805333, 51975385), the Natural Science Foundation of Guangdong Province (Grant No. 2017A030313309), the Science and Technology Innovation Commission Shenzhen (Grant Nos. JCYJ20170817094310049 and JSGG20170824111725200).

Acknowledgments: The authors are grateful to their colleagues for their essential contribution to the work.

Conflicts of Interest: The authors declare no conflicts of interest.

References

1. Hourmand, M.; Sarhan, A.A.D.; Sayuti, M. Micro-electrode fabrication processes for micro-EDM drilling and milling: A state-of-the-art review. *Int. J. Adv. Manuf. Technol.* **2017**, *91*, 1023–1056. [CrossRef]
2. Bilal, A.; Pervej Jahan, M.; Talamona, D.; Perveen, A. Electro-Discharge Machining of Ceramics: A Review. *Micromachines* **2019**, *10*, 10. [CrossRef] [PubMed]
3. Yu, Z.Y.; Masuzawa, T.; Fujino, M. Micro-EDM for Three-Dimensional Cavities-Development of Uniform Wear Method. *CIRP Ann.* **1998**, *47*, 169–172. [CrossRef]
4. Rajurkar, K.P.; Yu, Z.Y. 3D Micro-EDM Using CAD/CAM. *CIRP Ann.* **2000**, *49*, 127–130. [CrossRef]
5. Yang, F.; Qian, J.; Wang, J.; Reynaerts, D. Simulation and experimental analysis of alternating-current phenomenon in micro-EDM with a RC-type generator. *J. Mater. Process. Technol.* **2018**, *255*, 865–875. [CrossRef]
6. Tong, H.; Li, Y.; Wang, Y.; Yu, D. Servo scanning 3D micro-EDM based on macro/micro-dual-feed spindle. *Int. J. Mach. Tools Manuf.* **2008**, *48*, 858–869. [CrossRef]
7. Fu, X.Z.; Zhang, Q.H.; Gao, L.Y.; Liu, Q.Y.; Wang, K.; Zhang, Y.W. A novel micro-EDM-piezoelectric self-adaptive micro-EDM. *Int. J. Adv. Manuf. Technol.* **2016**, *85*, 817–824. [CrossRef]
8. Wang, J.; Yang, F.; Qian, J.; Reynaerts, D. Study of alternating current flow in micro-EDM through real-time pulse counting. *J. Mater. Process. Technol.* **2016**, *231*, 179–188. [CrossRef]
9. Wu, W.; Wu, X.Y.; Lei, J.G.; Xu, B.; Jiang, K.; Wu, Z.Z.; Yin, H.M.; Li, W. Fabrication of deep-narrow microgrooves by micro-EDM using rotary dentate disc foil electrodes in emulsion. *J. Micromech. Microeng.* **2019**, *29*, 035014. [CrossRef]
10. Schulze, V.; Weber, P.; Ruhs, C. Increase of process reliability in the micro-machining processes EDM-milling and laser ablation using on-machine sensors. *J. Mater. Process. Technol.* **2012**, *212*, 625–632. [CrossRef]
11. Wang, Y.K.; Chen, X.; Wang, Z.L.; Li, H.C.; Liu, H.Z. Fabrication of micro-rotating structure by micro reciprocated wire-EDM. *J. Micromech. Microeng.* **2016**, *26*, 115014. [CrossRef]
12. Sun, Y.; Gong, Y.D. Experimental study on the microelectrodes fabrication using low speed wire electrical discharge turning (LS-WEDT) combined with multiple cutting strategy. *J. Mater. Process. Technol.* **2017**, *250*, 121–131. [CrossRef]
13. Liu, Q.; Zhang, Q.; Zhang, M.; Yang, F. Study on the Discharge Characteristics of Single-Pulse Discharge in Micro-EDM. *Micromachines* **2020**, *11*, 55. [CrossRef] [PubMed]
14. Bellotti, M.; Qian, J.; Reynaerts, D. Process Fingerprint in Micro-EDM Drilling. *Micromachines* **2019**, *10*, 240. [CrossRef]
15. D'Urso, G.; Giardini, C.; Quarto, M.; Maccarini, G. Cost Index Model for the Process Performance Optimization of Micro-EDM Drilling on Tungsten Carbide. *Micromachines* **2017**, *8*, 251. [CrossRef]
16. Wyszynski, D.; Bizon, W.; Miernik, K. Electrodischarge Drilling of Microholes in c-BN. *Micromachines* **2020**, *11*, 179. [CrossRef]
17. Ahmed, N.; Pervez Mughal, M.; Shoaib, W.; Raza, S.F.; Alahmari, A.M. WEDM of Copper for the Fabrication of Large Surface-Area Micro-Channels: A Prerequisite for the High Heat-Transfer Rate. *Micromachines* **2020**, *11*, 173. [CrossRef]
18. Świercz, R.; Oniszczuk-Świercz, D.; Chmielewski, T. Multi-Response Optimization of Electrical Discharge Machining Using the Desirability Function. *Micromachines* **2019**, *10*, 72. [CrossRef]

19. Liu, Y.; Chang, H.; Zhang, W.C.; Ma, F.J.; Sha, Z.H.; Zhang, S.F. A Simulation Study of Debris Removal Process in Ultrasonic Vibration Assisted Electrical Discharge Machining (EDM) of Deep Holes. *Micromachines* **2018**, *9*, 378. [CrossRef]

20. Almacinha, J.A.; Lopes, A.M.; Rosa, P.; Marafona, J.D. How Hydrogen Dielectric Strength Forces the Work Voltage in the Electric Discharge Machining. *Micromachines* **2018**, *9*, 240. [CrossRef]

21. Wang, Y.K.; Wang, H.; Zhang, Y.X.; He, X.L.; Wang, Z.L.; Chi, G.X.; Chen, X.; Song, M.S. Micro Electrochemical Machining of Array Micro-Grooves Using In-Situ Disk Electrode Fabricated by Micro-WEDM. *Micromachines* **2020**, *11*, 66. [CrossRef] [PubMed]

22. Zhang, Y.; Wang, C.; Wang, Y.; Ji, L.; Tang, J.; Ni, Q. Effects of Helical Tube Electrode Structure on Mixed Machining Product Transfer in Micro-Machining Channel during Tube Electrode High-Speed Electrochemical Discharge Machining. *Micromachines* **2019**, *10*, 634. [CrossRef]

23. Wang, F.; Zhang, Y.; Liu, G.; Wang, Q. Improvement of processing quality based on VHF resonant micro-EDM pulse generator. *Int. J. Adv. Manuf. Technol.* **2019**, *104*, 3663–3677. [CrossRef]

24. Yu, Z.; Li, D.; Yang, J.; Zeng, Z.; Yang, X.; Li, J. Fabrication of micro punching mold for micro complex shape part by micro EDM. *Int. J. Adv. Manuf. Technol.* **2019**, *100*, 743–749. [CrossRef]

25. Huang, C.H.; Yang, A.B.; Hsu, C.Y. The optimization of micro EDM milling of Ti–6Al–4V using a grey Taguchi method and its improvement by electrode coating. *Int. J. Adv. Manuf. Technol.* **2018**, *96*, 3851–3859. [CrossRef]

26. Beigmoradi, S.; Ghoreishi, M.; Vahdati, M. Optimum design of vibratory electrode in micro-EDM process. *Int. J. Adv. Manuf. Technol.* **2018**, *95*, 3731–3744. [CrossRef]

27. Xu, B.; Wu, X.Y.; Lei, J.G.; Cheng, R.; Ruan, S.C.; Wang, Z.L. Laminated fabrication of 3D micro-electrode based on WEDM and thermal diffusion welding. *J. Mater. Process. Technol.* **2015**, *221*, 56–65. [CrossRef]

28. Xu, B.; Wu, X.Y.; Ma, J.; Liang, X.; Lei, J.G.; Wu, B.; Ruan, S.C.; Wang, Z.L. Micro-electrical discharge machining of 3D micro-molds from Pd40Cu30P20Ni10 metallic glass by using laminated 3D micro-electrodes. *J. Micromech. Microeng.* **2016**, *26*, 035004. [CrossRef]

29. Xu, B.; Guo, K.; Wu, X.Y.; Lei, J.G.; Liang, X.; Guo, D.J.; Ma, J.; Cheng, R. Applying a foil queue micro-electrode in micro-EDM to fabricate a 3D micro-structure. *J. Micromech. Microeng.* **2018**, *28*, 055008. [CrossRef]

30. Xu, B.; Guo, K.; Zhu, L.K.; Wu, X.Y.; Lei, J.G.; Zhao, H.; Tang, Y.; Liang, X. The wear of foil queue microelectrode in 3D micro-EDM. *Int. J. Adv. Manuf. Technol.* **2019**, *104*, 3107–3117. [CrossRef]

 *micromachines*

Article

Electrodischarge Drilling of Microholes in c-BN

Dominik Wyszynski [1,*] , **Wojciech Bizon** [1] **and Krzysztof Miernik** [2]

1   Faculty of Mechanical Engineering, Cracow University of Technology, al. Jana Pawla II 37, 31-864 Krakow, Poland; wojciech.bizon@pk.edu.pl
2   Faculty of Materials Science and Physics, Cracow University of Technology, al. Jana Pawla II 37, 31-864 Krakow, Poland; kmiernik@pk.edu.pl
*   Correspondence: dominik.wyszynski@pk.edu.pl; Tel.: +48-691-175-649

Received: 10 January 2020; Accepted: 7 February 2020; Published: 10 February 2020

Abstract: Cubic boron nitride (c-BN) is a "difficult-to-cut" material. High precision machining of this material is problematic because it is difficult to control the material removal rate and maintain acceptable accuracy. This paper describes an application of electrodischarge machining (EDM) for drilling micro holes in c-BN. The goal of this research was to determine a set of parameters and technical specifications for such a process. We used an isoenergetic transistor power supply with a microsecond voltage pulse generator and a tungsten tool electrode of diameter d = 381 μm. Each hole was drilled for 10 min. The holes did not exceed 410 μm in diameter and were at least 1000 μm deep. The process was carried out in a hydrocarbon dielectric liquid. We assess the quality of the holes from a qualitative and quantitative point of view. The results show that electrodischarge is a precise, accurate, and efficient method for machining c-BN.

Keywords: electrodischarge micromachining; drilling; cubic boron nitride

1. Introduction

Machining of very hard and high strength materials—called "difficult-to-cut" materials—has been a challenge for production engineers for decades [1]. The low efficiency of current machining processes and expensive machine tools make precise industrial-scale machining for this category of materials expensive. Boron nitride (BN) is among these difficult-to-cut materials and, due to its extraordinary properties, has been present in various technical applications for more than 150 years. These properties vary depending on the polymorph structure of BN. This chemical compound exists in an amorphous form (a-BN) and in its basic and the most stable soft hexagonal form (h-BN), which is commonly used as a lubricant. The cubic form of boron nitride c-BN, however, is one of the hardest materials on Earth [2]. Its hardness makes it very attractive in many applications, and c-BN is used more frequently than the other forms of boron nitride. Due to high thermal and chemical stability, c-BN is widely used in the manufacturing of cutting tools for ferrous alloys machining, where diamond tools are less durable due to carbonization and chemical solubility. There is also a wurtzite form of BN, which is structurally similar to c-BN. It is said to be 18% harder than diamond, but due to its rare occurrence in nature, this has not been scientifically verified.

Even if recent developments in material science have introduced an efficient way to create c-BN, which makes the material more viable for applications such as high-power electronics, transistors, and solid-state devices, they have not resolved the problem of the inefficient machining of c-BN [3]. The most frequently used technology for manufacturing parts of c-BN is sintering [4,5]. This process introduces some limitations, i.e., limited part size and complexity (internal curvilinear channels) as well as the high cost of tooling. The novelty of this research is to apply electrodischarge machining on a micro-scale in order to offer a cost-effective and versatile approach to forming c-BN parts. It would make it more attractive for a broad variety of engineering applications where expensive and limited (in terms of depth of the holes) laser drilling methods are applied.

2. Materials and Methods

Machining of difficult-to-cut materials such a c-BN requires non-traditional machining processes. These methods are considered unconventional because they do not require direct contact of the tool with the machined material. The energy necessary to remove the machined material is delivered by means of kinetics (i.e., abrasive water jet and ultrasound abrasive machining), electromagnetic radiation (i.e., laser beam machining [6]), or electric field (i.e., electrochemical and electrodischarge machining [7]).

In order to drill with acceptable accuracy and precision, we used electrodischarge machining which is effective given the partial electroconductivity of c-BN. EDM is an electrically induced thermal process, whereby the machined material is removed from the workpiece by energy from electrical discharges occurring between the working electrode tool and the workpiece electrode. The electrodes are immersed in a dielectric medium (air, deionized water, hydrocarbon liquids, etc.) Both the workpiece and the electrode tool material are removed by melting and evaporation coming from energy generated by electrical discharges or sparks in the inter-electrode gap. The role of the dielectric is to provide optimal conditions (heat exchange and flow) for discharge and to evacuate debris from the inter-electrode gap between the voltage pulses [8,9]. Figure 1 below presents a scheme of the electrodischarge process using a tubular electrode tool. Rotation is introduced to better clean the debris (eroded particles) from the inter-electrode gap.

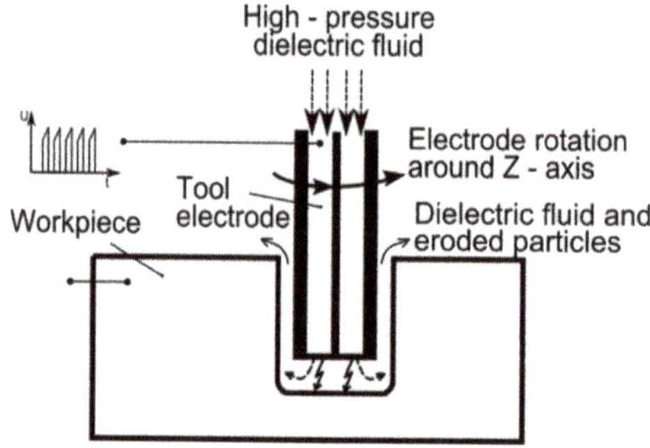

Figure 1. Scheme of EDM drilling [10].

The current research was motivated by needs voiced by manufacturers of cutting tools for aircraft parts' machining. The objective was to check the feasibility of the application of the EDM method for sinking or drilling channels in the c-BN layer of an insert for grooving, and to compare it to laser machining. The results of laser machining were not included in the current research.

For the purpose of the research we used a Sandvik Coromant CB20 grade cutting tool as a machined part to make blind holes in the c-BN layer by means of micro EDM drilling (see Figure 2).

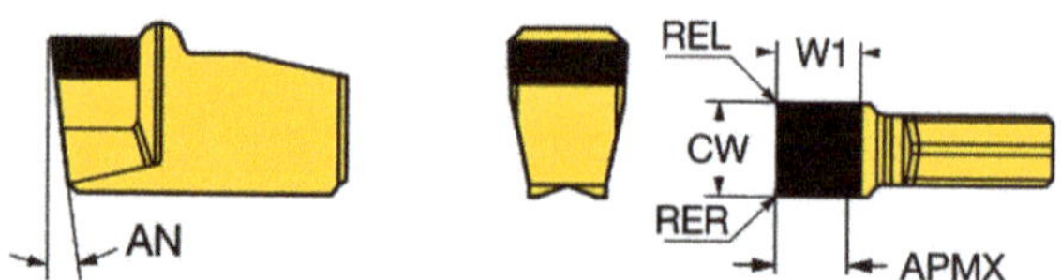

Figure 2. T-Max® Q-Cut insert for grooving (N151.2-600-50E-G CB20) [11].

We chose this cutting tool because it allowed us to deliver various cutting fluids directly to the machining zone (under the tip), which improved cutting efficiency. To this end, the top surface (black) of the insert was subjected to several electrodischarge drilling tests. The experimental part was preceded by an analysis of the authors' experience in electrodischarge machining of difficult-to-cut materials, preliminary machining tests, and confirmed with the results presented in [6]. Based on these preliminary tests, the range of the most important parameters was selected and the experiment was planned. The experiment was prepared in accordance with factorial design [12]. After preliminary tests, we have decided that the experiment plan should cover a relatively wide range of pulse-on time and symmetrical pulse-off durations (1 and 10 μs). The goal was to check the process indices for the shortest possible, pulse-on times and relatively longer ones on our pulse generator. We assumed a maximum of 10 μs pulse-on times in order to not overheat the inter-electrode gap that could result in excessive electrode tool wear and dielectric decomposition to graphite. The excessive appearance of conductive graphite corrupts the machining process. Detailed information about the design of the experiment is presented below in Tables 1–3. Table 4 below presents output process factors and measures.

Table 1. Input factors for the experiment.

Factors	Parameters	Values
Constant factors:	Voltage (V)	120
	Machining time (s)	600
	Material and diameter of electrode tool (μm)	Tungsten, $\phi = 381$
	Dielectric liquid	Exxol80 (hydrocarbon)
	Electrode tool rotation (rpm)*	250
	Pulse duty cycle D (%) $D = \frac{t_{on}}{t_{on}+t_{off}} \cdot 100\%$	50
	Threshold current I_t (A)	0.3
Disrupting factors:	Uneven dielectric liquid flow	
Variable process parameters:	Pulse-on time t_{on} (μs)	1; 10
	Pulse-off time t_{off} (μs)	1; 10
	Pulse period (μs)	2; 20
	Pulse frequency (kHz)	500; 50
	Working current I_w (A)	0.9, 1.35, 1.8

Table 2. The research plan—series one.

No.	Constant Values			Variable Parameter
	t_{on} (μs)	t_{off} (μs)	I_t (A)	I_w (A)
1	10	10	0.3	0.9
2	10	10	0.3	1.35
3	10	10	0.3	1.8

Table 3. The research plan—series two.

No.	Constant Values			Variable Parameter
	t_{on} (μ)	t_{off} (μs)	I_t (A)	I_w (A)
4	1	1	0.3	0.9
5	1	1	0.3	1.35
6	1	1	0.3	1.8

The following aspects of the process were measured and determined:

Table 4. Output process factors and measures.

Processing Factors
Hole diameter (µm)
Hole depth (µm)
Linear electrode tool wear (µm)
Average drilling speed (µm/min)
Material Removal Rate (µm^3/min)

The threshold current (I_t) is the minimum current value acceptable for feed regulator that should be maintained to support the discharges, while the working current (I_w) is the current value that is a reference for the feed regulator to be the default current during the drilling.

The holes were drilled with the micromachining machine prototype designed and built in the Institute of Production Engineering at the Cracow University of Technology in Krakow, Poland, presented below in Figure 3. The machine body was designed and manufactured of materials ensuring minimal thermal expansion and high stiffness (granite).

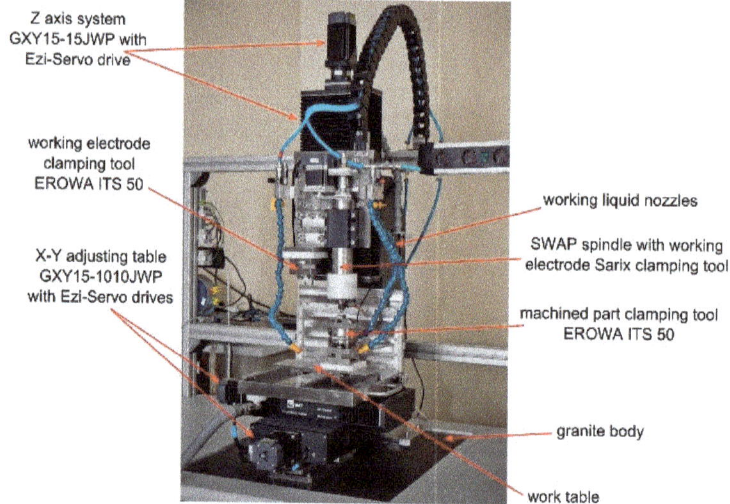

Figure 3. Electrochemical/electrodischarge hybrid micromachining machine prototype [13].

This hybrid micromachining machine prototype was designed and manufactured for micromachining involving pulse electrochemical machining and electrodischarge machining. The application of both the aforementioned methods in a sequential or synergic way enables obtaining most of the advantages of both methods. For the current study, the machine tool was used in the EDM work regime and equipped with a transistor isoenergetic voltage pulse generator and a power supply that enables setting rectangular voltage pulses at a range from 1 to 999 µs and an amplitude of 60–120 V. The chosen cylindrical tungsten electrode tool, which is produced by Balzer Technik in Switzerland, of $\phi = 381$ was clamped on a Sarix, Switzerland clamping tool. The sample was fixed with an EROWA ITS 50, Switzerland clamping tool. As the working electrode tool wears during machining due to electrical discharges, we used a high melting point T = 3410 °C tungsten electrode tool [14]. The phase diagram for c-BN, presented in Figure 4, shows that the temperature required to melt or evaporate the machined material is relatively high (more than 3000 °C). Application of the standard low melting point copper electrode could result in excessive electrode tool wear. Moreover, the tungsten electrode tool's Young modulus is higher and the electrode tool is less prone to plastic deformation during fixing in the clamping tool and homing. Unfortunately, no tungsten tubular electrode tool of this diameter is commercially available. Preliminary machining tests revealed also excessive tungsten electrode tool wear. In the current research positive (higher) electrical potential was applied to the machined part. The chosen polarity of the electrodes ensures maximal material removal rate and minimal electrode tool wear.

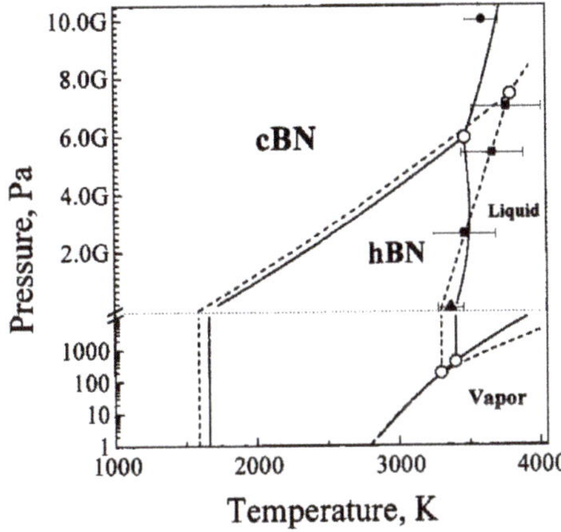

Figure 4. Phase p, T-diagram of boron nitride [15].

Initially, the electrode tool was not rotated during the sinking process (first two holes). Then the electrode tool was rotated for drilling in order to improve the removal of resolidified electroerosion products from the inter-electrode gap. The gap was flushed with fresh dielectric from the side. A scheme of the test stand is presented below in Figure 5.

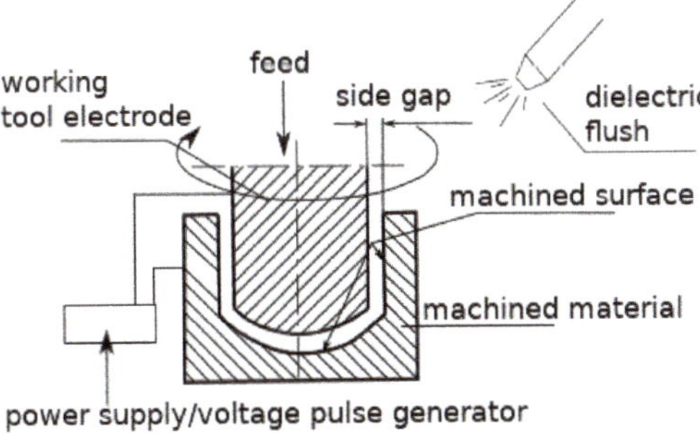

Figure 5. Scheme of the electrodischarge drilling process.

3. Results and Discussion

The goal of the work was to describe the possibility of application of the method for machining of cubic boron nitride. The research was designed and prepared to show the potential of the method and describe technological aspects. The measurements were taken with the use of an optical microscope Motic series K equipped with Instant Digital Microscopy camera Moticam 2300 (1/2" Live 3.0 Megapixels, Hongkong, China), and Motic Images Plus software (Hongkong, China). The scanning electron microscopy images were prepared by JEOL JSM-5500 Scanning Electron Microscope (Tokyo, Japan).

Two magnification levels (200× and 1000×) are displayed in order to give a comprehensive view of the machined holes' shape and edge quality. The noticeable shape inaccuracy (Figure 6a) was caused by carbon (graphite) deposited on the electrode tool surface. Relatively long voltage pulse (t_{on} = 10 µs), high working current value (I_w = 0.9 A), and insufficient dielectric flush in the inter-electrode gap caused hydrocarbon dielectric thermal decomposition and graphite deposition on the surface of the electrode tool. The electrically conductive graphite took over the role of the tungsten electrode tool at the deposited area. Nevertheless, the shape of the electrode tool was acceptably copied, which can be observed in the round shape of the hole in Figure 6. It would be difficult to measure the edge surface roughness Ra precisely, but it was estimated by digital image analysis to be lower than Ra < 5 µm. The Ra was evaluated based on known magnification of SEM (Scanning Electron Microscope) image and proportion of used marker. For example, if the 10 µm marker has 50 pixels, then counting the size of spatial amplitudes (of the valleys and peaks on the edge of the hole) in pixels gives rough information about the physical size and enables to evaluate Ra upon a mathematical formula. Digital image analysis can give approximated values of 2D roughness. The same phenomenon related to inter-electrode gap overheating and dielectric thermal decomposition was observed in the second sinking approach. This inaccuracy was eliminated in successive drilling tests by more intense dielectric flushing [16] and electrode tool rotation. The hole remained round and the edge sharp. The estimated surface roughness Ra was less than 10 µm.

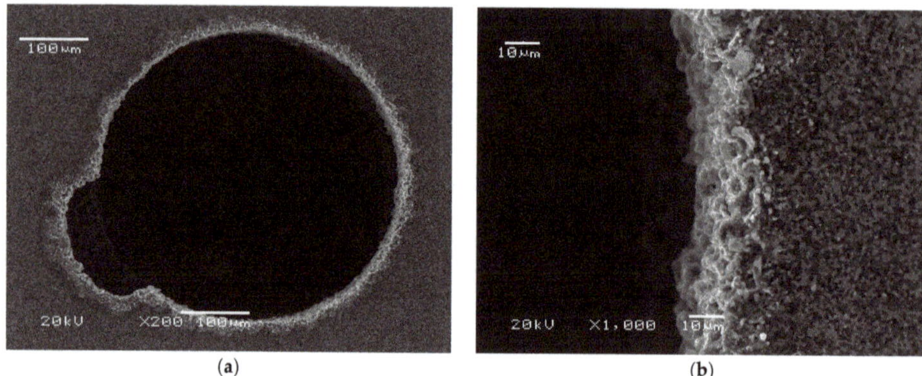

Figure 6. SEM images of hole no.1. $U = 120V$, $t_{on} = 10$ µs, $t_{off} = 10$ µs, threshold current $I_t = 0.3$ A, working current $I_w = 0.9$ A. (**a**) magnification 200×, (**b**) magnification 1000×.

Intensified dielectric flushing improved evacuation of the debris from the inter-electrode side gap (see Figure 7a). The graphite from high-temperature dielectric decomposition was not deposited on the electrode tool, and there was no deformation. The shape was properly reproduced from the cylindrical electrode tool.

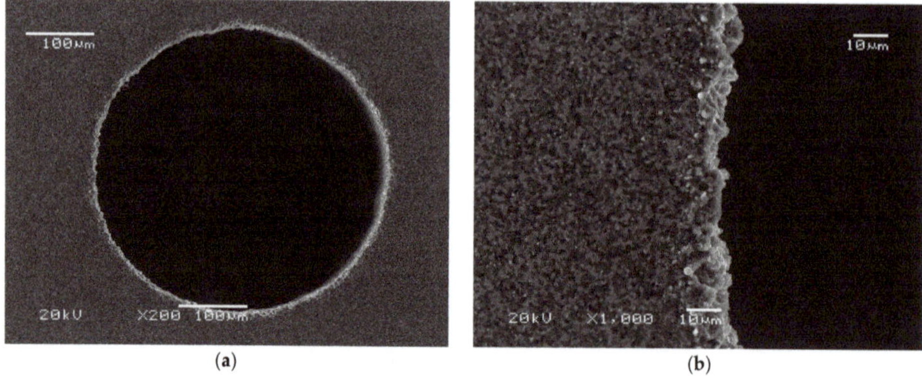

Figure 7. SEM images of hole no.4. $U = 120$ V, $t_{on} = 1$ µs, $t_{off} = 1$ µs, threshold current $I_t = 0.3$ A, working current $I_w = 0.9$ A. (**a**) magnification 200×, (**b**) magnification 1000×.

Figure 7 shows that, from the qualitative point of view, the shape of the hole is round. The SEM pictures of all obtained holes confirm this tendency. It would also be interesting to know the chemical composition of the hole edges. Unfortunately, energy-dispersive X-ray spectroscopy could not determine the chemical composition because the atomic mass of the examined compound was too small. Nevertheless, the SEM images (Figures 6 and 7) showed no visual changes of the material surface on the edge. This is consistent with limited heat-induced phase change in this area.

The depth of the holes was calculated with a stepper motor encoder. Reading values were reduced by the linear electrode tool wear and inter-electrode gap thickness.

The measured and calculated quantities are presented in Figure 8 below and summarized in Table 5.

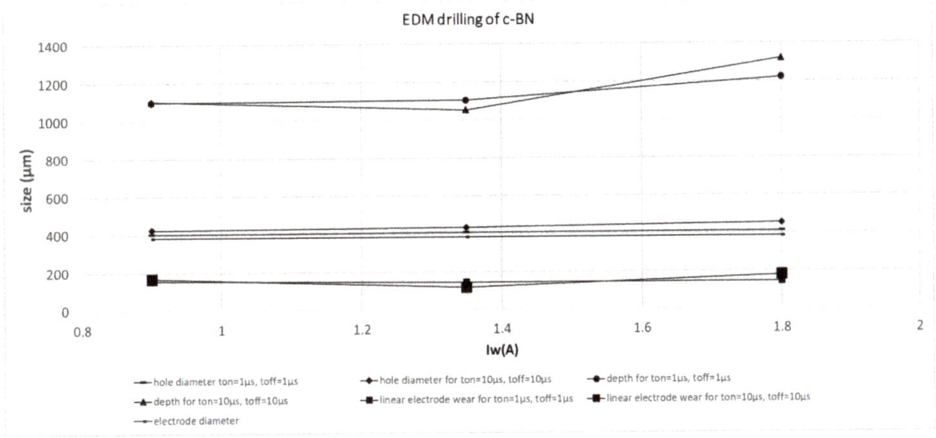

Figure 8. Results of electrodischarge drilling with the use of a tungsten electrode tool in a c-BN sample for various t_{on} and t_{off} values.

Figure 8 shows that the pulse-on and off times have a significant impact on hole depth for higher working current values while the impact of the increase in hole diameter could be neglected. The electrical current intensity is proportional to the number of charges that are transferred in a given time. As electrical potential, in this case, is constant and the power $P = U \times I$, the bigger the I value, the bigger the P value. P stands for the energy (work) necessary for material melting and evaporation over time of machining. As the machining time, in this case, is constant, the higher the P value, the deeper the hole. The energy in this case is consumed for drilling (in the z-axis direction) due to the feed regulator that compensates inter-electrode gap thickness and thus enables discharges. The side gap thickness is constant due to continuous working electrode feed and its circumferential wear. When the working current I_w is higher, higher linear electrode wear and hole diameters are also observed. It leads to a higher side gap size that can facilitate the removal of debris from the inter-electrode volume during the pulse-off time. This can improve electrical erosion conditions. Figure 9 shows the relation between average drilling speed and changes of working current I_w for various pulse-on and pulse-off times. Average drilling speed relates to the depth of the hole over the machining time. One can observe that the higher working current has an impact on average drilling speed. Therefore, it improves the material removal rate. The relation between working current I_w and MRR is presented in Figure 10.

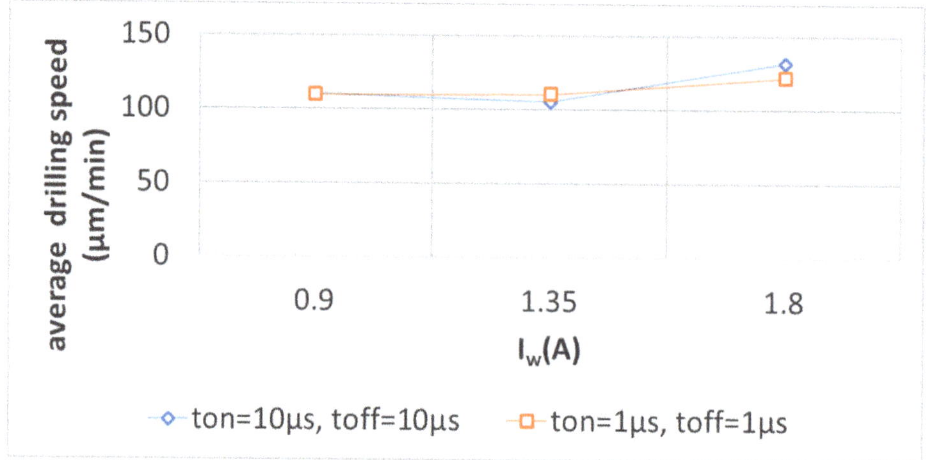

Figure 9. Average electrodischarge drilling speed for various t_{on} and t_{off} and working current I_w.

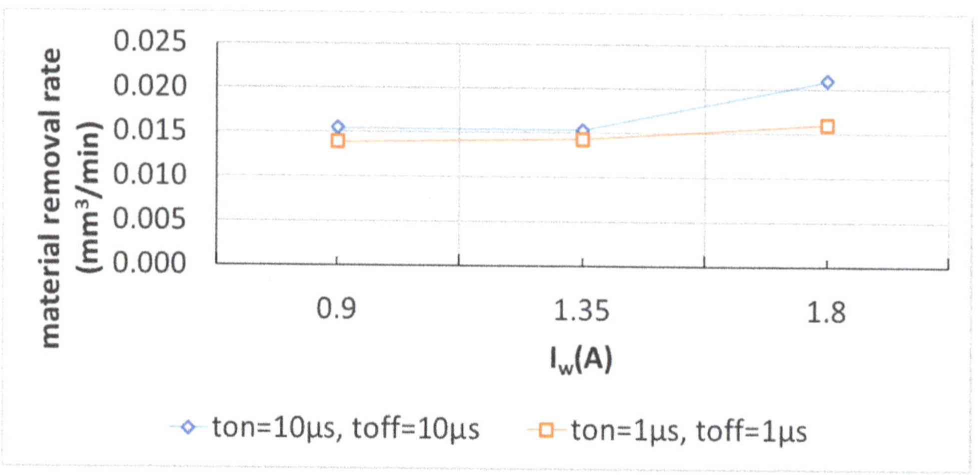

Figure 10. The material removal rate for various t_{on} and t_{off} and working current I_w.

Designer and metrology communities face a huge challenge regarding the measurement of micro-scale features [17]. Out-of-roundness depends on the size and is defined in so-called International Tolerances. If we take these norms into consideration, according to standard tolerance grades, our results fit IT8-IT9. This result needs to be explained. These values of tolerances refer to features ranging from 0 to 3150 mm in size. The dimensions of the holes prepared in the research are close to zero (micrometer scale). It is worth considering if the dimensions of holes presented in the current research are not too close to the tolerance value. For that reason we propose the formula for determining relative out-of-roundness in Equation (1) below in order to quantitatively address the roundness of the holes:

$$\text{RoR} = \frac{\text{Maximum inner diameter} - \text{Minimum inner diameter}}{\text{Nominal diameter}} \times 100\% \tag{1}$$

where:

maximum inner diameter = mean hole diameter + 1SD,
minimum hole diameter = mean hole diameter - 1SD,
nominal diameter = mean hole diameter.

Table 5. Mean and corrected standard deviations (SD) of the measured and calculated values (Bessel's correction).

Hole Number	Hole Diameter (µm) SD = 10	Hole Depth (µm) SD = 74	Linear Electrode Wear (µm) SD = 3	Average Drilling Speed (µm/min) SD = 7	Relative Electrode Wear (%) SD = 0.3	Side Gap (µm) SD = 10	MRR (mm³/min) SD = 0.002	Edge Ra (µm)	Relative Roundness
1	423	1102	167	110	15	43	0.016	< 5	4.7%
2	431	1053	118	105	11	50	0.015	< 10	4.6%
3	450	1319	178	132	13	69	0.021	< 10	4.4%
4	402	1097	157	110	14	21	0.014	< 5	4.9%
5	406	1105	145	111	13	25	0.014	< 5	4.9%
6	409	1220	144	122	12	27	0.016	< 5	4.8%

Relative electrode tool wear refers to the linear electrode tool wear in regard to the depths of the holes.

As the prepared holes were of small diameter (ca. 400 µm), of a relatively high depth (ca. 1000–1300 µm), and were blind holes, it is difficult to present the sidewall profile. It could be assumed that the surface roughness value is close to the one determined on the edge of the hole. The linear electrode tool wear was dependent on the depth of the hole and voltage pulse energy. The linear wear was measured and presented in Table 5 above (less than 200 µm). The circumference electrode tool wear is definitely low and dependent on the time that the electrode tool spent in the material during machining. In the area close to the face of the electrode tool the circumferential wear is usually higher than in the area close to the hole edge due to machining time. It has some impact on the hole diameter decreasing along the depth profile. The determined side gap value (the lowest for the less energetic pulses ca. 20–30 µm) allows assuming that the circumferential wear value is not high and does not introduce significant taper of the holes.

The best set of parameters for drilling are those that balance higher material removal rate, lower electrode tool wear, and smaller side gaps, and is the one that ensures good shape and accuracy and cost-effectiveness. Of those presented above in Tables 1–3, the best conditions for micro EDM drilling were achieved when the working current was the highest. Relatively high working current and short pulses improved machining quality. The resulting increase in hole diameter was negligible, while the material removal rate was significantly higher and the electrode tool wear was acceptable.

4. Conclusions

The tests prepared in the study proved that the micro EDM drilling process is stable, reliable and efficient. The obtained results allow formulating the conclusion that electrodischarge drilling offers c-BN tools' manufacturers a cheap, accurate, and precise alternative to other machining methods. Micro EDM drilling enables the manufacturing of channels for delivering various cutting fluids directly to the machining zone (under the tip) to improve cutting efficiency and extend the tip life of cutting tools [17]. Designer and metrology communities face a huge challenge regarding the measurement of micro-scale features [18]. The analysis of the surface geometrical structure features of the holes obtained in current research revealed that there is still a need to bridge the gap in measurement standards concerning the so-called mezzo scale (transient between millimeter and micrometer scale). Nevertheless, it could be concluded that micro electrodischarge drilling of c-BN can also be applied in many manufacturing applications i.e., semiconductor devices for harsh environments such as solar-blind UV sensors in space [19]. Due to c-BN's extraordinary thermal conductivity, it can also be applied in the manufacturing of heat sinks for semiconductor lasers and microwave devices. The resolution of micro electrodischarge drilling depends on minimal removed volumes. In order to improve the machining accuracy and precision, it is very important to apply adequate machine body design and materials limiting temperature expansion and ensuring its required stiffness. The other important factors are power supply, voltage pulse generators, as well as the properties of electrode tool materials and reliability of clamping systems. Also, the machined material homogeneity has significant importance for the obtained results. Of course, this approach is adequate for finishing operations or for machining on a micro-scale. This research was part of the Innolot project "Technologies of forming micro- and macro-geometry of the cutting tools, made of ultra-hard materials in the process of implementation of advanced laser techniques" founded by the Polish National Centre for Research and Development.

Author Contributions: Conceptualization, D.W.; methodology, D.W.; validation, D.W.; formal analysis, D.W.; investigation, D.W. and W.B.; resources, D.W.; data curation, D.W.; writing—original draft preparation, D.W.; writing—review and editing, D.W.; visualization, D.W. and K.M.; supervision, D.W.; project administration, D.W.; funding acquisition, D.W. All authors have read and agreed to the published version of the manuscript.

Funding: This research was funded by POLISH NATIONAL CENTRE FOR RESEARCH AND DEVELOPMENT, grant number POIR.01.02.00-00-0008/15-00. The APC was funded by CRACOW UNIVERSITY OF TECHNOLOGY.

Conflicts of Interest: The authors declare no conflict of interest. The funders had no role in the design of the study; in the collection, analyses, or interpretation of data; in the writing of the manuscript, or in the decision to publish the results.

References

1. Cook, M.; Bossom, P. Trends and recent developments in the material manufacture and cutting tool application of polycrystalline diamond and polycrystalline cubic boron nitride. *Int. J. Refract. Met. Hard Mater.* **2000**, *18*, 147–152. [CrossRef]

2. Dub, S.; Lytvyn, P.; Strelchuk, V.; Nikolenko, A.; Stubrov, Y.; Petrusha, I.; Taniguchi, T.; Ivakhnenko, S. Vickers Hardness of Diamond and cBN Single Crystals: AFM Approach. *Crystals* **2017**, *7*, 369. [CrossRef]

3. Narayan, J.; Bhaumik, A. Research Update: Direct conversion of h-BN into pure c-BN at ambient temperatures and pressures in air. *APL Mater.* **2016**, *4*, 20701. [CrossRef]

4. Fukunaga, O. Science and technology in the recent development of boron nitride materials. *J. Phys. Condens. Matter* **2002**, *14*, 10979–10982. [CrossRef]

5. Sumiya, H.; Uesaka, S.; Satoh, S. Mechanical properties of high purity polycrystalline cBN synthesized by direct conversion sintering method. *J. Mater. Sci.* **2000**, *35*, 1181–1186. [CrossRef]

6. Denkena, B.; Krödel, A.; Grove, T. On the pulsed laser ablation of polycrystalline cubic boron nitride—Influence of pulse duration and material properties on ablation characteristics. *J. Laser Appl.* **2019**, *31*, 022004. [CrossRef]

7. Jia, Y.H. Study on EDM Technics of Polycrystalline Cubic Boron Nitride Cutting Tool and PcBN Cutting Tool's Life. *Appl. Mech. Mater.* **2011**, *120*, 311–315. [CrossRef]

8. Schumacher, B.M. After 60 years of EDM the discharge process remains still disputed. *J. Mater. Process. Technol.* **2004**, *149*, 376–381. [CrossRef]

9. Kunieda, M.; Lauwers, B.; Rajurkar, K.; Schumacher, B. Advancing EDM through Fundamental Insight into the Process. *CIRP Ann.* **2005**, *54*, 64–87. [CrossRef]

10. Skoczypiec, S.; Machno, M.; Bizon, W. The capabilities of electrodischarge microdrilling of high aspect ratio holes in ceramic materials. *Manag. Prod. Eng. Rev.* **2016**, *6*, 10–20. [CrossRef]

11. Sandvik, C. N151.2-600-50E-G CB20. Available online: https://www.sandvik.coromant.com/en-gb/products/pages/productdetails.aspx?c=N151.2-600-50E-G%20%20%20%20CB20 (accessed on 29 September 2018).

12. NIST/SEMATECH e-Handbook of Statistical Methods. Available online: http://www.itl.nist.gov/div898/handbook/ (accessed on 4 February 2020).

13. Skoczypiec, S.; Ruszaj, A. A sequential electrochemical-electrodischarge process for micro part manufacturing. *Precis. Eng.* **2014**, *38*, 680–690. [CrossRef]

14. Rajurkar, K.P.; Sundaramb, M.M.; Malshec, A.P. Review of Electrochemical and Electrodischarge Machining. *Procedia CIRP* **2013**, *6*, 13–26. [CrossRef]

15. Solozhenko, V.L.; Turkevich, V.Z.; Holzapfel, W.B. Refined Phase Diagram of Boron Nitride. *J. Phys. Chem. B* **1999**, *103*, 2903–2905. [CrossRef]

16. Liu, Y.; Chang, H.; Zhang, W.; Ma, F.; Sha, Z.; Zhang, S. A Simulation Study of Debris Removal Process in Ultrasonic Vibration Assisted Electrical Discharge Machining (EDM) of Deep Holes †. *Micromachines* **2018**, *9*, 378. [CrossRef] [PubMed]

17. Reiter, M.; Brier, J.; Bleicher, F. Machining of Iron-Carbon Alloys by the Use of Poly-Crystalline Diamond Cutting Inserts with Internal Cooling. *J. Manuf. Mater. Process.* **2018**, *2*, 57. [CrossRef]

18. Soltani, A.; Talbi, A.; Mortet, V.; Benmoussa, A.; Zhang, W.J.; Gerbedoen, J.-C.; De Jaeger, J.-C.; Gokarna, A.; Haenen, K.; Wagner, P.; et al. Diamond and Cubic Boron Nitride: Properties, Growth and Applications. *AIP Conf. Proc.* **2010**, *1292*, 191.

19. Uhlmann, E.; Mullany, B.; Biermann, D.; Rajurkar, K.P.; Hausotte, T.; Brinksmeier, E. Process chains for high –precision components wit micro-scale features. *Manuf. Technol.* **2016**, *65*, 549–572. [CrossRef]

 *micromachines*

Review

Recent Advances and Perceptive Insights into Powder-Mixed Dielectric Fluid of EDM

Asarudheen Abdudeen, **Jaber E. Abu Qudeiri ***, **Ansar Kareem, Thanveer Ahammed and Aiman Ziout**

Mechanical Engineering Department, College of Engineering, United Arab Emirates University, Al Ain 15551, UAE; 201990133@uaeu.ac.ae (A.A.); 201990209@uaeu.ac.ae (A.K.); 201870099@uaeu.ac.ae (T.A.); ziout@uaeu.ac.ae (A.Z.)
* Correspondence: jqudeiri@uaeu.ac.ae

Received: 28 June 2020; Accepted: 29 July 2020; Published: 31 July 2020

Abstract: Electrical discharge machining (EDM) is an advanced machining method which removes metal by a series of recurring electrical discharges between an electrode and a conductive workpiece, submerged in a dielectric fluid. Even though EDM techniques are widely used to cut hard materials, low efficiency and high tool wear remain remarkable challenges in this process. Various studies, such as mixing different powders to dielectric fluids, are progressing to improve their efficiency. This paper reviews advances in the powder-mixed EDM process. Furthermore, studies about various powders used for the process and its comparison are carried out. This review looks at the objectives of achieving a more efficient metal removal rate, reduction in tool wear, and improved surface quality of the powder-mixed EDM process. Moreover, this paper helps researchers select suitable powders which are exhibiting better results and identifying different aspects of powder-mixed dielectric fluid of EDM.

Keywords: EDM; SR; TWR; PMEDM; MRR

1. Introduction

Among all the non-conventional machining processes, electrical discharge machining (EDM) is one of the major and popular machining processes. Electrical discharge machining is also known as spark machining, spark eroding, die sinking, and wire burning or wire erosion. This technique is usually used to machine hard materials and high-temperature resistant alloys which are electrically conductive. The principle of the EDM technique is the use of thermoelectric energy to erode conductive components through rapidly occurring sparks between the uncontacted electrode and the workpiece [1]. It is an electro-thermal non-traditional machining process, where electrical energy is used to generate electrical spark, and the material is removed from the workpiece by a series of rapidly requiring current discharges between two electrodes separated by a dielectric liquid. It can be used to machine difficult geometries in small batches or even on a job-shop basis [2]. A large body of research discusses EDM processes with different objectives [3–11].

In the case of development of super-tough electrically conductive materials like carbides, stainless steels, hastalloys, nitralloys, waspalloy, nomonics, etc., the demand for non-traditional manufacturing processes has become more relevant. These super tough materials, which have extensive applications, such as manufacturing of dies, automobile, and aerospace components, are very difficult to machine by conventional methods. To machine all the electrically conductive materials irrespective of their hardness and toughness, EDM processes extensively use thermal energy [2].

In non-conventional machining processes, EDM has tremendous potential on account of its versatility because of its applications in modern industries. The EDM process can also produce holes, external shapes, profiles or cavities in an electrically conductive work piece by means of controlled application of high-frequency electrical discharges by vaporizing or melting the work piece material in a particular area. The electrical discharges are the results of controlled pulses of direct current and occur between the tool electrode (cathode) and the work piece (anode) [12]. Generally, one of the largest uses of the EDM process is in tool-, die-, and mold-making [13]. In the powder-mixed EDM (PMEDM) process, electrically conductive powder is used in dielectric medium which reduces the insulating strength of dielectric fluid, and it increases the spark gap between tool and work piece which will result in an improvement of material removal rate (MRR) and surface finish [14,15]. The mostly used powders are aluminum, chromium, graphite, silicon, copper, and silicon carbide. Pulse on time, duty cycle, peak current, and concentration of the powder added to the dielectric fluid of EDM are the main variables to study the process of the performance in terms of MRR and surface roughness (SR) [16].

Material removal rate is defined as the volume of material removed over a unit period [17,18]. It is commonly expressed by the unit (mm^3/min). A high value of the discharge voltage, peak current, pulse duration, duty cycle, and the low values of pulse interval will result in a higher MRR. In addition to these abovementioned electrical parameters, non-electrical parameters and material properties also have their own significant influence on the MRR.

Tool wear ratio (TWR) can be defined as the ration of volume of materials removed from the tool electrode to that of work piece. It can be also called as electrode wear ratio (EWR). Tool wear ratio depends on electrode polarity and the electrode materials properties. Surface roughness is a very important parameter to consider in die-sinking EDM. In most die-sinking operations, separate finishing and roughing operations are carried out to complete the final product. It is represented by the average SR and measured in microns [19].

The aim of this paper was to provide comprehensive information and details of the PMEDM process. It is pertinent that there is great importance given to the review articles to bring out intricacies, parameter–response co-relations, and critical analysis of its response for better utilization of this process. This paper gives state-of-the-art current research studies conducted with PMEDM variants for machining different materials.

The paper is organized to begin with a brief introduction of PMEDM, its variants and introduces its working principle. It further discusses the optimization of process parameters and the types of EDM studies for different powder-mixed dielectrics in various EDM processes. Moreover, performance measures in PMEDM and materials are also discussed. Furthermore, effect of powders and their concentration, magnetic field-assisted PMEDM, modeling and simulation of PMEDM, combined and hybrid process with powder-mixed dielectric fluid are scrutinized in the paper.

In addition, other objective functions, such as analysis of performance of plain water, performance of water mixed with organic compounds, performance of commercial water-based dielectrics and their surface effects are also evaluated. The main text of this article provides a meticulous review on major areas of PMEDM research on different powder materials. It may help researchers find suitable dielectric fluid having better properties according to their needs. The last section of the paper draws conclusions and trends of the reviewed bodies of research are subsequently drawn.

2. Different Types of EDM Techniques

There are many types of EDM. This research categorized EDM processes under three main categories, namely, Sinker EDM, Wire EDM, and fast hole drilling EDM. Spark machining's working principle is based on the erosion of the material by frequent sparks between the workpiece and the device or tool that is immersed in a dielectric medium. A gap separates the workpiece and the electrode to establish a pulsed spark through which the dielectric fluid flows. Schematic representation of the basic working principle of EDM process is shown in Figure 1.

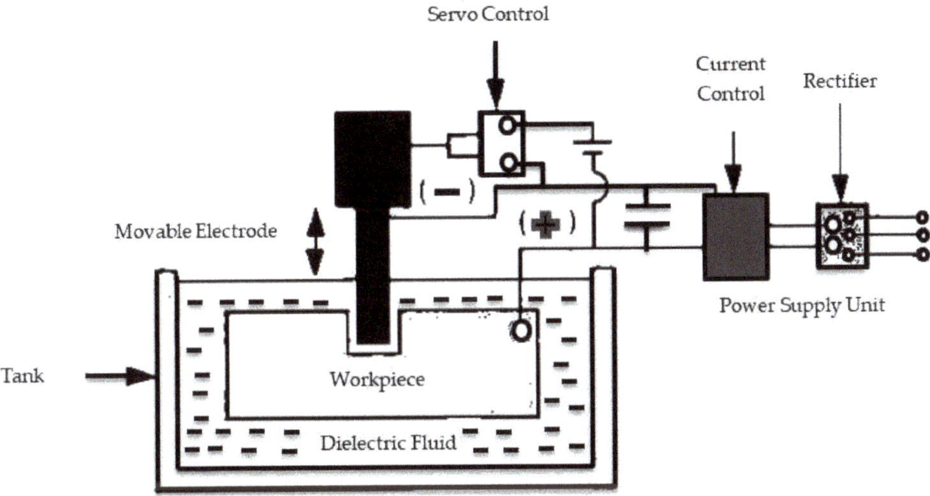

Figure 1. Schematic representation of the basic working principle of electrical discharge machining (EDM) process.

2.1. Wire EDM

One of the most emerging non-conventional manufacturing processes is the wire electrical discharge machining [20,21]. Widely this process is used to machine hard materials and intricate shapes which are not possible with conventional machining methods. It is more efficient and economical. In wire EDM, an electric spark is created between an electrode and the work piece [22,23]. The spark is the visible evidence of the flow of electricity. This electric spark creates intense heat of temperature ranges from 8000 to 12,000 degree Celsius, which melts almost anything. The spark is carefully controlled and localized so that it only affects the surface of the materials. The EDM process does not affect the heat treat below the surface of the work piece. In wire EDM, the spark always takes place in the dielectric of de-ionized water. The conductivity of the water is carefully controlled for making an excellent environment for the EDM process. Here, the water acts as a coolant and flushes away the eroded metal particles [24]. The working principle of wire EDM is shown in Figure 2.

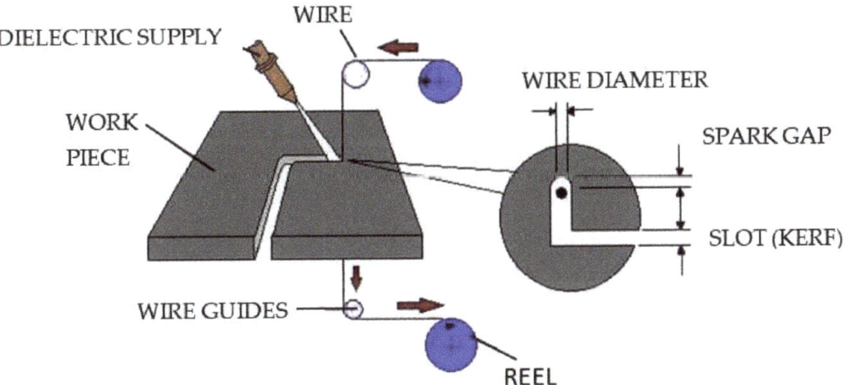

Figure 2. Working principle of wire EDM.

Figure 2. Working principle of wire EDM.

The wire EDM process has a wide range of applications such as in die making, electronics and automotive industries [25,26].

As per the literature, the main parameters are:

- Pulse-on time (Ton)
- Pulse-off time (Toff)
- Servo voltage (V)
- Peak current (I)
- Gap voltage (Vgap)
- Dielectric flow rate
- Wire feed rate
- Wire tension

2.2. Sinker EDM

It is also called cavity-type EDM or volume EDM. It consists of an electrode and work piece submerged in an insulating liquid, such as, more typically, oil, or, less frequently, other dielectric fluids. The electrode and work piece are connected to a suitable power supply [27]. The power supply generates an electrical voltage between the two parts. As the electrode approaches the work piece, the dielectric breakdown occurs in the fluid, which forms a plasma channel and small spark jumps. These jumping sparks usually strike on one at a time. The sinker EDM process uses an electrically charged electrode that is configured to a specific geometry to burn the electrode's geometry into a metal component. This process is commonly used in dies and tool manufacturing [28]. A schematic diagram of a sinker EDM is shown in Figure 3.

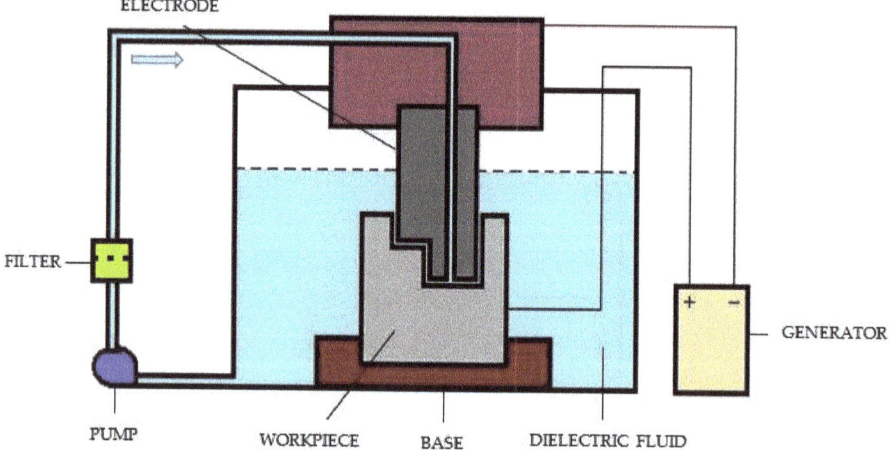

Figure 3. Schematic diagram of sinker EDM.

Here, the main components are power supply, dielectric system, electrode, and the servo system. So, this is the schematic which explains the principle of sparking. The work piece is usually connected to the positive terminal of the power supply. Here, the shape of the tool is different, and the same shape will be replicated as well on the work piece. The sparking takes place through the different zones or the through different points which are closer to work piece. Due to the fact of this sparking, the plasma formation zone will create the bubbles and high pressure which subsequently collapse and erode the work piece. Whenever the sparking takes place, erosion of the work piece will also occur. Thus, we have seen in the case of EDM, small erosion will take place on the electrode as well. Sinker EDM is one of the advanced methods for machining electrically conductive materials [29–33].

2.3. Fast Hole Drilling EDM

Figure 4 shows fast hole drilling EDM. This EDM was designed for fast, accurate, and deep hole manufacturing. By concept, it is similar to sinker EDM process but the electrode is a rotating tube conveying pressurized jet of dielectric fluid. It can make deep holes in about a minute, and it is a good way to machine holes in materials which is too hard for twist drilling machining. This EDM drilling process is mostly used in aerospace industries for producing cooling holes in aero blades and other components which requires cooling. It is also used in industrial gas turbine blades, in molds and dies, and in bearings.

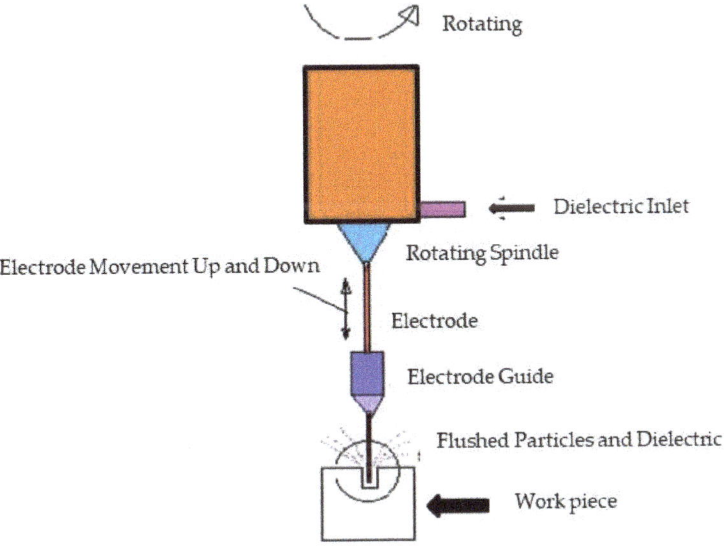

Figure 4. Fast hole drilling EDM.

The abovementioned EDM methods was about machining using dielectric fluid. Various powders can be mixed to improve their efficiency.

3. Powder-Mixed EDM

To improve the ionization and ease in the frequency of spark between the cathode (tool) and work piece, fine-grained conductive powder is added into dielectric fluid in the EDM process [34,35]. In the PMEDM process, the suitable material in the powder form is mixed into the dielectric fluid tank [36] for the better circulation of the dielectric fluid by the mixing system which includes aluminum, graphite, copper, chromium, silicon carbide, etc. The presence of this powder makes the process mechanism substantially different from the conventional EDM process [37,38]. Here a spark gap was provided by the additive particle which ranged in between 25–50 micrometers. The voltage applied was in between 80 and 320V, and the electrical field ranged in between 105–105 V/m [39]. A large body of research discussed the PMEDM processes with different objectives [40–50].

Principle of Powder-Mixed EDM

The influence of the powder in the PMEDM mainly depends on the powder parameters, that is, powder material, particle size, and particle concentration. There are many powder materials mixed with dielectric fluid including aluminum (20%), chromium (0.9%), graphite (5%), silicon (0.03%) or silicon carbide (0.3%). The previous studies used partial sizes of 10–25 µm, also the particle concentrations are about 6 g/L. These fine powders mixed to the dielectric fluid in the EDM, resulted in the reduced insulating strength and the generation of more spark [24]. Thus, table MRR and surface quality can be expected. The powder particles are energized and moved in crisscross directions. Under the action of applied electric force, these particles arrange in a suitable form at different locations in the sparking area [51]. Tool steel, alloy steel, and especially nickel-based super alloy Inconel-800 has been commonly used as workpieces by various researchers [45,52,53]. Other commonly used workpieces in PMEDM are H-11 Die Steel [54], AISI D2 Die Steel [41], Hastelloy [55], H 13 steel [56], W300 Die Steel [57], EN 31 Steel [58], AISI H-11 [59], Stavax [60], etc. The schematic representation of the principle of PMEDM is shown in Figure 5.

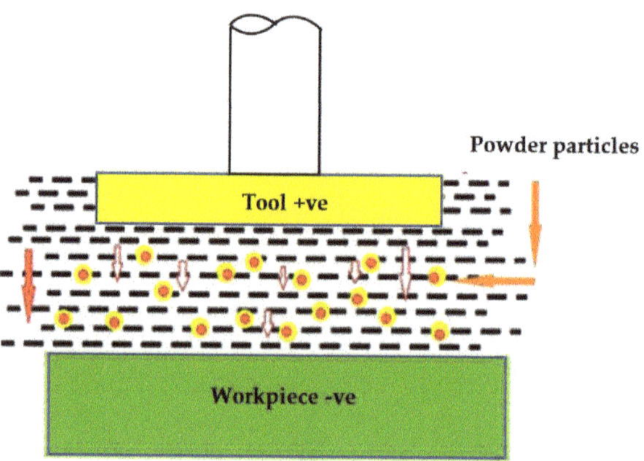

Figure 5. Schematic representation of the principle of powder-mixed EDM.

One of the main issues faced while researching PMEDM is that the literatures available on this topic is limited. However, it can be easily found from the available literature that considerable efforts have been directed to improve the quality of the surface by suspending the powder particles in the dielectric fluid used in the EDM. There were few works reported for the improvement of machining efficiency of powder-mixed EDM.

By the addition of powder particles to the electric discharge machining process, the dielectric fluid will modify some process variables, and it creates the condition to achieve a higher surface quality. Here, the improvement in the polishing performance of conventional EDM and the analysis were carried out by varying the silicon powder concentration and flushing flow rate over a set of different process areas. It was found that, effects in the final surface after machining was evaluated by surface morphologic analysis and measured through some surface quality indicators [61]. Jeswani (1981) [62] conducted experiments with the addition of 4 gm/L of fine graphite powder in the dielectric and found that MRR improved by 60% and the electrode wear ratio reduced by about 15%. And the further investigations show high efficiency application of water mixture and glycerin for reducing disadvantages in EDM sinking using pure water.

Purna Chandra Sekhar et al. [40] explained elaborately about their research work. That saying, the material removal mechanism of the EDM process was very complex, and also the theoretical modeling of this process was very difficult. Powder-mixed dielectric is a promising research area. Only a few studies have touched on the introduction of using nano powders into EDM process. The author reported that most of the research work carried out with aluminum, silicon, and graphite powders [63–66] and some with other types of powders. They are Cr, Ni, Mo, Ti, etc. [67–69]. The Impact of such machining on MRR, SR, and TWR was studied by the most available research works on the powder-mixed dielectric fluids [40]. Generally, the tool electrodes used in PMEDM can be copper, tungsten, brass, aluminum, and graphite [46,60,70–73].

When cryotreated electrode was used, MRR, TWR, and SR decreased by 12%, 24%, and 13.3%, respectively, and when SiC powder was used, MRR increased by 23.2% and TWR and Ra decreased by about 25% and 14.2%, respectively [74]. Moreover, the machining performance can be improved using graphene nanofluid as the dielectric [75]. Studies on plasma channel expansion, nano-powder-mixed sinking and milling in micro-EDM were also carried out [76,77].

4. Various Powders Used in EDM

Powders mixed with dielectric fluid in EDM can be categorized into six different types, namely, aluminum, silicon, chromium, graphite, silicon carbide, and nickel micro powder. Each of these types has its own characteristics which makes it suitable to be used for different machining conditions. In order to meet the required conditions, these powders have many properties such as MRR, SR, TWR, etc. General composition of widely and recently used powder types (materials) for EDM process considered for each type are summarized in Table 1 and their physical properties are shown in Table 2.

Table 1. Details of general composition of various powders used by researchers for different EDM processes.

Powder Type	Composition (weight %)	Parameters Studied
Silicon	0.03	material removal rate (MRR), surface roughness (SR) [70]
Aluminum	20	MRR, SR, Tool wear ratio (TWR) [78]
Silicon	0.3	MRR [41]
Chromium	9	MRR, TWR [69]
Nickel micro powder	6	MRR [79]

Table 2. Commonly used powders in powder-mixed EDM (PMEDM) and their physical properties [80].

Material	Density (g/cm^3)	Electrical Resistivity (μΩ-cm)	Thermal Conductivity (W/m-K)
Aluminum (Al)	2.70	2.89	236
Graphite C	1.26	103	3000
Chromium (Cr)	7.16	2.6	95
Copper (Cu)	8.96	1.71	401
Silicon (Si)	2.33	2325	168
Nickel (Ni)	8.91	9.5	94
Silicon Carbide (SiC)	3.22	1013	300
Titanium (Ti)	4.72	47	22
Tungsten (W)	19.25	5.3	182
Alumina (Al$_2$O$_3$)	3.98	103	25.1
Boron Carbide (B$_4$C)	2.52	5.5×105	27.9
Carbon nano tubes (CNTs)	2.0	50	4000
Molybdenum Disulfide (MoS$_2$)	5.06	106	138

The material removal rate can be increased by mixing powder with the dielectric fluid as compared with ordinary EDM process. Also, it can be increased by the increasing of the peak current. Peak current in a higher value will produce a rougher surface in the EDM process. That means different mixing powders in different input parameters have different responses and results. Review of optimization of process parameters are given in Table 3.

Table 3. Review on optimization of process parameters.

Work Material	Powder	Input Parameter	Response Variables	Optimization Technique	Results	Reference
En-31	Silicon	Peak current, Pulse on, Duty cycle, Powder concentration	MRR, SR	Response surface methodology (RSM)	Powder concentration and peak current were the most influential parameters	[70]
Inconel 718	Aluminum	Voltage, Discharge current, Duty cycle, Powder concentration	MRR, SE, WR	One variable at a time	Size and particle concentration significantly affect machining efficiency	[81]
AISI-D2 Die steel	Silicon	Peak current, Pulse on time, Pulse off time, Powder concentration, Grain, nozzle flushing	MR	Taguchi method	The peak current and concentration of silicon powder mostly influences the machining rate	[41]
EN-8	Chromium	Current, Tool angle, Powder concentration, Duty cycle	MRR, TWR	RSM	The most significant parameters affecting MRR are powder concentration and current, whereas, current and electrode angle greatly influences TWR	[69]
EN-19	Nickel micro powder	Peak current, Duty cycle, Electrode angle, Powder concentration	MRR, TWR	RSM	ANOVA results revealed that the current was the most dominant factor affecting both MRR and TWR increases with increase in current and powder concentration	[79]
AISI 1045 Steel	Aluminum	Current, Voltage, Pulse on time, Duty factor constant	MRR, SR	Taguchi method	As the concentration of aluminum powder and grain size in EDM oil increases, surface roughness starts decreasing. MRR and surface roughness are equally important. With the increase in concentration of aluminum powder and grain size MRR and surface finish of AISI 1045 Steel increases	[63]
W300 Die Steel	Aluminum	Peak current, Pulse on time, Powder concentration, and polarity	MRR, EWR, SR, WLT	Signal-to-noise (S/N) ratio and the analysis of variance (ANOVA)	Polarity plays an important role in PMEDM. High MRR is obtained in positive polarity, whereas better surface quality (surface roughness and white layer thickness) is achieved in negative polarity. Distilled water can be used as dielectric fluid instead of hydrocarbon oil and, moreover, the performance can be improved by the addition of aluminum powder	[57]

Table 3. *Cont.*

Work Material	Powder	Input Parameter	Response Variables	Optimization Technique	Results	Reference
EN 31 Steel	Silicon	Pulse on time, Duty cycle and Peak current, Powder material, Powder size, Powder concentration, Dielectric type, Peak voltage, Pulse off time, Polarity, Inter electrode gap (IEG)	MRR. TWR, WR, SR	RSM	MRR and SR roughness have been measured for each setting. The use of powder-mixed dielectric promotes the reduction of surface roughness and enhances material removal rate	[80]
SKD-11	Aluminum chromium copper and silicon carbide powders concentration	Pulse on time, Peak current	MRR, TWR, SR	RSM	The discharge gap distance and material removal rate increased as powder granularity was increased. Aluminum produced the largest discharge gap enlargement and silicon carbide produced the smallest.	[82]
AISI D2 Die steel	Chromium	Peak current, Pulse on time, Pulse off time, Powder concentration	MRR, TWR, SR	Taguchi, Anova	With the increase in current and pulse-on time, MRR increases. Due to the increased concentration of chromium powder, MRR tends to decrease. TWR is mainly affected by current. With the increase in current, TWR increases. Also, TWR tends to decrease with the increase in chromium powder concentration. surface roughness is higher with the increase in pulse-off time	[83]
AISI D3 Die Steel	Aluminum Powder	Peak current, Pulse on time	MRR, TWR, SR	Central composite design (CCD) of response surface methodology (RSM)	Maximum MRR is obtained at a high peak current of 14 Amp, higher Ton of 150 µs, and high concentration of Al powder 6 g/L. Low TWR and SR are made with low peak current of 2 Amp, lower ton of 50 µs and higher concentration of Al powder of 6 g/L.	[84]

5. Relevant Studies on PMEDM

In this section, the studies dealing with the PMEDM process are reviewed based on the following classification scheme:

Type of EDM process: The relevant studies classified according to the EDM processes used powders mixed with the dielectric fluid are as follows:

- D-S—Die sinking EDM;
- WEDM—Wire EDM;
- FHD—Fast hole drilling.

From the authors' perspective, reviewing the literature related to the PMEDM technique shows that most of the studies discussed the powder-mixed for D-S EDM technologies such as Reference [85] While few studies discussed the powder mixed for wire cut such as Reference [16], and for fast hole drilling there are very few researchers studying the effect of PMEDM and its process parameters [86]. But there are relevant researches available related with PMEDM using other EDM processes or by modifying the above three processes [87].

Objective functions:

The following objective functions are reported in the current relevant literature:

Objective 1: Performance measures in PMEDM;

Objective 2: Powder materials;

Objective 3: Effect of powders and their concentration;

Objective 4: Magnetic field-assisted PMEDM;

Objective 5: Other objective functions.

Using this classification scheme, Table 4 chronologically lists the studies for powder-mixed dielectric fluid in different EDM process.

Table 4. List of the studies for different powder-mixed dielectric in various EDM processes.

No.	Author, Year	EDM Process				Objectives				
		D-S EDM	W EDM	FHD EDM	Other EDM Processes	1	2	3	4	5
1	[2]				x	√	√			
2	[55]				x	√	√	√		
3	[16]		x			√	√			
4	[88]		x							√
5	[51]	x	x			√				
6	[61]				x	√	√	√		√
7	[62]	x				√	√	√		
8	[40]					√	√	√		√
9	[70]	x				√	√			
10	[78]				x	√	√	√		√
11	[41]	x					√	√		
12	[69]				x	√	√	√		
13	[79]				x	√	√	√		
14	[19]				x					√
15	[89]				x			√		
16	[86]			x						√
17	[90]				x		√	√		
18	[91]	x				√	√	√		
19	[87]				x					√

Table 4. *Cont.*

No.	Author, Year	EDM Process				Objectives				
		D-S EDM	W EDM	FHD EDM	Other EDM Processes	1	2	3	4	5
20	[92]				x	√	√	√		
21	[93]				x					√
22	[37]				x		√	√		√
23	[94]				x		√	√		
24	[95]				x					√
25	[96]	x								√
26	[97]				x					√
27	[98]	x								√
28	[99]	x								
29	[100]				x					√
30	[101]				x		√	√		√
31	[45]				x	√		√		√
32	[102]				x	√	√	√		
33	[12]				x		√			√
34	[68]				x					√
35	[103]				x					√
36	[47]				x					√
37	[104]				x		√			
38	[105]				x					√
39	[40]				x	√	√	√		√
40	[42]				x			√		
41	[106]				x	√				
42	[17]	x								√

5.1. Performance Measures in PMEDM

One of the important performance measures in EDM is MRR. Many research studies explored a number of ways to improve the SR and MRR in EDM. In order to obtain an optimum combination of performance measures for different work–tool interfaces, in major bulk of research studies introduced the optimization of process parameters. The reports generally agree that the SR has decreased and improved surface finish, with lower pulsed current and pulse-on time values and relatively higher pulse-off time. High-quality SR cannot be achieved when pulses of long duration are used during finishing process [107]. Several researchers tried innovative ways to get better performance parameters such as MRR improvement as well. Previous studies focused on reducing the TWR because the wear of the electrode tool affects the electrode tool profile and leads to a lower precision [108–110]. While referring the available literature references on the process, a need is felt to summarize all the results and conclusions made by different researchers [19]. There are four techniques to improve performance variables:

- By electrode design;
- By controlling process parameters;
- By EDM variations;
- By powder-mixed dielectric.

The effect of first three techniques mentioned above can be found in [1,12,75]. This paper concentrates on the effect of powder-mixed dielectric on the performance variables. Khan et al. [89] used a work piece of stainless-steel AISI 304 material and an electrode of copper material, they found that addition of 0.3 mg/L of Al_2O_3 in dielectric fluid will results in the improvement of SR and the MRR will increase. Here, the electrical discharge seems to be broken into many small discharges due to the presence of powder, which will result in the formation of many small cavities with smaller depths of the surface of the work piece. This produces smoother surface. As the result of the increasing of Ton, the heat energy is absorbed during a longer time making the surface less stressed that produces a better surface. In addition, the increase of MRR results in the increasing of pulse on time.

A Kumar et al. [111] studied the performance of EDM process for machining Inconel 825 alloy by mixing Al_2O_3 nanopowder in deionized water. The experimental investigation revealed that by setting optimal combinations of process parameters, the maximum MRR of 47 mg/min and minimum SR of 1.487 μm will be, respectively, 44 and 51 percent higher compared to conventional EDM process. S Patel et.al says that rotary tool is a recent innovation of PMEDM. Authors used aluminum oxide (Al_2O_3) powder with a particle concentration of 0.5–1.5 gm/L into the dielectric to study improved machining performance and found tremendous changes [81]. Hence, it can be noted that performance can be improved by adding various powders in dielectric fluids [112,113].

5.2. Powder Materials

In the PMEDM, the dielectric fluid mixed with the powder of different materials. Thus, the floating particles impede the burning process by creating a higher discharge probability and lowering the breakdown strength of the insulating dielectric fluid [101]. Material removal rate, SR, increase and reduces the TWR value along with the improvement of sparking efficiency and the gap distance enlarged by the conductive powders and dispersed the discharges more randomly throughout the surface. Here the micro cracks are reduced and the thickness of the recast layer becomes smaller. Thus, the corrosion resistance of the machined surface may subsequently improve [45].

From the study of surface modification technique using EDM with TiN-mixed fluid. A Muttamara and J Mesee [102] introduced a surface modification technique using EDM with TiN-mixed fluid. The study came to some points that the re-solidified layer containing TiCN can be generated on EDMed surface. The authors revealed that, the surface after EDM with TiN-mixed kerosene is rougher than that in EDM with kerosene-type fluid. The modified layer thickness grew up to 57 micrometers. On this thick layer there were few micro cracks and voids detected. Hardness of the modified surface with TiN mixed kerosene came to 980 HV which is harder than that EDMed with pure kerosene. On the machined surface, a hard layer containing TiCN can be formed. Smaller hardness value with a little higher SR value can be obtained by mixing TiN powder with dielectric kerosene [102].

For the machining of Inconel 718 (Nickel-based super alloys), copper tungsten in cylindrical shape was used as electrode tool. Here the nano alumina is used as the additive dielectric. It is used because of its high thermal conductivity. Due to the nano alumina having high thermal conductivity, more heat is distributed and dissipated to the surface of work piece to limit the size of craters produced. Mainly nano alumina has combined the effects of low electrical resistivity and low density. Low electrical resistivity creates a high spark and high thermal conductivity takes more heat away [70,114].

For the deposition in the powder-suspended EDM, the concentration of powder in the gap between the electrode and the work piece must be very high. To stabilize the machining process, the electrode moves reciprocally. Due to the reciprocating motion of thick electrode, the powder is flushed from the gap. During the downward motion of the electrode, the powder concentration is low, whereas in the conventional powder-suspended EDM, the powder hardly adheres on the work piece. To keep the powder concentration high, then powder should always provide into the gap [68]. Most researchers used concentrations of powder below 20 g/L and powders of micro-size ranging from 1 to 55 μm or powders of nano-size ranging from 20 to 150 nm [38].

Nano powder-mixed EDM is one of the recent advancements in this field. Certain studies were conducted for finding various properties of adding nano particles in the process for improving the efficiency [17,115–119]. Similarly experiments using graphene oxide flakes and scrolls are carried out for achieving related objectives [120].

5.3. Effect of Powders and Their Concentration

The concentration of powder-mixed with the dielectric fluid also discussed in many studies. For example, Sugunakar et.al [91] found that the MRR increases with the mixing of different powder particles into the dielectric medium. Due to the increasing of graphite powder from 0 to 9 g/L, the MRR further increases. Thereby, MRR decrease with the supplementary addition of graphite powder from 9 to 14 g/L, whereas insignificant increase in MRR is noticed when increasing in Al powder from 0 to 405 g/L and it decrease with increase in Al powder from 4.5 to 9 g/L. Here, we can observe that a considerable increase in MRR with increase in Al powder substance from 9 to 14 g/L. However significant increase in MRR was observed with increasing in combination of Al and graphite powders (1:1 ratio) from 0 to 4.5 g/L and then it decreases further increasing in combination of Al and graphite powders (1:1 ratio) from 4.5 to 14 g/L [91].

By improving material removal rate and optimizing various machining parameters in EDM, Vivek Kumar and Prakash Kumar made some summary points. They noted that MRR of electric discharge machining can be improved using optimization of various factors which are discharge Current, Ton, Toff and voltage. Discharge current has most significant parameters for MRR. Moderately, material removal rate effected by voltage and Ton time. Here the MRR significantly influenced by Toff [87].

The material removal rate is low at low current values as the discharge current is increased, the MRR also increases. The pulse time increases as MRR increases [92]. All factors have significant effect in varying degrees on the EDM performance. Pulse current is one of the most significant factors affecting the material removal rate, dimensional accuracy, and surface integrity of drilled hole. Comparatively the better electrode material is copper, because it gives better surface finish, high MRR, and less electrode wear than Al [121]. As per the optimization of process parameters of surfactant and graphite powder-mixed dielectric EDM through Taguchi–Grey relational analysis, it was found that the significance of each process parameter in sequence is, peak current. It is the most significant factor than surfactant concentration and graphite powder concentration based on largest delta (Δ) value which was found from the test response. The optimal combination of these process parameters was obtained [122]. According to Lamichhane et al. [123] the momentous parameters enhancing MRR (maximum 19.01 mg/minimum) were I, Ton along with the addition of hydroxyapatite nanoparticles (HAp) nano-powder in the dielectric fluid. Adding HAp also improves SR (0.340 μm) of machined surface. Hence, it shows the smallest particles produced the best surface finish while increasing the recast layer thickness. Suitable particle concentration tends to efficiency of the process and improved stability. Shinde et al. [124] used aluminum, silicon, and silicon carbide fine abrasive powders with particle concentration of 2 gm/L and size of 44 m were added into the SEO25 (spark erosion oil) dielectric liquid of electrical discharge machine. The finest concentration for maximizing the performance depends on powder characteristics.

5.4. Magnetic Field Assisted PMEDM

Magnetic field was found to be more effective in case of lower current values for desirable MRR and TWR. Here, the magnetic field resulted in an increase in the OC with strengths at different levels of current. For the SR, magnetic field resulted in the deterioration in surface finish for higher current settings, while it resulted in better surface finish at low current. It enhances the surface hardness by the support of magnetic field strength [103]. In another research, Al–metal matrix composites (MMCs) hybrid ED machining was studied in the magnetic field integrated in the traditional EDM. The experimental results witnessed a decrease in the micro hardness values and a decrease in recast layer thickness, followed by a major effect on MRR and surface finish when machining higher spark energy in the magnetic field. The experimental results brought consistency to the method and an excellent correspondence with experimental verifications [125]. The input processing parameters, magnetic field strength, pulse-on/off length, peak current, electrode variant, and workpiece were evaluated to determine their after-effects on the microhardness (MH) response and recast layer formation during Al–SiC composite machining. The experimental results indicate a 22% decline in surface microhardness and a thinner recast layer formation in the magnetic field coupled with higher spark energy [126]. Thus, a higher spark energy formed along with the magnetic field can decline the surface hardness. Bains et al. [126] studied the individual effect of machining parameters, namely, peak current, pulse on time, pulse off time, powder concentration, and magnetic field on material removal rate and tool wear rate. The effect of peak current on material removal rate and tool wear rate, followed by pulse on time, concentration of powder, and magnetic field was found to be dominant [127].

5.5. Other Objective Functions

Modeling and simulation of PMEDM process was also discussed. For the simulation process, a work piece model was generated using a numerical simulation like ANSYS software. This model was considered an AISI D2 die steel and the tool material was copper. The most common tool used is AISI D2 steel for the EDM applications, and it is best machined by copper electrode. Using finite element method, the model was tested under various conditions. For this required chemical composition, mechanical properties and the thermal properties of the material were given. For the analysis, process parameters for PMEDM like voltage, current, heat input to work piece, radius of spark, pulse duration, pulse on time, types of flushing, polarity, powder type and size, frequency constant, powder concentration, electrode lift, tool electrode diameter are considered. And the FEM model is validated by comparing it with the predicted theoretical results with experimental data.

The model developed for the study of PMEDM, can be further used to obtain the distributions of the temperature over the work piece, distributions of residual stress, flow of metal, and also to find out the cracks on the PMEDM work piece. So, the model can be used as an industrial tool to predict the evolution of temperature, stress, strain on the machined work piece [42].

A number of studies were performed using models and simulation approaches to achieve specific objective functions such as one investigation [128] which reported the development of a piezoelectric servo system to improve the gap control in dry EDM. In another investigation [129], an optimization system was proposed to generate the optimum EDM process parameters. Han et al. [130] used a simulated method developed for wire EDM. Additionally, Yadav et al. [131] developed a model to reflect the EDM process's thermal stress and Tsai et al. [132] a semi-empirical model for surface finishing of machined workpieces was developed in EDM. Furthermore, Wang et al. [133] developed a semi-empirical MRR model in the EDM process. Williams and Rajurkar [134] created an EDM wire process model which was developed to research the characteristics of the machined surfaces. Cogun and Savsar [135] introduced a statistical model for studying time-lapse variations in the EDM process.

Salah et al. [136] conducted numerical modelling of the temperature distribution caused by the EDM process. From these results the MRR and total roughness were deduced and compared with experimental observations for stainless-steel type AISI316L. Some other researchers, such as Gao et al. [137], experimented in such a way that a vibration model was set for the workpiece or the tool (wire). Srivastava and Pandey [138] concentrated on kerf analysis in micro-WEDM and lateral wire vibration and breakdown distance determined for a micro-WEDM process. They also advanced a lateral vibration model for the wire which they used to measure the maximum wire vibration amplitude. These authors also conducted numerous experiments with different machining parameters on stainless steel. Rafał Swiercz et al. [139] says PMEDM with reduced graphene oxide flakes causes more stable electrical discharges with less energy and resulted in a uniform thickness of the recast layer. Studies using graphenes are recent in this field. Different characteristics like erosion characteristics, surface characteristics, surface integrity, and machining parameters were studied using grapheme oxide [140–142].

Also, the PMEDM was combined and hybrid with other processes as presented in Reference [56]. The aim of this combination was to control the process stability and to compare the pulse shapes in different processes. There was an electronic circuit connected to a computer display which recorded the current and voltage pulses. In this hybrid machining mode, the tool electrode will vibrate by ultrasonic frequency and SiO_2 nano powder was added to dielectric fluid. Due to the application of ultrasonic vibration on the tool electrode, an ultrasonic head of 20 kHz frequency of vibration and 20 W power mounted on EDM machine were obtained. Properties of typical dielectrics used in PMEDM are given in the Table 5.

Table 5. Properties of typical dielectrics used in PMEDM.

Dielectric Name	Specific Heat (J/kg-K)	Thermal Conductivity (W/m-K)	Breakdown Strength (kV/mm)	Flash Point (°C)
Deionized water	4200	0.62	65–70	Not applicable
Kerosene	2100	0.14	24	37–65
Mineral oil	1860	0.13	10–15	160
Silicon oil	1510	0.15	10–15	300

Using more advanced dielectric fluids and powders in EDM, the MRR and other properties of the machined materials will improve, and it makes the machining process very simple [98]. Besides that, health and atmosphere effects should be considered. The use of eco-friendly dielectric fluid will help to minimize the harmful environmental effects. Also, the use of water-based dielectric in die sink EDM influence the following factors [94].

- Performance of plain water;
- Performance of water mixed with organic compounds;
- Performance of commercial water-based dielectrics;
- Surface effects.

Effect on productivity by using deionized water or tap water may result in higher levels of material removal rate in some special cases like, in the time of using of brass electrode at negative polarity. In the effect on surface integrity, the deionized water usually has oxides on the machined surface and lower values of SR, while hydrocarbon oil has a contaminated appearance with carbon atoms inside the craters [95]. Most of the sinking processes use hydro-carbon oil like kerosene-based oil as the working fluids. Kerosene-based working fluids are used considering other factors such as health, safety, and environment. However, kerosene is inflammable and therefore undesirable, as the possibility of fire hazard has always been of great concern in sinking EDM [96]. To overcome the pollution and harmful vapor throughout the EDM machining process using the hydro-carbon dielectric, the EDM process is further refined and a novel method was proposed. Its name was given as dry-EDM (DEDM). In this process, air or compressed is passed through a thin walled tubular pipe, whenever the compressed gas cools the inter electrode gap and relived the debris from the machining zone [97].

Earlier studies also assessed the surface integrity and stresses of machined tool steel surfaces after EDM processes. For example, in one study [143], the impact of EDM process parameters on the 3D surface topography of tool steel was explored, while in another study [144], the crystallographic and metallurgical properties of the white layer in the Böhler W300 ferrite steel EDM were documented. In another investigation [145], residual stress was measured in material SKD11. Effect of EDM machining parameters on the surface characteristics and damage to machining of AISI D2 tool steel material [146] was studied. Other works surface alterations of tool steel were studied during the EDM process for AISI H13 and AISI D6 [147,148]. Some studies [149] looked at the effect of process parameters on the fatigue life of the AISI D6 steel tool. Changes in the chemical composition of the re-solidified layers of the electrodes and workpieces were also studied during the T215Cr12 die steel EDM [3]. Several experiments centered on resulting residual stresses in the machined surface, for example, Reference [150].

Both MRR and EWR decrease with increase in the temperature. Within the higher temperature, the droplets can align as much as easier under the action of the electrostatic force due to the smaller viscosity of the emulsion. And the better flushing condition can slightly improve the EDM performance. However, the improvement of the flushing condition cannot improve further performance of the EDM [151]. When hydro-carbon oil-based working fluid was used, there were several aerosols or gases generated due to the decomposition of the hydro-carbon oil. Besides the emission in the air, the accumulation of decomposition substance in the working fluid also increases the process complexity of waste fluid itself [99,100].

In consideration of accretion EDM mechanism, titanium powder suspended working oil is used as the dielectric. Here the titanium carbide will deposit when both discharge energy and power density are small. But, in the case of discharge power density exceeding the critical power density for the removal of carbon steel, then it melts and will evaporate at low discharge energy. Due to the consequent process, the titanium powder was not deposited, and the carbon steel was simply removed from the removal region [152]. The experimental setup is shown in Figure 6.

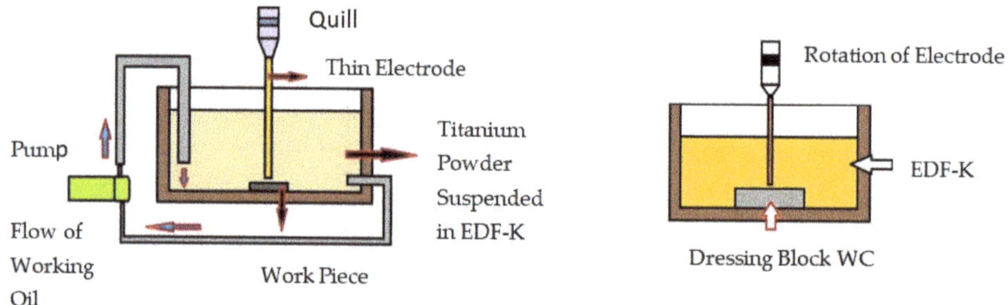

Figure 6. Experimental setup for accretion mechanism.

Besides its machining purpose, according to Furutani et al. [47] EDM is also used for the deposition process. In EDM the working fluid suspended with silicon powder will increase the SR because of the dispersion of discharge decreases the frictional coefficient. In EDM with MoS_2 suspended in the working oil, the MoS_2 could be deposited on the metals which have low melting point than MoS_2. Dispersion of the lubricant duration occurs due to the uneven concentration of MoS2. Gap length expansion caused the improvement of roughness [104].

6. Discussion

Reviewing the publication related to dielectric fluid in the EDM process showed that the majority of the studies investigated the effect of variation of powder concentration in different process parameters, such as voltage, current, pulse rate, pulse on time, and it has been observed that for low current values the MRR is very low. But the MRR increased with the increase in the discharge current. For reverse polarity, the spark energy was low at low discharge current and pulse on time. Other bodies of researches have been conducted for improving the machining efficiency and for reducing the tool wear ratio. These studies point out the advanced methods for achieving better surface finish, reducing the tool wear ratio, and increasing MRR. Moreover, this method has tremendous application in the field of mold and die making and nano size hole drilling.

PMEDM is really useful for:

- Machining of widely available advanced materials like MMCs, insulating ceramics like TiO_2 which have been successfully machined by dispersing various powders into EDM dielectric;
- Production and machining of micro products and sophisticated micro mechanical elements.

But, the setup has the following drawbacks too:

- Electrically conductive material can only be machined;

- For achieving more accuracy in machining, it consumes more time;
- Health problems are occurring due to the usage of some dielectric fluids;
- Materials like water hardened die steel, molybdenum high-speed tool steel have not been tried as work material.

7. Conclusions

The reviews on the state of art, studies on the dielectric fluid in EDM processes lead to the following conclusions.

- According to the general agreement on the results, different conductive powders, with different concentrations can improve the material removal rate in EDM
- Peak current, pulse on and off time, duty cycle, voltage, discharge current, tool angle, powder concentration, nozzle flushing, and grain are the input parameters that improve the material removal rate and changing of SR with the different configuration of the above input parameters
- According to the major observations by the researchers, the increase in SR and improvement in MRR were observed in the mixing of powders into the dielectric medium
- This review reveals that, with a specific increase in concentration of powder in the dielectric fluid, the MRR and SR will increase. Increasing the concentration of powder in the dielectric fluid beyond the certain optimum concentration of particles in the dielectric, short-circuit discharges occur, and, as a consequence, MRR decreases. From this analysis, it can be concluded that PMEDM holds a brilliant promise in the application of EDM, in particular, regarding process efficiency and work piece surface quality. Therefore, an extensive study is required to understand the machining mechanics and other aspects of PMEDM and will be performed as future research.

Author Contributions: Conceptualization, A.A. and J.E.A.Q.; methodology, A.A. and J.E.A.Q.; software, A.A.; validation, A.Z. and T.A.; formal analysis, A.A. and J.E.A.Q.; investigation, A.A. and A.K.; resources, A.A. and J.E.A.Q., A.A. and A.Z.; writing—original draft preparation, A.A. and J.E.A.Q.; writing—review and editing, A.A. and J.E.A.Q., and A.K.; visualization, A.A. and A.Z.; supervision, A.A. and J.E.A.Q.; project administration, J.E.A.Q. and A.Z.; funding acquisition, J.E.A.Q. All authors have read and agreed to the published version of the manuscript.

Funding: This research was funded by the Research Affairs Office at UAE University, grant number 31N396.

Conflicts of Interest: The authors declare no conflict of interest.

References

1. Qudeiri, J.E.A.; Mourad, A.H.I.; Ziout, A.; Abidi, M.H.; Elkaseer, A. Electric discharge machining of titanium and its alloys: Review. *Int. J. Adv. Manuf. Technol.* **2018**, *11*, 84–111. [CrossRef]
2. Kansal, H.K.; Singh, S.; Kumar, P. Technology and research developments in powder mixed electric discharge machining (PMEDM). *J. Mater. Process. Technol.* **2007**, *184*, 32–41. [CrossRef]
3. Soni, J.S.; Chakraverti, G. Experimental investigation on migration of material during EDM of die steel (T215 Cr12). *J. Mater. Process. Technol.* **1996**, *56*, 439–451. [CrossRef]
4. Soni, J.S. Microanalysis of debris formed during rotary EDM of titanium alloy (Ti 6A1 4V) and die steel (T 215 Cr12). *Wear* **1994**, *177*, 71–79. [CrossRef]
5. Dewangan, S.K. Experimental Investigation of Machining Parameters for EDM Using U-shaped Electrode of AISI P20 Tool Steel. *Natl. Inst. Technol. Rourkela* **2010**, *93*, 1–4.
6. Nallusamy, S. Analysis of MRR and TWR on OHNS die steel with different electrodes using electrical discharge machining. *Int. J. Eng. Res. Afr.* **2016**, *22*, 112–120. [CrossRef]
7. Tripathy, S.; Tripathy, D.K. An approach for increasing the micro-hardness in electrical discharge machining by adding conductive powder to the dielectric. *Mater. Today Proc.* **2017**, *4*, 1215–1224. [CrossRef]
8. Khedkar, N.K.; Singh, T.P.; Jatti, V.K.S. Surface improvement of H13 hot die steel material by EDM method using silicon carbide powder-mixed dielectric. *Int. J. Appl. Eng. Res.* **2014**, *9*, 4809–4816.
9. Laad, M.; Jatti, V.S.; Jadhav, P.P. Investigation into application of electrical discharge machining as a surface treatment process. *WSEAS Trans. Appl. Theor. Mech.* **2014**, *9*, 245–251.

10. Iqbal, A.K.M.A.; Khan, A.A. Optimization of process parameters on EDM milling of stainless steel AISI 304. *Adv. Mater. Res.* **2011**, *264*, 979–984. [CrossRef]

11. Lorincz, J. Drilling Small, Deep Holes with Precision EDM. *Manuf. Eng. Media* **2013**. Available online: https://www.sme.org/drilling-small-deep-holes-precision-edm (accessed on 28 June 2020).

12. Hourmand, M.; Sarhan, A.A.D.; Sayuti, M. Micro-electrode fabrication processes for micro-EDM drilling and milling: A state-of-the-art review. *Int. J. Adv. Manuf. Technol.* **2017**, *91*, 1023–1056. [CrossRef]

13. Qudeiri, J.E.A.; Zaiout, A.; Mourad, A.H.I.; Abidi, M.H.; Elkaseer, A. Principles and Characteristics of Different EDM Processes in Machining Tool and Die Steels. *Appl. Sci.* **2020**, *10*, 2082. [CrossRef]

14. Wong, Y.S.; Lim, L.C.; Rahuman, I.; Tee, W.M. Near-mirror-finish phenomenon in EDM using powder-mixed dielectric. *J. Mater. Process. Technol.* **1998**, *79*, 30–40. [CrossRef]

15. Talla, G.; Gangopadhayay, S.; Biswas, C.K. State of the art in powder-mixed electric discharge machining: A review. *Proc. Inst. Mech. Eng. Part B J. Eng. Manuf.* **2017**, *231*, 2511–2526. [CrossRef]

16. Saliya, J. Comparative Study for the effect of Powder mixed dielectric on Performance of Wire EDM‖. *Int. J. Eng. Dev. Res.* **2014**, *2*, 3671–3673.

17. Mohanty, S.; Mishra, A.; Nanda, B.K.; Routara, B.C. Multi-objective parametric optimization of nano powder mixed electrical discharge machining of AlSiCp using response surface methodology and particle swarm optimization. *Alexandria Eng. J.* **2018**, *57*, 609–619. [CrossRef]

18. Backer, S.; Mathew, C.; George, S.K. Optimization of MRR and TWR on EDM by Using Taguchi's Method and ANOVA. *Int. J. Innov. Res. Adv. Eng. (IJIRAE)* **2014**, *1*, 106–112.

19. Ojha, K.; Garg, R.K.; Singh, K.K. MRR improvement in sinking electrical discharge machining: A review. *J. Miner. Mater. Charact. Eng.* **2010**, *9*, 709. [CrossRef]

20. Roy, A.; Narendra, N.; Nedelcu, D. Experimental investigation on variation of output responses of as cast TiNiCu shape memory alloys using wire EDM. *Int. J. Modern Manuf. Technol.* **2017**, *9*, 90–101.

21. Ho, K.H.; Newman, S.T.; Rahimifard, S.; Allen, R.D. State of the art in wire electrical discharge machining (WEDM). *Int. J. Mach. Tools Manuf.* **2004**, *44*, 1247–1259. [CrossRef]

22. Mahapatra, S.S.; Patnaik, A. Optimization of wire electrical discharge machining (WEDM) process parameters using Taguchi method. *Int. J. Adv. Manuf. Technol.* **2007**, *34*, 911–925. [CrossRef]

23. Kumar, S.; Singh, R.; Singh, T.P.; Sethi, B.L. Surface modification by electrical discharge machining: A review. *J. Mater. Process. Technol.* **2009**, *209*, 3675–3687. [CrossRef]

24. Rao, M.S.; Venkaiah, N. *Review on Wire-Cut EDM Process*; World Academy of Research in Science and Engineering: Hyderabad, India, 2013.

25. Pradhan, M.K. Estimating the effect of process parameters on MRR, TWR and radial overcut of EDMed AISI D2 tool steel by RSM and GRA coupled with PCA. *Int. J. Adv. Manuf. Technol.* **2013**, *68*, 591–605. [CrossRef]

26. Yan, M.T.; Lai, Y.P. Surface quality improvement of wire-EDM using a fine-finish power supply. *Int. J. Mach. Tools Manuf.* **2007**, *47*, 1686–1694. [CrossRef]

27. Wang, C.C.; Yan, B.H.; Chow, H.M.; Suzuki, Y. Cutting austempered ductile iron using an EDM sinker. *J. Mater. Process. Technol.* **1999**, *88*, 83–89. [CrossRef]

28. Kandpal, B.C.; kumar, J.; Singh, H. Machining of Aluminium Metal Matrix Composites with Electrical Discharge Machining—A Review. *Mater. Today Proc.* **2015**, *2*, 1665–1671. [CrossRef]

29. Kumar, U.A.; Laxminarayana, P. Optimization of Electrode Tool Wear in micro holes machining by Die Sinker EDM using Taguchi Approach*. *Mater. Today Proc.* **2018**, *5*, 1824–1831. [CrossRef]

30. Kumar, U.A.; Laxminarayana, P.; Aravindan, N. Study of surface morphology on micro machined surfaces of AISI 316 by Die Sinker EDM. *Mater. Today Proc.* **2017**, *4*, 1285–1292. [CrossRef]

31. Razak, D.M.; Syahrullail, S.; Nuraliza, N.; Azli, Y.; Sapawe, S. Surface modification of biomaterial embedded with pits using die sinker machine. *Sci. Iran.* **2017**, *24*, 1901–1911. [CrossRef]

32. Isvilanonda, V.; Ramulu, M.; Laxminarayana, P.; Briggs, T. Effect of die sinker EDM and AWJ machining processes on flexural properties of hybrid titanium laminates. In Proceedings of the 55th International SAMPE Symposium, Seattle, WA, USA, 17 May 2010.

33. Singh, G.; Bhui, A.S.; Lamichhane, Y.; Mukhiya, P.; Kumar, P.; Thapa, B. Machining performance and influence of process parameters on stainless steel 316L using die-sinker EDM with Cu tool. *Mater. Today Proc.* **2019**, *18*, 2468–2476. [CrossRef]

34. Kolli, M.; Kumar, A. Effect of Boron Carbide Powder Mixed into Dielectric Fluid on Electrical Discharge Machining of Titanium Alloy. *Procedia Mater. Sci.* **2014**, *5*, 1957–1965. [CrossRef]

35. Sanghani, C.R.; Acharya, G.D. Effect of Various Dielectric Fluids on Performance of EDM: A Review. *Trends Mech. Eng. Technol.* **2016**, *6*, 55–71.

36. Modi, M.; Agarwal, G. Effect of aluminium and chromium powder mixed dielectric fluid on electrical discharge machining effectiveness. *Adv. Prod. Eng. Manag.* **2019**, *14*, 323–332. [CrossRef]

37. Reddy, V.V.; Kumar, A.; Valli, P.M.; Reddy, C.S. Influence of surfactant and graphite powder concentration on electrical discharge machining of PH17-4 stainless steel. *J. Braz. Soc. Mech. Sci. Eng.* **2015**, *37*, 641–655. [CrossRef]

38. Marashi, H.; Jafarlou, D.M.; Sarhan, A.A.D.; Hamdi, M. State of the art in powder mixed dielectric for EDM applications. *Precis. Eng.* **2016**, *46*, 11–33. [CrossRef]

39. Choudhary, S.K.; Jadoun, R.S. Current research issue, trend & applications of powder mixed dielectric electric discharge machining (PM-EDM): A review. *Int. J. Eng. Sci. Res. Technol.* **2014**, *3*, 335–358.

40. Sekhar, B.; Sekhar, B.P.C.; Radhika, S.; Kumar, D.S. Powder Mixed Dielectric in EDM—An overview. *Int. J. Res.* **2015**, *2*, 720–726.

41. Kansal, H.K.; Singh, S.; Kumar, P. Effect of silicon powder mixed EDM on machining rate of AISI D2 die steel. *J. Manuf. Process.* **2007**, *9*, 13–22. [CrossRef]

42. Kansal, H.K.; Singh, S.; Kumar, P. Numerical simulation of powder mixed electric discharge machining (PMEDM) using finite element method. *Math. Comput. Model.* **2008**, *47*, 1217–1237. [CrossRef]

43. Kumar, S.; Batra, U. Surface modification of die steel materials by EDM method using tungsten powder-mixed dielectric. *J. Manuf. Process.* **2012**, *14*, 35–40. [CrossRef]

44. Grigoryev, E.G.; Olevsky, E.A. Thermal processes during high-voltage electric discharge consolidation of powder materials. *Scr. Mater.* **2012**, *66*, 662–665. [CrossRef]

45. Kansal, H.K.; Singh, S.; Kumar, P. Parametric optimization of powder mixed electrical discharge machining by response surface methodology. *J. Mater. Process. Technol.* **2005**, *169*, 427–436. [CrossRef]

46. Zhao, W.S.; Meng, Q.G.; Wang, Z.L. The application of research on powder mixed EDM in rough machining. *J. Mater. Process. Technol.* **2002**, *129*, 30–33. [CrossRef]

47. Furutani, K.; Sato, H.; Mieda, Y. Consideration of accretion mechanism by electrical discharge machining with titanium powder suspended in working oil. In Proceedings of the 20th Annual ASPE Meeting ASPE 2005, Norfolk, VA, USA, 9–14 October 2005.

48. Furutania, K.; Saneto, A.; Takezawa, H.; Mohri, N.; Miyake, H. Accretion of titanium carbide by electrical discharge machining with powder suspended in working fluid. *Precis. Eng.* **2001**, *25*, 138–144. [CrossRef]

49. Prabu, M.; Ramadoss, G.; Senthilkumar, C.; Boopathi, R.; Magibalan, S. Experimental investigation on effect of graphite powder suspended dielectric in electric discharge machining of AL-TIB2 composites. *J. Chem. Pharm. Sci.* **2015**, *974*, 2115.

50. Singh, J.; Sharma, R.K. Study the surface modification and characterization for powder mixed electrical discharge machining of tungsten carbide. *Indian J. Eng. Mater. Sci.* **2016**, *23*, 321–335.

51. Viswanth, V.S.; Ramanujam, R.; Rajyalakshmi, G. A review of research scope on sustainable and eco-friendly electrical discharge machining (E-EDM). *Mater. Today Proc.* **2018**, *5*, 12525–12533. [CrossRef]

52. Kumar, S.; Dhingra, A.K.; Kumar, S. Parametric optimization of powder mixed electrical discharge machining for nickel-based superalloy inconel-800 using response surface methodology. *Mech. Adv. Mater. Mod. Process.* **2017**, *3*, 7. [CrossRef]

53. Sharma, S.; Kumar, A.; Beri, N.; Kumar, D. Effect of aluminium powder addition in dielectric during electric discharge machining of hastelloy on machining performance using reverse polarity. *Int. J. Adv. Eng. Technol.* **2010**, *1*, 13–24.

54. Kansal, H.K.; Singh, S.; Kumar, P. Performance parameters optimization (multi-characteristics) of powder mixed electric discharge machining (PMEDM) through Taguchi's method and utility concept. *Indian J. Eng. Mater. Sci.* **2006**, *13*, 209–216.

55. Singh, P.; Kumar, A.; Beri, N.; Kumar, V. Some experimental investigation on aluminum powder mixed EDM on machining performance of hastelloy steel. *Int. J. Adv. Eng. Technol.* **2010**, *1*, 28–45.

56. Singh, G.; Singh, P.; Tejpal, G.; Singh, B. Effect of machining parameters on surface roughness of H13 Steel in EDM process using powder mixed fluid. *Int. J. Adv. Eng. Res. Stud.* **2012**, *2*, 148–150.

57. Syed, K.H.; Palaniyandi, K. Performance of electrical discharge machining using aluminium powder suspended distilled water. *Turk. J. Eng. Environ. Sci.* **2012**, *36*, 195–207.

58. Padhee, S.; Nayak, N.; Panda, S.K.; Dhal, P.R.; Mahapatra, S.S. Multi-objective parametric optimization of powder mixed electro-discharge machining using response surface methodology and non-dominated sorting genetic algorithm. *Sadhana* **2012**, *37*, 223–240. [CrossRef]
59. Mir, M.J.; Sheikh, K.; Singh, B.; Malhotra, N. Modeling and analysis of machining parameters for surface roughness in powder mixed EDM using RSM approach. *Int. J. Eng. Sci. Technol.* **2012**, *4*, 45–52. [CrossRef]
60. Razak, M.A.; Abdul-Rani, A.M.; Nanimina, A.M. Improving EDM efficiency with silicon carbide powder-mixed dielectric fluid. *Int. J. Mater. Mech. Manuf.* **2015**, *3*, 40–43. [CrossRef]
61. Peças, P.; Henriques, E. Effect of the powder concentration and dielectric flow in the surface morphology in electrical discharge machining with powder-mixed dielectric (PMD-EDM). *Int. J. Adv. Manuf. Technol.* **2008**, *37*, 1120–1132. [CrossRef]
62. Jeswani, M.L. Effect of the addition of graphite powder to kerosene used as the dielectric fluid in electrical discharge machining. *Wear* **1981**, *70*, 133–139. [CrossRef]
63. Goyal, S.; Singh, R.K. Parametric study of powder mixed EDM and optimization of MRR & surface roughness. *Int. J. Sci. Eng. Technol.* **2014**, *3*, 56–62.
64. Khundrakpam, N.S.; Singh, H.; Kumar, S.; Brar, G.S. Investigation and modeling of silicon powder mixed EDM using response surface method. *Int. J. Curr. Eng. Technol.* **2014**, *4*, 1022–1026.
65. Muniu, J.M.; Ikua, B.W.; Nyaanga, D.M.; Gicharu, S.N. Study on Effects of Powder-Mixed Dielectric Fluids on Electrical Discharge Machining Processes. *Int. J. Eng. Res. Technol.* **2013**, *2*, 2449–2456.
66. Mathpathi, U.; Jeevraj, S.K.; Ramola, I.C. Analysis of material removal rate with powder mixed dielectric. *IJCAE* **2013**, *4*, 316–332.
67. Yih-Fong, T.; Fu-Chen, C. Investigation into some surface characteristics of electrical discharge machined SKD-11 using powder-suspension dielectric oil. *J. Mater. Process. Technol.* **2005**, *170*, 385–391. [CrossRef]
68. Furutani, K.; Sato, H.; Suzuki, M. Influence of electrical conditions on performance of electrical discharge machining with powder suspended in working oil for titanium carbide deposition process. *Int. J. Adv. Manuf. Technol.* **2009**, *40*, 1093–1101. [CrossRef]
69. Ojha, K.; Garg, R.K.; Singh, K.K. Parametric optimization of PMEDM process using chromium powder mixed dielectric and triangular shape electrodes. *J. Miner. Mater. Charact. Eng.* **2011**, *10*, 1087. [CrossRef]
70. Paul, B.K.; Sahu, S.K.; Jadam, T.; Datta, S.; Dhupal, D.; Mahapatra, S.S. Effects of addition of copper powder in the dielectric media (EDM Oil) on Electro-discharge machining performance of inconel 718 super alloys. *Mater. Today Proc.* **2018**, *5*, 17618–17626. [CrossRef]
71. Pecas, P.; Henriques, E. Influence of silicon powder-mixed dielectric on conventional electrical discharge machining. *Int. J. Mach. Tools Manuf.* **2003**, *43*, 1465–1471. [CrossRef]
72. Klocke, F.; Lung, D.; Antonoglou, G.; Thomaidis, D. The effects of powder suspended dielectrics on the thermal influenced zone by electrodischarge machining with small discharge energies. *J. Mater. Process. Technol.* **2004**, *149*, 191–197. [CrossRef]
73. Gurule, N.B.; Nandurkar, K.N. Effect of tool rotation on material removal rate during powder mixed electric discharge machining of die steel. *Int. J. Emerg. Technol. Adv. Eng.* **2012**, *2*, 328–332.
74. Bhaumik, M.; Maity, K. Effect of deep cryotreated tungsten carbide electrode and SiC powder on EDM performance of AISI 304. *Part. Sci. Technol.* **2019**, *37*, 981–992. [CrossRef]
75. Paswan, K.; Pramanik, A.; Chattopadhyaya, S. Machining performance of Inconel 718 using graphene nanofluid in EDM. *Mater. Manuf. Process.* **2020**, *35*, 33–42. [CrossRef]
76. Chu, X.; Zhu, K.; Wang, C.; Hu, Z.; Zhang, Y. A Study on Plasma Channel Expansion in Micro-EDM. *Mater. Manuf. Process.* **2016**, *31*, 381–390. [CrossRef]
77. Jahan, M.P.; Rahman, M.; Wong, Y.S. Study on the nano-powder-mixed sinking and milling micro-EDM of WC-Co. *Int. J. Adv. Manuf. Technol.* **2011**, *53*, 167–180. [CrossRef]
78. Kumar, A.; Maheshwari, S.; Sharma, C.; Beri, N. Analysis of machining characteristics in additive mixed electric discharge machining of nickel-based super alloy Inconel 718. *Mater. Manuf. Process.* **2011**, *26*, 1011–1018. [CrossRef]
79. Garg, R.K.; Ojha, K. Parametric optimization of PMEDM process with nickel micro powder suspended dielectric and varying triangular shapes electrodes on EN-19 steel. *J. Eng. Appl. Sci.* **2011**, *6*, 152–156. [CrossRef]
80. Rajkumar, H.; Vishwakamra, M. Performance parameters characteristics of PMEDM: A review. *Int. J. Appl. Eng. Res.* **2018**, *13*, 5281–5290.

81. Patel, S.; Thesiya, D.; Rajurkar, A. Aluminium powder mixed rotary electric discharge machining (PMEDM) on Inconel 718. *Aust. J. Mech. Eng.* **2018**, *16*, 21–30. [CrossRef]

82. Tzeng, Y.-F.; Lee, C.-Y. Effects of powder characteristics on electrodischarge machining efficiency. *Int. J. Adv. Manuf. Technol.* **2001**, *17*, 586–592. [CrossRef]

83. Abrol, A.; Sharma, S. Effect of chromium powder mixed dielectric on performance characteristic of AISI D2 die steel using EDM. *Int. J. Res. Eng. Technol.* **2015**, *4*, 344–356.

84. Korada, S.; Prasad, L.V.R.S.V.; Chilamkurti, S.G. Employing Sic Nano-Powder Dielectric to Enhance Machinability of Aisi D3 Steel in Electrical Spark Machining. *Int. J. Modern Manuf. Technol.* **2019**, *XI*, 3.

85. Malhotra, N.; Singh, H.; Rani, S. Improvements in performance of EDM—A review. *Conf. Proc.-IEEE Southeastcon* **2008**. [CrossRef]

86. Manivannan, R.; Kumar, M.P. Improving the machining performance characteristics of the μEDM drilling process by the online cryogenic cooling approach. *Mater. Manuf. Process.* **2018**, *33*, 390–396. [CrossRef]

87. Kumar, V.; Kumar, P. Improving Material Removal Rate and Optimizing Variousmachining Parameters in EDM. *Int. J. Eng. Sci.* **2017**, *6*, 64–68. [CrossRef]

88. Mahajan, R.; Krishna, H.; Singh, A.K.; Ghadai, R.K. A review on Copper and its alloys used as electrode in EDM. In Proceedings of the IOP Conference Series: Materials Science and Engineering, Volume 377, International Conference on Mechanical, Materials and Renewable Energy, Sikkim, India, 8–10 December 2017. [CrossRef]

89. Khan, A.A.; Mohiuddin, A.K.M.; Latif, M.A.A. Improvement of MRR and surface roughness during electrical discharge machining (EDM) using aluminum oxide powder mixed dielectric fluid. *IOP Conf. Ser. Mater. Sci. Eng.* **2018**, *290*, 012063. [CrossRef]

90. Gudur, S.; Potdar, V.V.; Gudur, S. A Review on Effect of Aluminum & Silicon Powder Mixed EDM on Response Variables of Various Materials. *Int. J. Innov. Res. Sci. Eng. Technol.* **2014**, *3*, 17937–17945. [CrossRef]

91. Sugunakar, A.; Kumar, A.; Markandeya, R. Effect of Powder Mixed Dielectric Fluid on MRR And SR During Electrical Discharge Machining of RENE 80. *OSR J. Mech. Civ. Eng.* **2017**, *14*, 43–50. [CrossRef]

92. Goyal, P. Enhancement of MRR in EDM by composite material electrode on die steel. *Int. J. Sci. Eng. Technol. Res.* **2014**, *3*, 2640–2643.

93. Mhatre, M.S.; Sapkal, S.U.; Pawade, R.S. Electro discharge machining characteristics of Ti-6Al-4V alloy: A grey relational optimization. *Procedia Mater. Sci.* **2014**, *5*, 2014–2022. [CrossRef]

94. Leão, F.N.; Pashby, I.R. A review on the use of environmentally-friendly dielectric fluids in electrical discharge machining. *J. Mater. Process. Technol.* **2004**, *149*, 341–346. [CrossRef]

95. Singh, N.K.; Singh, K.K. Review on recent development in environmental-friendly EDM techniques. *Adv. Manuf. Sci. Technol.* **2015**, *39*, 17–37.

96. Zhang, Y.; Liu, Y.; Ji, R.; Cai, B.; Shen, Y. Sinking EDM in water-in-oil emulsion. *Int. J. Adv. Manuf. Technol.* **2013**, *65*, 705–716. [CrossRef]

97. Singh, N.K.; Pandey, P.M.; Singh, K.K.; Sharma, M.K. Steps towards green manufacturing through EDM process: A review. *Cogent Eng.* **2016**, *3*, 1272662. [CrossRef]

98. Zhang, Y.; Liu, Y.; Shen, Y.; Ji, R.; Li, Z.; Zheng, C. Investigation on the influence of the dielectrics on the material removal characteristics of EDM. *J. Mater. Process. Technol.* **2014**, *214*, 1052–1061. [CrossRef]

99. Paramashivan, S.S.; Mathew, J.; Mahadevan, S. Mathematical modeling of aerosol emission from die sinking electrical discharge machining process. *Appl. Math. Model.* **2012**, *36*, 1493–1503. [CrossRef]

100. Evertz, S.; Dott, W.; Eisentraeger, A. Electrical discharge machining: Occupational hygienic characterization using emission-based monitoring. *Int. J. Hyg. Environ. Health* **2006**, *209*, 423–434. [CrossRef] [PubMed]

101. Singh, S.; Bhardwaj, A. Review to EDM by using water and powder-mixed dielectric fluid. *J. Miner. Mater. Charact. Eng.* **2011**, *10*, 199. [CrossRef]

102. Muttamara, A.; Mesee, J. Effect of TiN powder mixed in Electrical Discharge Machining. *IOP Conf. Ser. Mater. Sci. Eng.* **2016**, *157*, 12021. [CrossRef]

103. Bhatt, G.; Batish, A.; Bhattacharya, A. Experimental investigation of magnetic field assisted powder mixed electric discharge machining. *Part. Sci. Technol.* **2015**, *33*, 246–256. [CrossRef]

104. Furutani, K.; Shimizu, Y. Experimental analysis of deposition process of lubricant surface by electrical discharge machining with molybdenum disulfide powder suspended in working oil. In Proceedings of the 18th Annual Meet of the American Society of Precision Engineering, Portland, OR, USA, 28 October 2003; pp. 547–550.

105. Kim, Y.S.; Chu, C.N. The Effects of Graphite Powder on Tool Wear in Micro Electrical Discharge Machining. *Procedia CIRP.* **2018**, *68*, 553–558. [CrossRef]

106. Khosrozadeh, B.; Shabgard, M. Effects of hybrid electrical discharge machining processes on surface integrity and residual stresses of Ti-6Al-4V titanium alloy. *Int. J. Adv. Manuf. Technol.* **2017**, *93*, 1999–2011. [CrossRef]

107. Qudeiri, J.E.A.; Saleh, A.; Ziout, A.; Mourad, A.H.I.; Abidi, M.H.; Elkaseer, A. Advanced electric discharge machining of stainless steels: Assessment of the state of the art, gaps and future prospect. *Materials* **2019**, *12*, 907. [CrossRef] [PubMed]

108. Marafona, J.; Wykes, C. New method of optimising material removal rate using EDM with copper-tungsten electrodes. *Int. J. Mach. Tools Manuf.* **2000**, *40*, 153–164. [CrossRef]

109. Mohri, N.; Suzuki, M.; Furuya, M.; Saito, N.; Kobayashi, A. Electrode Wear Process in Electrical Discharge Machinings. *CIRP Ann.-Manuf. Technol.* **1995**, *44*, 165–168. [CrossRef]

110. Staelens, F.; Kruth, J.P. A Computer Integrated Machining Strategy for Planetary EDM. *CIRP Ann.-Manuf. Technol.* **1989**, *38*, 187–190. [CrossRef]

111. Kumar, A.; Mandal, A.; Dixit, A.R.; Das, A.K. Performance evaluation of Al2O3 nano powder mixed dielectric for electric discharge machining of Inconel 825. *Mater. Manuf. Process.* **2018**, *33*, 986–995. [CrossRef]

112. Hosni, N.A.J.; Lajis, M.A. Multi-response optimization of the machining characteristics in electrical discharge machining (EDM) using span-20 surfactant and chromium (Cr) powder mixed. *Materwiss. Werksttech.* **2019**, *50*, 329–335. [CrossRef]

113. Lajis, M.A.; Hosni, N.A.J. The influences of various mixed dielectric fluids on the performance electrical discharge machining of AISI D2 hardened steel. *Materwiss. Werksttech.* **2018**, *49*, 413–419. [CrossRef]

114. Sivaprakasam, P.; Hariharan, P.; Gowri, S. Experimental investigations on nano powder mixed Micro-Wire EDM process of inconel-718 alloy. *Meas. J. Int. Meas. Confed.* **2019**, *147*, 106844. [CrossRef]

115. Marashi, H.; Sarhan, A.A.D.; Hamdi, M. Employing Ti nano-powder dielectric to enhance surface characteristics in electrical discharge machining of AISI D2 steel. *Appl. Surf. Sci.* **2015**, *357*, 892–907. [CrossRef]

116. Hosni, N.A.J.; Lajis, M.A. The influence of Span-20 surfactant and micro-/nano-Chromium (Cr) Powder Mixed Electrical Discharge Machining (PMEDM) on the surface characteristics of AISI D2 hardened steel. *IOP Conf. Ser. Mater. Sci. Eng.* **2018**, *342*, 1–11. [CrossRef]

117. Hosni, N.A.J.; Lajis, M.A. Experimental investigation and economic analysis of surfactant (Span-20) in powder mixed electrical discharge machining (PMEDM) of AISI D2 hardened steel. *Mach. Sci. Technol.* **2020**, *24*, 398–424. [CrossRef]

118. Sethuramalingam, P.; Vinayagam, B.K. Adaptive neuro fuzzy inference system modelling of multi-objective optimisation of electrical discharge machining process using single-wall carbon nanotubes. *Aust. J. Mech. Eng.* **2015**, *13*, 97–117. [CrossRef]

119. Abbas, M.A.; Lajis, M.A.; Abbas, D.R.; Merzah, O.M.; Kadhim, M.H.; Shamran, A.A. Influence of additive materials on the roughness of AISI D2 steel in electrical discharge machining (EDM) environment. *Materwiss. Werksttech.* **2020**, *51*, 719–724. [CrossRef]

120. Rani, J.R.; Thangavel, R.; Oh, S.I.; Lee, Y.S.; Jang, J.H. An ultra-high-energy density supercapacitor; fabrication based on thiol-functionalized graphene oxide scrolls. *Nanomaterials* **2019**, *9*, 148. [CrossRef]

121. Somashekhar, K.P.; Ramachandran, N.; Mathew, J. Optimization of material removal rate in micro-EDM using artificial neural network and genetic algorithms. *Mater. Manuf. Process.* **2010**, *25*, 467–475. [CrossRef]

122. Reddy, V.V.; Reddy, C.S.; Valli, P.M. Optimization of Process Parameters of Surfactant and Graphite Powder Mixed Dielectric EDM through Taguchi-Grey Relational Analysis. *Bonfring Int. J. Ind. Eng. Manag. Sci.* **2015**, *5*, 175. [CrossRef]

123. Lamichhane, Y.; Singh, G.; Bhui, A.S.; Mukhiya, P.; Kumar, P.; Thapa, B. Surface modification of 316L SS with HAp nano-particles using PMEDM for enhanced biocompatibility. *Mater. Today Proc.* **2019**, *15*, 336–343. [CrossRef]

124. Shinde, R.; Patil, N.; Raut, D.; Pawade, R.; Brahmaknakr, P. *Experimental Investigations into Powder-Mixed Electrical Discharge Machining (PMEDM) of HCHCr D2 Die Steel*; Atlantis Press: Paris, France, 2017. [CrossRef]

125. Bains, P.S.; Sidhu, S.S.; Payal, H.S.; Kaur, S. Magnetic field influence on surface modifications in powder mixed EDM. *Silicon* **2019**, *11*, 415–423. [CrossRef]

126. Bains, P.S.; Sidhu, S.S.; Payal, H.S. Magnetic field assisted EDM: New horizons for improved surface properties. *Silicon* **2018**, *10*, 1275–1282. [CrossRef]

127. Rouniyar, A.K.; Shandilya, P. Fabrication and experimental investigation of magnetic field assisted powder mixed electrical discharge machining on machining of aluminum 6061 alloy. *Proc. Inst. Mech. Eng. Part B J. Eng. Manuf.* **2019**, *233*, 2283–2291. [CrossRef]
128. Kunieda, M.; Takaya, T.; Nakano, S. Improvement of dry EDM characteristics using piezoelectric actuator. *CIRP Ann.-Manuf. Technol.* **2004**, *53*, 183–186. [CrossRef]
129. Fenggou, C.; Dayong, Y. The study of high efficiency and intelligent optimization system in EDM sinking process. *J. Mater. Process. Technol.* **2004**, *149*, 83–87. [CrossRef]
130. Han, F.; Kunieda, M.; Sendai, T.; Imai, Y. High precision simulation of WEDM using parametric programming. *CIRP Ann.-Manuf. Technol.* **2002**, *51*, 165–168. [CrossRef]
131. Yadav, V.; Jain, V.K.; Dixit, P.M. Thermal stresses due to electrical discharge machining. *Int. J. Mach. Tools Manuf.* **2002**, *42*, 877–888. [CrossRef]
132. Tsai, K.M.; Wang, P.J. Semi-empirical model of surface finish on electrical discharge machining. *Int. J. Mach. Tools Manuf.* **2001**, *41*, 1455–1477. [CrossRef]
133. Wang, P.J.; Tsai, K.M. Semi-empirical model on work removal and tool wear in electrical discharge machining. *J. Mater. Process. Technol.* **2001**, *114*, 1–17. [CrossRef]
134. Williams, R.E.; Rajurkar, K.P. Study of wire electrical discharge machined surface characteristics. *J. Mater. Process. Technol.* **1991**, *28*, 127–138. [CrossRef]
135. Cogun, C.; Savsar, M. Statistical modelling of properties of discharge pulses in electric discharge machining. *Int. J. Mach. Tools Manuf.* **1990**, *30*, 467–474. [CrossRef]
136. Salah, N.B.; Ghanem, F.; Atig, K.B. Numerical study of thermal aspects of electric discharge machining process. *Int. J. Mach. Tools Manuf.* **2006**, *46*, 908–911. [CrossRef]
137. Gao, C.; Liu, Z. A study of ultrasonically aided micro-electrical-discharge machining by the application of workpiece vibration. *J. Mater. Process. Technol.* **2003**, *139*, 226–228. [CrossRef]
138. Srivastava, V.; Pandey, P.M. Performance evaluation of electrical discharge machining (EDM) process using cryogenically cooled electrode. *Mater. Manuf. Process.* **2012**, *27*, 683–688. [CrossRef]
139. Świercz, R.; Oniszczuk-świercz, D. The effects of reduced graphene oxide flakes in the dielectric on electrical discharge machining. *Nanomaterials* **2019**, *9*, 335. [CrossRef] [PubMed]
140. Kavimani, V.; Prakash, K.S.; Thankachan, T. Influence of machining parameters on wire electrical discharge machining performance of reduced graphene oxide/magnesium composite and its surface integrity characteristics. *Compos. Part B Eng.* **2019**, *167*, 621–630. [CrossRef]
141. Świercz, R.; Oniszczuk-Świercz, D. Investigation of the Influence of Reduced Graphene Oxide Flakes in the Dielectric on Surface Characteristics and Material Removal Rate in EDM. *Materials* **2019**, *12*, 943. [CrossRef] [PubMed]
142. Wang, X.; Yi, S.; Guo, H.; Li, C.; Ding, S. Erosion characteristics of electrical discharge machining using graphene powder in deionized water as dielectric. *Int. J. Adv. Manuf. Technol.* **2020**, *108*, 357–368. [CrossRef]
143. parameters on 3D surface topography. *J. Mater. Process. Technol.* **2004**, *148*, 155–164. [CrossRef]
144. Cusanelli, G.; Hessler-Wyser, A.; Bobard, F.; Demellayer, R.; Perez, R.; Flükiger, R. Microstructure at submicron scale of the white layer produced by EDM technique. *J. Mater. Process. Technol.* **2004**, *149*, 289–295. [CrossRef]
145. Lee, H.T.; Rehbach, W.P.; Hsu, F.C.; Tai, T.Y.; Hsu, E. The study of EDM hole-drilling method for measuring residual stress in SKD11 tool steel. *J. Mater. Process. Technol.* **2004**, *149*, 88–93. [CrossRef]
146. Guu, Y.H.; Hocheng, H.; Chou, C.Y.; Deng, C.S. Effect of electrical discharge machining on surface characteristics and machining damage of AISI D2 tool steel. *Mater. Sci. Eng. A* **2003**, *358*, 37–43. [CrossRef]
147. Simão, J.; Aspinwall, D.; El-Menshawy, F.; Meadows, K. Surface alloying using PM composite electrode materials when electrical discharge texturing hardened AISI D2. *J. Mater. Process. Technol.* **2002**, *127*, 211–216. [CrossRef]
148. Simao, J.; Lee, H.G.; Aspinwall, D.K.; Dewes, R.C.; Aspinwall, E.M. Workpiece surface modification using electrical discharge machining. *Int. J. Mach. Tools Manuf.* **2003**, *43*, 121–128. [CrossRef]
149. Zeid, O.A.A. On the effect of electrodischarge machining parameters on the fatigue life of AISI D6 tool steel. *J. Mater. Process. Technol.* **1997**, *68*, 27–32. [CrossRef]
150. Younis, M.A.; Abbas, M.S.; Gouda, M.A.; Mahmoud, F.H.; Allah, S.A.A. Effect of electrode material on electrical discharge machining of tool steel surface. *Ain Shams Eng. J.* **2015**, *6*, 977–986. [CrossRef]

151. Liu, Y.; Zhang, Y.; Ji, R.; Cai, B.; Wang, F.; Tian, X.; Dong, X. Experimental characterization of sinking electrical discharge machining using water in oil emulsion as dielectric. *Mater. Manuf. Process.* **2013**, *28*, 355–363. [CrossRef]

152. Assarzadeh, S.; Ghoreishi, M.; Shariyyat, M. Response surface methodology approach to process modeling and optimization of powder mixed electrical discharge machining (PMEDM). In Proceedings of the 16th International Symposium on Electromachining, Shanghai, China, 19–23 April 2010; pp. 19–23.

MDPI
St. Alban-Anlage 66
4052 Basel
Switzerland
Tel. +41 61 683 77 34
Fax +41 61 302 89 18
www.mdpi.com

Micromachines Editorial Office
E-mail: micromachines@mdpi.com
www.mdpi.com/journal/micromachines